Composites in Structural Engineering and Architecture

Composites in Structural Engineering and Architecture

Thomas Keller

EPFL PRESS

Published with the support of the Swiss National Science Foundation

General Management: Lucas Giossi
Editorial and Sales Management: Sylvain Collette and May Yang
Communications Manager: Manon Reber
Production Manager: Christophe Borlat
Editorial: Alice Micheau-Thiébaud and Jean Rime
Graphic design: Kim Nanette
Accounting: Daniela Castan
Logistics: Emile Razafimanjaka

Cover illustration: Tara Habibi, EPFL-CCLab
Credit: Thomas Keller, EPFL

EPFL is an imprint owned by the Presses polytechniques et universitaires romandes,
a Swiss academic publishing company whose main purpose is to publish
the teaching and research works of the Ecole polytechnique fédérale de Lausanne (EPFL).
PPUR
EPFL – Rolex Learning Center,
CM Station 10, CH-1015 Lausanne
info@epflpress.org
tel.: +41 21 693 21 30

www.epflpress.org

First edtion in English 2024
© EPFL Press
ISBN 978-2-88915-647-4 for the print edition (in gray)
ISBN 978-2-8323-2268-0 for the ebook edition (PDF), DOI: 10.55430/6225TKVA01

This text is under Creative Commons license:

it requires you, if you use this writing, to cite the author, the source
and the original publisher, without modifications to the text or
of the extract and without commercial use.

PREFACE

COMPOSITES – OR HOW THINGS BEGAN

Fibre-polymer composites – henceforth referred to as "composites" – are construction materials that can offer unique properties in structural engineering and architecture when compared to conventional structural materials, such as reinforced concrete, steel, or timber. In addition to excellent mechanical properties, their physical properties – including low thermal conductivity, transparency, and colour, among others – allow structural, building physics, and architectural functions to be integrated into individual members of buildings, as illustrated in **Fig. P.1** [1]. This integration of functions may simultaneously fulfil requirements regarding structural safety, serviceability, aesthetics, economy, and sustainability. However, alongside these opportunities, composites may also exhibit some limitations in their structural application, which can be caused by their anisotropy or viscoelasticity, or in terms of ductility, fatigue, fire resistance, and durability.

As the title specifies, structural engineering and architecture are addressed in this monograph. The inclusion of architecture is crucial since, in building construction, architects normally choose the material, and, in most cases, this material selection is not primarily based on the benefits offered by superior mechanical properties. Priority may be given to criteria such as versatility in shaping, texture, transparency, and colour. Together with a discussion on structural design, aspects related to architectural design therefore also play an important role in the monograph.

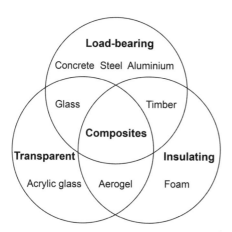

Fig. P.1 Functions and their integration offered by composite materials, and comparison to conventional materials (diagram shown during inaugural lecture at EPFL in 2000)

The content of the monograph is based on my more than 25 years of experience as a pioneer, scientist, designer, and lecturer in the field of composites, applied in structural engineering and architecture. To be more specific, I acquired the knowledge to write this document from (i) the scientific work I carried out with my team at the Composite Construction Laboratory, CCLab at EPFL, which I established in 2000, (ii) the design and construction of several composite structures in collaboration with industry, since 1996, (iii) my work in the Project Team of the European Technical Specification *Design of fibre-polymer composite structures* (CEN/TS 19101), published in 2022, and (iv) cross-disciplinary thinking, developed during my collaboration with architects and my teaching activities, the latter involving the disciplines of structural engineering, building physics, and architecture. Looking back on this experience, I found it convincing and meaningful to organise the main content of the monograph according to the above-mentioned limitations and opportunities of composites, and then demonstrate how the former can be overcome and the latter explored, with the primary aim being to derive paths for the successful use of composites in structural engineering and architecture.

My intentions were always to link research, design, and execution, given that they cross-pollinate each other. New scientific questions have always arisen from the design and construction of structures, and the research findings obtained are then implemented in later, more advanced designs. This interaction between research and design and execution is reflected in the monograph; one example is the Pontresina Bridge, one of the first composite bridges in Europe, which was built in Switzerland in 1996 and which I had the opportunity to design. This two-span truss bridge comprises composite profiles which mimic steel sections, in accordance with the state-of-the-art knowledge of the time. One year later, I was lucky enough to be commissioned for the structural design of the Eyecatcher Building, which is still the tallest building in the world with a primary composite structure. The building was initially designed with steel frames; however, after having seen the Pontresina Bridge, the architect contacted me to discuss a possible change from steel to composite profiles, which, due to their low thermal conductivity, would allow for the structure's integration into the building envelope and thus its use as an architectural element. While designing these two composite structures, I realised that simply substituting isotropic steel with anisotropic composites is neither material-tailored nor economic.

These experiences laid the foundation for the CCLab and its main research axes, which were (and still are) directed towards material-tailored structural forms, including function integration and connection techniques for composite structures. Instead of bolted connections (again mimicking steel), material-tailored adhesively bonded connections were already implemented in one span of the Pontresina Bridge, and research into sandwich structures was initiated, which allow for a much better combination and tailoring of the materials. Complex architectural forms and function integration were developed, as implemented in the double-curved web-core sandwich roof of the Novartis Campus Entrance Building in 2006. Furthermore, rapid replacement of deteriorated bridges and their widening are possible, as performed in the case of the Avançon Bridge in 2012, where a (heavy) one-lane concrete bridge was replaced by a (lightweight) two-lane composite-balsa wood sandwich bridge deck, adhesively

Preface

bonded onto two steel girders, with a significant reduction in traffic obstruction compared to a conventional concrete replacement.

The field of composites is vast, ranging from automotive, aerospace, and maritime applications to – more recently – civil engineering structures. It is difficult, however, to transfer research findings from one field to another as the boundary conditions are very different. The design service life, for instance, varies significantly, from approximately 10 to 30 to 100 years in the cases of cars, airplanes, and bridges, respectively. Another typical example is the adhesive layer thickness in connections, which can vary from 0.1 mm in aerospace to 30 mm in bridge construction, in the latter case to compensate for tolerances. A further aim of this monograph is thus to extract, from the vast field of composite material science, the knowledge that is relevant in the field of structural engineering and architecture, and to present this knowledge in a consistent framework. By doing so, it also becomes clear that exploring the potential opportunities of composites requires a close collaboration between structural engineers, architects, and manufacturers. Against this background, the target audience of the monograph is manifold, ranging from students to scientists, practical engineers, and architects.

The monograph is, on the one hand, a critical review of the state of the art in the field, but on the other hand, it contains a significant number of original analyses and unpublished elements. Furthermore, areas that still require research concerning the design and use of composite structures are identified. The document also provides the opportunity to revise or re-interpret findings from my own earlier works, based on more recent insights.

When I designed the Pontresina Bridge in 1996, composites were expensive and the design was mainly based on testing, but I was told that costs would soon come down following more widespread application, which would be fostered by (i) the fact that composites are durable and do not require any protection and maintenance (in contrast to reinforced concrete or steel) and (ii) the development of design codes. A survey which I performed in 2001 on the prospects of using composites for bridges pointed to a bright future for composite highway bridges, as shown in **Fig. P.2**. The survey also demonstrated the expected transition from short-term applications, which were already in place at that time, to a much larger field with new types emerging in the medium term, and a focus on these new types in the long term.

Today, about 25 years later, it is clear that (i) material costs are still relatively high, (ii) composites often require protective coating against environmental conditions, just as steel profiles do, (iii) the above-mentioned European Technical Specification has only just been published (in 2022), and (iv) composite structures are mostly still in the phase of pilot applications (except strengthening of existing structures with composites). The development is thus much slower and more complicated than predicted. On the other hand, the history of structural engineering and architecture reveals that it has always taken several decades for new materials to become widely adopted (as will also be pointed out in the monograph). The development in the case of composites is thus still in line with these experiences.

Composites will not replace conventional construction materials (as was occasionally suggested 25 years ago), but they do have the potential to establish themselves

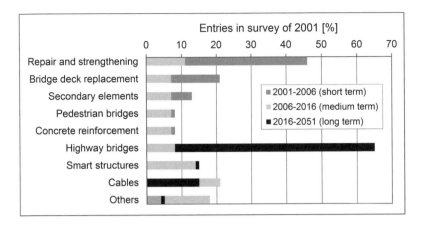

Fig. P.2 Survey in 2001: Expected types of composite applications in bridge construction in short term (0–5 years), medium term (5–15 years), and long term (15–50 years), derived from [2]

on an equal footing, as discussed in this document. In hybrid construction, they can also contribute to a better use of conventional structural materials, by combining the latter with composites. Furthermore, and most interestingly, composites can offer unique options in design and thus significantly broaden the current design space – and may thus contribute to extend "*i limiti dell'immaginazione*", according to Leonardo Da Vinci:

"*I limiti dell'immaginazione dell'uomo sono dati dai materiali di cui dispone*"

Thomas Keller
July 2024

REFERENCES

[1] T. Keller, "Fibre reinforced polymer materials in building construction", in *IABSE Symposium Towards a Better Built Environment*, Melbourne, 2002.

[2] T. Keller, *Use of fibre reinforced polymers in bridge construction*. Structural Engineering Documents, SED 7, International Association for Bridge and Structural Engineering, IABSE, Zurich, 2003.

ACKNOWLEDGEMENTS

I would like to take the opportunity to sincerely thank all those who made it possible for this monograph to see the light of the day, including colleagues, friends, and institutions.

Since studying civil engineering at the Swiss Federal Institute of Technology in Zurich (ETH), I have been lucky enough to meet several outstanding personalities, who guided my way. From my doctoral supervisor, Professor Christian Menn at ETH, I learned the principles of conceptual design of structures and understood the significance of conceptual design in structures that are economical but also aesthetically convincing. Architect and engineer Santiago Calatrava was my first employer after my studies and further raised my interests in architecture and its relationship to structural engineering. After I returned to ETH as a part-time assistant professor at the Department of Architecture, Professor Otto Künzle gave me the opportunity to design the Pontresina Bridge, which was the seed from which my interest and career in composite construction grew. Luckily, Professor Urs Meier at EMPA was not far from ETH, and I could always – and still can – rely on his advice and thus gain confidence on my way. I was also fortunate enough to meet Professor Marie-Anne Erki from the Royal Military College of Canada, who helped me in the design of the adhesive connections of the Pontresina Bridge and thus paved the way for what is now an important research axis at the CCLab. Finally, Professor Jean-Claude Badoux, president of EPFL at the time (1998), offered me the position I hold to this day, which allowed me to establish the CCLab and develop my activities. I owe a deep debt of gratitude to all these colleagues.

The two backbones of the CCLab have always been my PhD students (see list below) and the experimental platform. What I know today about composites and what is written in this monograph originates, to a large extent, from the work I do with my PhD students, who have come to the CCLab from all over the world and who continue to inspire me. Nothing is more satisfying than seeing them today in important positions in industry or as professors in academia, spread out across the world once again. The support provided by the technicians of the experimental platform in developing original and complex experiments has always been outstanding and contributed to the success of many projects. My seniors, Julia de Castro and Anastasios Vassilopoulos have supported me in supervising the students and in teaching. Furthermore, I have always been able to count on the support from industry and funding institutions, such as Fiberline, Sika, Scobalit, Martin Marietta Composites, 3A Composites, Blue-Office Architecture, and the Swiss National Science Foundation, Swiss Innovation

Promotion Agency, and Federal Roads Office. I would like to sincerely thank all of them for their commitment and support.

With regard to the monograph itself, its quality is particularly due to the thorough reviews of three of my peers. The discussions with professors Yu Bai (Monash University), João Ramôa Correia (University of Lisbon), and Wendel Sebastian (University College London) were always very fruitful and stimulating. A former and two of my current PhD students reviewed the topics in which they are experts today, that is, Abdolvahid Movahedirad for the chapters on viscoelasticity and fatigue, Congzhe Wang for the chapter on fracture, and Tara Habibi who gave support in drafting and reviewed the sections on composites in architecture and sculpture, and who designed the book cover. These colleagues and co-workers also deserve my heartfelt thanks and appreciation.

Last but not least, I would like to thank the publisher, EPFL Press, for the outstanding support and services, and the Swiss National Science Foundation, for the funding of the Open Access (obtained after a thorough scientific review of the monograph).

LIST OF MY PHD STUDENTS

Tara Habibi, 2020-2024

Congzhe Wang, 2019-2023

Gisele Góes Cintra, 2019-2023

Ghazaleh Eslami, 2019-2023

Niloufar Vahedi, 2018-2022

Lulu Liu, 2016-2021

Abdolvahid Movahedirad, 2015-2019

Aida Cameselle Molares, 2015-2019

Sonia Yanes Armas, 2013-2017

Maria Savvilotidou, 2013-2017

Haifeng Fan, 2012-2017

Myrsini Angelidi, 2012-2017

Kyriaki Goulouti, 2013-2016

Mário De Jesus Garrido, 2011-2016

Wei Sun, 2010-2015

Robert Koppitz, 2010-2015

Carlos Pascual Agulló, 2010-2014

Michael Osei-Antwi, 2010-2014

Moslem Shahverdi, 2009-2013

Roohollah Sarfaraz Khabbaz, 2008-2012

Omar Moussa, 2008-2011

Behzad Dehghan Manshadi, 2008-2011

Ye Zhang, 2004-2010

Yu Bai, 2004-2009

Martin Schollmayer, 2002-2009

João Ramôa Correia, 2004-2008

Erika Schaumann, 2004-2008

Florian Riebel, 2003-2006

Craig Tracy, 2001-2005

Julia de Castro, 2000-2005

Herbert Gürtler, 2001-2004

Till Vallée, 2000-2004

Sean Dooley, 1999-2004

WHO AM I

I obtained a Civil Engineering degree from the Swiss Federal Institute of Technology (ETH) Zurich in 1983. Subsequently, I worked at the architecture and engineering office of Santiago Calatrava, where, in 1987, I developed the structural concept for the Alamillo Bridge in Seville. Back at ETH, I received my doctoral degree in 1992, under the supervision of Prof. Christian Menn. In 1996, I was appointed as part-time assistant professor at the Department of Architecture of ETH, in 1998 as part-time associate professor, and in 2007 as full professor of structural engineering at the Swiss Federal Institute of Technology (EPFL) Lausanne. From 1996 to 2004, I also worked as a practical engineer and was a partner in a civil engineering company.

Today, I am the head of the Composite Construction Laboratory (CCLab) at EPFL, which I established in 2000. My research work is focused on the development of material-tailored applications of composites in structural engineering and architecture. I was a founding member of the International Institute for FRP in Construction (IIFC) and a member of the CEN/TC 250 Project Team to establish the European Technical Specification *Design of fibre-polymer composite structures*, CEN/TS 19101, published in 2022.

As a practical engineer, I designed the first composite pedestrian bridge in Switzerland (Pontresina Bridge, 1997), which is also one of the first composite bridges in Europe. Furthermore, I was responsible for the structural design of the five-storey Eyecatcher Building (Basel, 1998), which is still the tallest building in the world with a primary composite structure. I also contributed to the design of the free-form multifunctional composite sandwich roof of the Novartis Campus Entrance Building (Basel, 2006) and the hybrid Avançon (vehicular) Bridge with an adhesively bonded composite-balsa sandwich deck (Bex, 2012).

Fig. P.3 Steel-concrete Alamillo Bridge, Seville, 1992, first cable-stayed bridge without back-cables (pylon weight balances girder weight), 200 m main span **(left)**, and five-storey composite-timber Eyecatcher Building, Basel, 1998 (with author on third floor) **(right)**

TABLE OF CONTENTS

Preface: Composites – Or how things began 5

Acknowledgements 9

Who am I 11

1 Composites – Compared...19
    1.1   Terminology...19
    1.2   Comparison across materials 20
    1.3   Scope and content 28
    1.4   References..29

2 Composites – Compact 31
    2.1   Introduction...31
    2.2   Materials 33
          2.2.1   Fibres and resins...33
          2.2.2   Sandwich core materials...35
          2.2.3   Adhesives 40
    2.3   Components and members 42
          2.3.1   Components..42
          2.3.2   One-dimensional members..45
          2.3.3   Two-dimensional members 47
    2.4   Connections and joints..50
          2.4.1   Overview 50
          2.4.2   Bolted lap shear connections and joints 53
          2.4.3   Adhesively bonded lap shear connections and joints.................53
          2.4.4   Hybrid bonded/bolted lap shear connections and joints.............57
    2.5   Manufacturing and installation 59
          2.5.1   Hand lay-up...59
          2.5.2   Pultrusion 60
          2.5.3   Vacuum infusion..62
          2.5.4   Prepreg moulding 64
          2.5.5   Installation 64
    2.6   Composites in bridge construction...64
          2.6.1   Introduction 64
          2.6.2   Vehicular bridge decks..66
          2.6.3   Pedestrian bridges 71

2.7	Composites in architecture	74
	2.7.1	Introduction	74
	2.7.2	Initiation phase of composites in architecture	76
	2.7.3	Laminated and profile structures	80
	2.7.4	Sandwich structures	81
	2.7.5	Design aspects in architecture	88
2.8	Composites in sculpture	90
2.9	References	95

3 Composites – Overcoming limitations 101

3.1	Introduction	101
3.2	Anisotropy	103
	3.2.1	Introduction	103
	3.2.2	Anisotropy in composite and core materials	103
	3.2.3	Effects of anisotropy on structural response	107
	3.2.4	Conclusions	110
3.3	Imperfections and tolerances	111
	3.3.1	Introduction	111
	3.3.2	Types and consequences of imperfections in composites	112
	3.3.3	Tolerances	114
	3.3.4	Conclusions	115
3.4	Viscoelasticity	116
	3.4.1	Introduction	116
	3.4.2	Engineering and phenomenological models	117
	3.4.3	Glass transition temperature	119
	3.4.4	Time and temperature dependence	122
	3.4.5	Creep, recovery, and relaxation	125
	3.4.6	Moisture effects	129
	3.4.7	Conclusions	131
3.5	Curing and physical ageing	131
	3.5.1	Introduction	131
	3.5.2	Processes of curing, physical ageing, and moisture uptake	132
	3.5.3	Early-age development of adhesive properties	134
	3.5.4	Long-term development of adhesive properties	143
	3.5.5	Conclusions	144
3.6	Ductility	145
	3.6.1	Introduction	145
	3.6.2	Energy-based definition and pseudo-ductility	146
	3.6.3	Examples of pseudo-ductility	150
	3.6.4	Conclusions	157
3.7	Fracture	157
	3.7.1	Introduction	157
	3.7.2	One-dimensional crack propagation	160
	3.7.3	Two-dimensional crack propagation	165
	3.7.4	Multiple cracks	173

	3.7.5	Conclusions	176
3.8	Fatigue		177
	3.8.1	Introduction	177
	3.8.2	Basic representation of fatigue resistance	178
	3.8.3	Mean stress dependence	180
	3.8.4	Creep-fatigue interaction	181
	3.8.5	Variable amplitude loading	183
	3.8.6	Temperature and moisture dependence	186
	3.8.7	Fatigue modelling vs fatigue design	186
	3.8.8	Fatigue-tailored detailing	187
	3.8.9	Fatigue load model for composite bridges	192
	3.8.10	Conclusions	195
3.9	Fire		196
	3.9.1	Introduction	196
	3.9.2	Thermophysical and thermomechanical properties	198
	3.9.3	Fire resistance behaviour	208
	3.9.4	Fire dynamic simulation of indoor fires	221
	3.9.5	Conclusions	224
3.10	Durability		226
	3.10.1	Introduction	226
	3.10.2	Long-term surveys of Pontresina Bridge and Eyecatcher Building	227
	3.10.3	Further long-term studies	240
	3.10.4	Detailing for durability	242
	3.10.5	Conclusions	244
3.11	Sustainability		245
	3.11.1	Introduction	245
	3.11.2	Sustainability assessment of composites	246
	3.11.3	Results from case studies	248
	3.11.4	Bio-based composites	250
	3.11.5	Conclusions	250
3.12	References		251

4   Composites – Exploring opportunities 263

4.1	Introduction		263
4.2	Lessons learnt from nature		265
	4.2.1	Introduction	265
	4.2.2	Materials of natural structures	266
	4.2.3	Design principles of nature	267
	4.2.4	Structural design concepts of nature	270
	4.2.5	Conclusions	274
4.3	Structural form		274
	4.3.1	Introduction	274
	4.3.2	Language of composites	276
	4.3.3	Material-tailored structural form	277

		4.3.4	Material-tailored structural system	283

4.3.4 Material-tailored structural system 283
4.3.5 Conclusions 291
4.4 Transparency 292
4.4.1 Introduction 292
4.4.2 Scientific background 292
4.4.3 Laminate case study 295
4.4.4 Building integration...300
4.4.5 Conclusions 302
4.5 Function integration 302
4.5.1 Introduction 302
4.5.2 Thermal insulation...305
4.5.3 Thermal activation...311
4.5.4 Solar cell encapsulation...313
4.5.5 Colour, light, and texture...316
4.5.6 Conclusions 318
4.6 Hybrid construction 319
4.6.1 Introduction 319
4.6.2 Hybrid composite construction 320
4.6.3 Composite action...325
4.6.4 Conclusions 326
4.7 Modular construction 327
4.7.1 Introduction 327
4.7.2 Modular composite construction..................................327
4.7.3 Fields of application 328
4.7.4 Conclusions 330
4.8 References...331

5 Composites – How to design...335
5.1 Introduction..335
5.2 Conceptual design..337
5.2.1 Member composition and function integration 337
5.2.2 Material selection and protection systems...................337
5.2.3 Structural system and redundancy...............................338
5.2.4 Concepts of joints and details......................................338
5.2.5 Manufacturing and installation....................................339
5.2.6 Preliminary dimensioning and cost estimate...............340
5.3 Detailed fatigue and adhesive joint resistance verifications 341
5.3.1 Verification concept...341
5.3.2 Fatigue resistance verification.....................................342
5.3.3 Adhesive joint resistance verification 344
5.4 Execution 345
5.5 Maintenance and rehabilitation..346
5.6 References...348

6	**Composites – How to proceed**		**351**
	6.1	Introduction	351
	6.2	Lessons learnt and conclusions	351
		6.2.1 Composites – Compact	351
		6.2.2 Composites – Overcoming limitations	352
		6.2.3 Composites – Exploring opportunities	354
		6.2.4 Composites – How to design	356
	6.3	Research needs	356
	6.4	Education	359
	6.5	References	359

Postscript: Composites – Or why things (don't) get built **361**

    References 363

Appendix **365**

    A.0   Introduction 365

    A.1   Verdasio Bridge 366

    A.2   Pontresina Bridge 368

    A.3   Eyecatcher Building 373

    A.4   Novartis Building 380

    A.5   Avançon Bridge 386

    A.6   Flower Sculpture 393

    A.7   TCy Sculpture 395

    B.1   Dock Tower 397

    B.2   RLC Building 400

    B.3   CLP Building 402

    B.4   TSCB Bridge 405

    B.5   1K Bridge 409

Image & video credits **411**

Index **421**

CHAPTER 1

COMPOSITES – COMPARED

1.1   TERMINOLOGY

The materials to which this monograph refers are designated "Fibre-Polymer Composites", or just "Composites" for short, in accordance with the terms used in the new European Technical Specification CEN/TS 19101 *Design of fibre-polymer composite structures* [1]. During the development of this specification, a decision was made to replace the denominations that had been commonly used for these materials up until then, i.e., "Fibre-Reinforced Polymers" or just "FRPs", and to no longer use acronyms. Indeed, in modern composites, the fibres do not reinforce the polymer matrix but are the main load-bearing constituents, which are supported by the matrix. Furthermore, denominations and acronyms such as "FRP" do not attribute a recognisable identity to the material, as for instance "Timber" does – and "Composites" will now also do. To specify the fibre type, the terms "glass fibre composites" or "carbon fibre composites" are applied, instead of the acronyms "GFRP" or "CFRP".

The term "Composite" already has other uses in structural design, the most significant ones being (i) "Composite steel and concrete structures" in [2] and (ii) "composite action", also see [2]. In the first case, "composite" is used as an adjective, i.e., in the sense of "hybrid", as defined below. In the second case, the term designates a shared structural action effect, e.g., of a bridge deck acting as a top chord of a steel girder. These uses of the term are, however, clearly distinct from the new material denomination "Composites".

Another often-used term that requires specification is "hybrid". In this document, the term is used in the two cases of "hybrid structures" (or members) and "hybrid joints" (between members), again in accordance with [1]. The first term designates load-bearing structures (or members) composed of composite members (or components) and members (components) of other materials, such as steel, concrete, or timber; an example is a composite bridge deck bonded to steel girders. "Hybrid joints" is used for joints composed of combined adhesively bonded and bolted connections, which may share the joint load (depending on the relative connection stiffness).

To make distinctions between composites and steel, reinforced concrete, and timber materials, the latter three materials are subsumed under "conventional materials" in the following chapters.

1.2 COMPARISON ACROSS MATERIALS

As the denomination "Fibre-Polymer Composites" already indicates, these materials are composed of two constituents, i.e., fibres as a linear and flexible constituent (such as glass, basalt, aramid, or carbon fibres), which are embedded in a polymeric matrix (normally polyester, vinylester, or epoxy resins). The polymeric matrix represents the volumetric constituent, which basically provides the shape of a composite member, as shown in **Fig. 1.1** (left), taking the example of a composite profile with channel section. Core materials of sandwich panels can also be subsumed under "Composites", as depicted in Fig. 1.1 (right), while adhesives are the third type of polymeric materials used in the context of composite structures.

Materials composed of combined linear and volumetric constituents typically exhibit anisotropic or orthotropic behaviour, i.e., their properties are direction-dependent and differ along three orthogonal axes in the latter case. Materials with direction-independent properties are designated isotropic. Crucial for the composite action between fibres and matrix is the interfacial bond strength, which mainly depends on the physical and chemical compatibility of the two constituents and an appropriate manufacturing process.

Two-constituent compositions of linear and volumetric constituents can also be found in conventional structural materials, such as reinforced concrete and timber, see **Table 1.1**. The linear constituents in reinforced concrete and timber are steel rebars and cellulose fibres, while the volumetric constituents are the concrete itself (which is a composite of aggregates and a paste), and combined hemicellulose and lignin, respectively. Timber cellulose, hemicellulose, and lignin all are polymers.

The structural roles of the linear and volumetric constituents in reinforced concrete, timber, and composites are partially identical but also differ, see Table 1.1 and **Fig. 1.2**. All three types of linear constituents, i.e., steel rebars, timber cellulose, and composite fibres, normally take the tension forces. The volume fraction (vol%) of steel

Fig. 1.1 Composite profile (C300×90×15 mm, from Eyecatcher Building manufacturing, see Appendix A.3) with orthotropic glass fibre architecture embedded in a polyester matrix **(left)**; sandwich panel section (of Avançon Bridge deck, Appendix A.5) composed of glass fibre composite face sheets embedded in a vinylester matrix (22 mm thick) and balsa wood core (241 mm thick) **(right)**

1. Composites – Compared

Table 1.1 Constituents of composite materials used in engineering structures and their structural roles [3]

Composite material	Linear constituent			Volumetric constituent	
	Material	Structural role	Vol%	Material	Structural role
Reinforced concrete	Steel rebars	Tension (primary role)	1–3 [3] (80–240 kg/m^3)	Concrete	Compression (primary role)
Timber	Cellulose fibres	Tension or compression	40–50 [4]	Hemicellulose and lignin	Stabilisation of compressed fibres
Fibre-polymer composites	Glass, basalt, carbon fibres	Tension or compression	30–60 [1]	Polyester, vinylester, epoxy	Stabilisation of compressed fibres

rebars is much smaller than the similar fractions of cellulose and composite fibres. The transmission of compression differs, however. In reinforced concrete, compression forces are primarily transmitted by the concrete itself, the contribution of rebars is marginal in most cases, mainly due to the small cross-sectional area, see Table 1.1 (vol%). In timber and composites, compression is transmitted by the fibres. However, this transmission needs the collaboration of the volumetric constituents, which must stabilise the flexible fibres against buckling. The corresponding structural system is shown in Fig. 1.2 (right), where the fibres are supported by transverse springs, which represent this stabilising effect. It also becomes clear that if the stabilising effect is lost, e.g., under elevated temperatures above the glass transition temperature of the polymeric constituent, a member under bending or compression can exhibit an instability failure, e.g., a wrinkling failure in the case of a compressed sandwich panel face sheet, see Section 3.4 on viscoelasticity. A further stabilising effect occurs in timber through the geometrical stiffness of the quasi-hexagonal structure of the cell walls. In all cases, furthermore, the volumetric constituents ensure the load transmission between the linear constituents. An additional function of the volumetric constituents in reinforced concrete and composites is the protection of the rebars and fibres from environmental degradation, thus ensuring their durability.

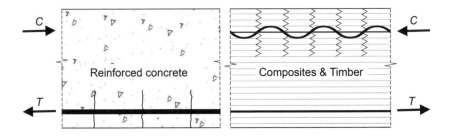

Fig. 1.2 Roles of linear and volumetric constituents in reinforced concrete (**left**) and timber and composites (**right**) (volumetric constituents represented by springs that stabilise fibres, T = tension, C = compression) [3]

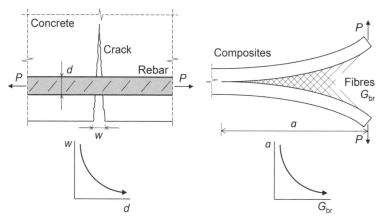

Fig. 1.3 Comparison of concrete crack width (w) vs steel rebar diameter (d) **(left)** and composite crack length (a) vs fracture resistance of crack bridging fibres (G_{br}) (assumptions: large scale bridging, bridging length ≅ crack length) **(right)** [3]

A further similarity exists between reinforced concrete and composites, which concerns the role of the linear constituents in increasing the resistance against fracture, as illustrated in **Fig. 1.3**. Since the tensile strength of concrete is almost zero, concrete structures are normally cracked in tension zones and the cracks are bridged by steel rebars, which transmit the tension forces. An important metric in this respect is the crack width (w), which should be limited to not affect durability through corrosion

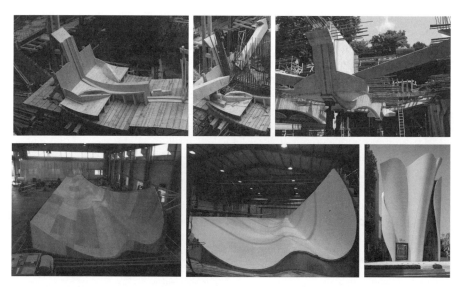

Fig. 1.4 Reinforced concrete "Bird" structure of Stadelhofen Train Station in Zurich, 1988 (architect S. Calatrava): formwork, added reinforcement and prestressed prefabricated concrete tension members, and final structure (length ~15 m) **(top left to right)**; composite envelope of House of Dior in Seoul, 2015 (architect C. de Portzamparc): mould, cured envelope panel (20×6 m), finalised building **(bottom left to right)**

Fig. 1.5 Arrangement of stiff concrete steel rebars to follow complex formwork shape **(left)** vs flexible glass fibre composite fabric easily draped and cut to follow complex mould shape **(right)**

of the steel rebars. This limitation can be achieved by increasing the rebar diameter (or reducing the rebar spacing), and thus decreasing the rebar stress and strain, see Fig. 1.3 (left). In composites, one of the most critical failure modes is delamination fracture between composite plies, see Section 3.7 on fracture. The fracture resistance can be significantly increased by an appropriate fibre architecture, which allows tensioned fibres to bridge the crack and thus reduce the crack length and prevent unstable crack propagation.

A final similarity to be mentioned between reinforced concrete and composites is based on the liquid nature of the volumetric constituent during manufacturing. This liquid condition requires formwork in concrete and moulds in composite structures, as the examples in **Fig. 1.4** depict. Also nicely illustrated in these cases is the potential of both materials to conceive structures of complex double-curved shapes, which are basically defined by these formworks or moulds. The linear constituents, rebars or fibres, must follow these shapes near the surface in all directions, which is, however, much easier in the case of thin and flexibles fibres than for much thicker and stiffer rebars, as illustrated in **Fig. 1.5**. Fibres and corresponding fabrics can be draped easily and even cut by scissors during their lay-up. Furthermore, the positioning of formwork and rebars, and pouring and hardening of the concrete normally occurs on the construction site, which is very time consuming, while entire composite superstructures can be prefabricated and, due to their low weight, rapidly installed on site, as shown in **Fig. 1.6**.

When comparing timber and composites, both have a similar structural behaviour, i.e., the structural roles of linear and volumetric constituents are similar, as discussed above. Timber can thus also be designated a "natural composite" while fibre-polymers are "engineered composites". The effect of this difference on structural members can be seen in **Fig. 1.7**. If the bi-directional behaviour of a member is targeted, timber is normally cross-laminated into panels, while in the composite case, simple bi-directional woven fabrics can be used in composite plies. The difference in the thickness of these layers is significant, since one layer of timber is normally 20–40 mm thick due to the integrated nature of the material and for manufacturing reasons, while the fabric and corresponding ply thickness is in the order of 1 mm. Assuming an identical member thickness, the achievable effective depth in the two directions and thus the overall

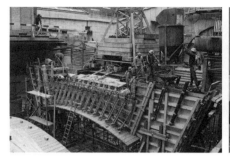 

Fig. 1.6 Cumbersome concrete pouring into encased frame of railway bridge over lower shop floor of Stadelhofen Train Station, Zurich **(left)** vs rapid installation of prefabricated superstructure of Friedberg Bridge (Germany), with pultruded deck system adhesively bonded on two steel girders **(right)**

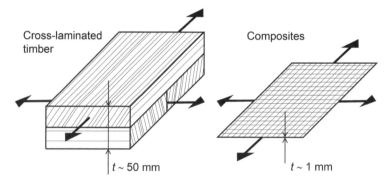

Fig. 1.7 Typical thicknesses of bi-directional cross-laminated timber (CLT) layers **(left)** and composite ply **(right)** [3]

structural efficiency is therefore significantly smaller in the timber than in the composite case. Furthermore, in the case of both materials, the joining of members is made difficult by the material anisotropy, as force directions change at these locations; effective joining is thus easier with isotropic materials. The material strength of timber and composite members can thus often not be fully used since the joints determine the cross-sectional dimensions (as further explained in Section 2.4).

A structural material quite different from the materials discussed above is steel, which is an isotropic and homogeneous material, and thus not consisting of constituents. Although there are some similarities, differences from composites are significant. The most obvious similarity concerns the cross-sectional shapes of steel and composite profiles, which are basically identical, as shown in **Fig. 1.8**. Why this simple material substitution of steel by composites is done, which is not material-tailored from the structural point of view, is discussed in Section 4.3.

As has been shown above, composite laminates can be conceived with a very low thickness, as is often also the case in flanges and webs of steel profiles. Accordingly, as in the steel case, composite laminates and profiles, depending on their slenderness,

1. Composites – Compared

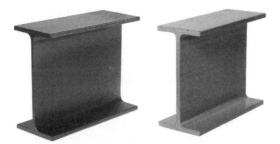

Fig. 1.8 I-section profiles made of steel (painted) **(left)**; glass fibre composites (as produced by pultrusion) **(right)**, both 160 mm high

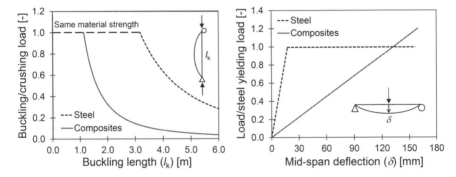

Fig. 1.9 Comparison of Euler buckling load (normalised with crushing load, i.e., material strength) vs buckling length (l_k) of steel and composite column **(left)**; load (normalised with steel yielding load) vs mid-span deflection (δ) of steel and composite beam (simplified curves) **(right)** (assumptions: same I cross sections 300×150×12/8 mm, beam span = 5.0 m, steel grade S235, E_{comp} = 28 GPa (glass fibres) identical steel and composite compressive strengths of 240 MPa [3]

are sensitive to local or global buckling under compression loading. The situation is even more critical for composites, as their stiffness is significantly lower than that of steel if glass fibres are used. Buckling loads are thus lower, as shown in **Fig. 1.9** (left), where the Euler buckling loads of a steel and a glass fibre composite column are compared, as a function of the buckling length. The Euler buckling curves are limited by the crushing load, where the full cross sections reach their compressive material strength, which has been assumed to be similar in steel and composites in Fig. 1.9 (left).

The most significant difference between the two materials is that steel exhibits a ductile behaviour, while composites basically, but not always, have a brittle response. Profiles, as depicted in Fig. 1.8, used as beams under bending, behave as outlined in Fig. 1.9 (right). A steel beam is initially much stiffer than a glass fibre composite beam of identical cross section, i.e., the mid-span deflections are much smaller under the same load. If the yield strength of steel is reached, however, the deformations grow rapidly without any significant increase in load. The response of a composite beam remains linear, however, up to a brittle failure. The advantages of a ductile nonlinear response, as exhibited by steel, and how this response can also be achieved with composites, are discussed in Section 3.6 on ductility.

Composite fibres exhibit high strength and low density compared to steel, for instance, as shown in **Table 1.2**. The specific strength, i.e., the strength per density, is a metric for a material's suitability for lightweight construction, as explained in Section 4.3. The specific strength vs specific elastic modulus relationships of composite fibres and laminates, concrete, steel, and timber, are shown in **Fig. 1.10** (which is based on Table 1.2). Composite fibres exhibit significantly higher specific values when compared to the other materials, while timber performs as well as steel, and both perform much better than concrete. The superior composite fibre values, however, decrease if composite laminates are considered because of the low strength and modulus of the matrix; but carbon fibre composites still exhibit an excellent performance in lightweight construction.

These examples also demonstrate that "Composites" are not a single and well-defined material, as is the case for steel, for instance. Composites, when further including core materials in the case of sandwich panels, allow for the conception of an almost infinite number of material configurations, which can be optimally tailored to a specific application. A somewhat similar flexibility can only be offered by reinforced concrete, where the rebar arrangement can be varied. In the cases of steel and timber, only different grades or species are available, and the adaptability is thus

Table 1.2 Comparison of typical properties of composite fibres and profiles and conventional materials (in longitudinal direction if orthotropic) [1, 5–7]

Material	Tensile strength [MPa]	Elastic modulus [GPa]	Density [kg/m^3]	Coefficient of linear thermal expansion $\times 10^{-6}$ [K^{-1}]	Thermal conductivity [W/mK]
E-glass fibres	2 500	74	2 500	5	1.0
Glass fibre composites	350	28	1 800	10	0.4
Basalt fibres	3 000	90	2 700	8	0.04
Carbon fibres	4 400–4 900	230–450	1 800	-0.4 to -0.8	24–105
Carbon fibre composites	1 800	140	1 600	1	5
Aramid fibres	3 600	130	1 450	-2	0.04
Flax fibres	1 100	65	1 500	-8	0.05
Steel S235	360	210	7 850	10	50
Concrete C35/45	3 (-40[1])	30	2 300	10	2.5
Timber	10	11	350	3	0.2

[1] Compressive strength

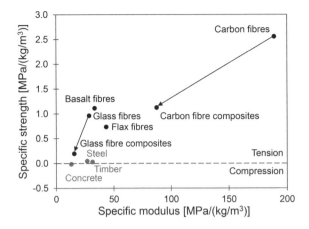

Fig. 1.10 Specific strength vs specific modulus (i.e., strength/density vs modulus/density) of fibres, laminates (composites), and conventional materials, indicating suitability for lightweight construction (tensile strength, except concrete with compressive strength) [3]

much reduced. Engineered wood-based panels can alleviate this disadvantage to a certain extent, although limitations, such as those discussed above in the case of CLT (Fig. 1.7), still exist.

The possible tailoring of composite materials also allows for the imposition of flows of forces through structures, by grading the stiffness accordingly, since the forces always follow the paths of highest stiffness. Such flows of forces can also be visualised and used as an architectural feature.

Hybrid-composite structures (as defined above) may outperform all-composite structures in terms of structural efficiency and material consumption, as discussed in Section 4.6. The fact that the coefficients of thermal expansion of concrete, steel, and glass fibre composites are almost identical, as also shown in Table 1.2, represents a significant advantage in the combination of these materials in hybrid structures or members. As can also be seen, carbon, aramid, and natural (flax) fibres normally exhibit a negative coefficient, while carbon fibre composites may not notably deform under temperature variations.

Composites can be used in sandwich panels, as already mentioned above, see Fig. 1.1 (right). The structural behaviour of sandwich panels and profiles is very similar, as shown in **Fig. 1.11** for the structural bending of an I-beam. A bending moment (M) is taken by tension (T) and compression (C) forces in the profile flanges, while the same forces apply in the composite face sheets in the sandwich panel case. A shear force (V), which applies to the web of the profile, is taken by the core of the sandwich panel. Since the core is much wider than the thin steel web, lower strength materials, such as polymeric foams or balsa wood, can be used as core materials to transmit the shear forces. Profiles and sandwich panels can be applied in structural systems such as beams, columns, frames, trusses, slabs, plates, and shells.

Core materials such as polymeric foams and balsa wood furthermore have very low thermal conductivities (see Table 4.6). Together with the low thermal conductivity

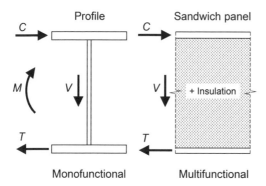

Fig. 1.11 Structural roles of components of monofunctional I-profile and multifunctional sandwich panel with additional insulation function (M = bending moment, V = shear force in web/core, T = tension force in flange/face sheet, C = compression force in flange/face sheet) [3]

of glass and basalt fibre composites (Table 1.2), composite sandwich panels may also provide a high thermal insulation capacity. In building construction, the outer thermally insulating envelope and inner load-bearing structure – which are separated in concrete and steel structures – can thus be merged into a single-layer structural envelope, which can furthermore be conceived with a complex shape. This multifunctional and material-tailored use of composites may provide new opportunities in architecture, and regarding sustainability and economy, as discussed in Section 4.5 and [8].

1.3 SCOPE AND CONTENT

The scope of this monograph covers stand-alone all-composite and hybrid-composite structures, primarily used for bridges and buildings, but also for other structural engineering applications, such as off-shore platforms, as an example. More details about the scope, regarding considered composite materials, components, joints, members, and manufacturing processes, are specified in Section 2.1. Outside of the scope are (i) strengthening of existing structures with composites, (ii) composite rebars used for the reinforcement of concrete, and (iii) membrane structures.

The aspects discussed in the monograph are also basically applicable to composite wind turbine blades. Such blades are (i) of a large scale, as building roofs and bridges can be, and (ii) composed of composite laminates, sandwich panels, and adhesive bond lines, and manufactured by similar processes, as discussed in this document. Consulting design standards for wind turbine blades, for example [9], reveals that the primary differences between bridge and building structures and such blades lie on the loading and not on the material or resistance sides.

The content of this monograph focuses – within the vast field of composite material science – on the knowledge and understanding that are relevant in the field of structural engineering and architecture, as emphasised in the Preface. The monograph is composed of six chapters and an appendix. The first two chapters introduce the content, define the scope, and provide a compact summary of the state-of-the-art knowledge in the field. They are followed by the two principal chapters, namely (i) the third chapter, which focuses on limitations of composites and measures of how they can be overcome, and (ii) the fourth chapter, which presents opportunities offered by composites and shows how they can be explored, as is also explained in the Preface. In the fifth chapter, relevant aspects of structural design are discussed, based on the previously mentioned European Technical Specification [1]. Conclusions are drawn in the sixth chapter, and axes of required future research are outlined. The appendix provides detailed information about the projects described in the above-defined chapters.

1.4 REFERENCES

[1]   *CEN/TS 19101: Design of fibre-polymer composite structures*. European Committee for Standardization, Brussels, 2022.

[2]   *EN 1994-1-1: Design of composite steel and concrete structures – Part 1-1: General rules and rules for buildings*. European Committee for Standardization, Brussels, 2004.

[3]   Original contribution.

[4]   C. Plomion, G. Leprovost, and A. Stokes, "Wood formation in trees", *Am Soc Plant Biol*, vol. 127, December, pp. 1513–1523, 2001, doi: 10.1104/pp.010816.

[5]   S. Amroune, A. Belaadi, M. Bourchak, A. Makhlouf, and H. Satha, "Statistical and experimental analysis of the mechanical properties of flax fibers", *J. Nat. Fibers*, vol. 19, no. 4, pp. 1387–1401, 2022, doi: 10.1080/15440478.2020.1775751.

[6]   K. L. Pickering, M. G. A. Efendy, and T. M. Le, "A review of recent developments in natural fibre composites and their mechanical performance", *Compos. Part A Appl. Sci. Manuf.*, vol. 83, pp. 98–112, 2016, doi: 10.1016/j.compositesa.2015.08.038.

[7]   J. Thomason, L. Yang, and F. Gentles, "Characterisation of the anisotropic thermoelastic properties of natural fibres for composite reinforcement", *Fibers*, vol. 5, no. 4, 2017, doi: 10.3390/fib5040036.

[8]   T. Keller, "Fibre reinforced polymer materials in building construction", in *IABSE Symposium Towards a Better Built Environment*, Melbourne, 2002.

[9]   *IEC 61400-5: Wind energy generation systems – Part 5: Wind turbine blades*. International Electrotechnical Commission, 2020.

CHAPTER 2

COMPOSITES – COMPACT

2.1 INTRODUCTION

Chapter 2 provides a concise state-of-the-art overview of composites in structural engineering and architecture. Following the European Technical Specification, CEN/ TS 19101 [1], the discussed subjects are materials, components, connections and joints, and members. Manufacturing methods are subsequently addressed, and overviews are given about the use of composites in bridge construction and architecture, with "architecture" being primarily related to the architectural and technical design of buildings. The two latter topics are complemented by a discussion on the application of composites in sculpture, which impressively demonstrates the formal and artistic potential of these materials. Concise conclusions concerning the state of the art are drawn in Section 6.2.

Only subject elements that have shown their relevance in structural engineering and architecture are considered; **Table 2.1** provides a corresponding overview. Concerning materials and considering the 50–100 year-long design service lives of buildings and bridges [2], natural fibres and biopolymers are not discussed in depth

Table 2.1   Considered and excluded subject elements

Subjects	Considered elements	Excluded elements
Materials	Glass, basalt, carbon, aramid, flax fibres	Other natural fibres
	Polyester, vinylester, epoxy thermosets	Phenolic thermosets, thermoplastics, biopolymers
	Epoxy, polyurethane, acrylic adhesives	Film adhesives
	PUR, PET, PVC foams, balsa wood cores	Honeycomb cores
Components and members	Ply, laminate, sandwich core components	Strengthening strips and sheets, concrete rebars
	Cables, profiles, sandwich panel members	Membrane structures
Connections and joints	Adhesively bonded, bolted, hybrid bonded-bolted joints	Snap fit joints
Manufacturing and installation	Hand lay-up, pultrusion, vacuum infusion, prepreg moulding	Filament winding, 3D printing

due to their limited outdoor durability, see Section 3.11.4. Nevertheless, in order to compare the properties of engineered and natural fibres, flax fibres are taken into account. Phenolic resins are only marginally mentioned since they only improve the fire reaction but not fire resistance behaviour. Thermoplastic resins are normally not used in structural engineering, mainly due to limited temperature resistance (softening) and susceptibility to creep, both caused by the missing cross-links between the molecular chains. Adhesives are applied as paste with a certain thickness to join structural members; film adhesives are thus not taken into account. Honeycomb cores are not addressed since they are only rarely used (see Section 2.6.2) and do not normally offer advantages over foam or balsa cores. The additional weight gain that they may provide is not relevant in most cases, and they cannot offer an insulation function in buildings (as foams do) or resist heavy traffic loads on bridges as high-density end-grain balsa can [3].

As already stated in Section 1.3, the monograph does not cover strengthening of existing structures and composite reinforcing bars for concrete. Furthermore, membrane structures are not addressed; information about this type of structure can be found in [4]. Concerning joints, snap fit joints are not discussed due to their limited structural application. The most relevant manufacturing processes for members of bridge and building structures are currently hand lay-up, pultrusion, vacuum infusion, and prepreg moulding; the much less frequently applied filament winding is not addressed. 3D printing is also not further discussed, since interlaminar fracture resistance (which often limits the design of composite members) is significantly lower than in members manufactured by current processes (see Section 3.7.1). Until this fact changes and significant volume fractions of bidirectional and continuous fibres can be implemented, 3D printing is not considered relevant to primary structural applications in the foreseeable future.

Only basic knowledge is summarised in Chapter 2. More specific and detailed information about each subject, i.e., concerning limitations, opportunities, and design, is given in Chapters 3 to 5, respectively. The information included in Chapter 2 is linked to Chapters 3 to 5 through Allocation Tables at the end of each section of Chapter 2. In these Allocation Tables, the topics of supplementary information are listed, together with the related section numbers of the respective Chapters 3 to 5, as shown in the schema in **Fig. 2.1**.

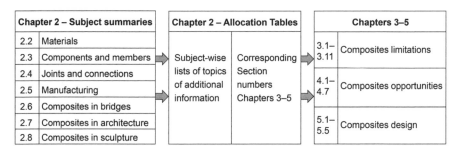

Fig. 2.1 Organisation of information: Chapter 2 compact state-of-the-art summaries by subject (**left**), and Chapter 2 Allocation Tables (**middle**), with topics of supplementary detailed information in respective sections of Chapters 3 to 5 (**right**)

2.2 MATERIALS

2.2.1 Fibres and resins

This organisation of information allows for, on the one hand, a compact and subject-wise state-of-the-art overview in Chapter 2 and, on the other, a discussion of relevant limitations, opportunities, and design aspects in Chapters 3 to 5, in detail and across the subjects of Chapter 2. The advantages of this organisation can be demonstrated by means of ductility, as an example, which is discussed in Chapter 3 (Section 3.6) across subjects, i.e., materials, components, joints, members, and structures, thus showing how ductility deficiency at the material level can be compensated for by system ductility at the member or structural levels, including pseudo-ductile joints.

2.2 MATERIALS

2.2.1 Fibres and resins

Relevant fibres in composite engineering structures are currently E-glass and carbon fibres, and to an increasing and decreasing extent basalt and aramid fibres, respectively. While carbon fibres exhibit the highest strength and stiffness properties, basalt fibres slightly outperform E-glass fibres at a lower level, see Table 1.2. Basalt fibres have other advantages over glass fibres, however, such as (i) a significantly higher creep rupture strength (Fig. 3.26, right), (ii) a better resistance against an alkaline environment (e.g., if embedded as rebars in concrete), (iii) less sensitivity to moisture, and (iv) less energy demand (Table 3.12). Aramid fibres are synthetic fibres and were used as cables in some pedestrian bridges in the early 1990s. However, they exhibit low compression properties, low creep rupture strength, and high sensitivity to elevated temperatures, and thus were not further used in structural engineering [5]. An exception are thermal break components, as described in Section 4.5.2 (Fig. 4.52), where aramid fibres were initially selected due to their very low thermal conductivity (see Tables 1.2 and 4.6). With the appearance of basalt fibres in the interim, which have similar low thermal conductivity values but exhibit a better mechanical performance, aramid fibres no longer offer such advantages in structural engineering. The mechanical properties of natural fibres (e.g., flax fibres) are inferior to those of the above-mentioned fibres (Table 1.2); the thermal conductivity is, however, in the same range as in the cases of aramid and basalt fibres. The coefficients of thermal expansion of all these fibres are discussed in Section 1.2.

While glass and basalt fibres are isotropic materials, carbon and natural fibres are strongly anisotropic [1, 6]. Graphitic planes with strong bonds are arranged in the microstructure of carbon fibres, in parallel to the fibre direction, providing excellent axial tension properties. In the transverse direction, however, these planes are held together only by van der Waals bonds and can thus slide with respect to each other. Accordingly, the transverse and interlaminar shear properties of carbon fibres are relatively low [1, 7]. Natural fibres, on the other hand, exhibit a similar anisotropy to wood, since they are composed of the same basic materials, i.e., cellulose microfibrils aligned more or less in the fibre direction, which are embedded into a hemicellulose and lignin matrix [8, 9].

A thin coating is normally applied to fibres, denominated "sizing", in order to protect them during processing and to ensure the fibre-matrix bond strength. Since individual fibres cannot be handled, they are normally bundled into rovings or processed into continuous filament mats (CFM) or randomly oriented chopped strand mats (CSM). In order to create a bidirectional load-bearing behaviour, woven or non-woven fabrics are manufactured from rovings, as discussed in Section 3.2.2. In some cases, mats and fabrics are stitched together. Fibre mats do not markedly improve the mechanical laminate properties, but they can, however, significantly increase fracture resistance through fibre bridging, as discussed in Section 3.7.

Polymers are composed of longer molecular chains of repetitive groups of atoms. Two basic types of polymers can be distinguished, i.e., thermoset and thermoplastic polymers. The main difference between them is that in thermosets the molecular chains are cross-linked by primary (covalent) bonds, while in thermoplastics only much weaker secondary (van der Waals and hydrogen) bonds exist between the chains, whose intensity, furthermore, decreases as the distance between the chains increases. Thermosets are thus thermally more stable but decompose at high temperatures, while thermoplastics show a comparably early softening and subsequently melt. Cross-linking generally strengthens the elastic over the viscous component of the viscoelastic response and thus also decreases the time dependence, i.e., the sensitivity to creep, for instance, as further discussed in Section 3.4. Due to these reasons, thermoset polymers are preferred in structural engineering in most cases. A specific case to mention are thermoset elastomers, such as acrylic adhesives, for instance, which have a low cross-link density and are often in the rubbery state at ambient temperature, see Sections 2.2.3 and 3.4.

The most commonly used thermoset resins in composite structures are unsaturated polyester, vinylester, and epoxy. Phenolic resins may be used if the fire reaction behaviour is of importance, see Section 3.9.1. Typical mechanical and physical properties of the first three resins are listed in **Table 2.2**; further properties are given in [1]. As can be seen, the mechanical properties of the different resins do not differ significantly. They influence the mechanical properties of composites mainly in matrix-dominated but much less in fibre-dominated cases, see Section 3.2.3. The glass transition temperatures listed in Table 2.2 depend significantly on the curing degree and moisture content, see Section 3.4.3.

Table 2.2 Indicative values for selected mechanical and physical properties of thermoset resins, based on [1, 10]

Property	Polyester	Vinylester	Epoxy
Elastic modulus [GPa]	3.4–4.0	3.3–3.5	3.0–4.0
Tensile strength [MPa]	40–90	75–82	35–130
Density [kg/m^3]	1 200	1 150	1 200
Glass transition temperature [°C]	40–110	40–120	40–300
Coefficient of linear thermal expansion $\times 10^{-6}$[K^{-1}]	30–200	30–50	50–110
Thermal conductivity [W/mK]	0.20	0.18–0.20	0.10–0.20

2. Composites – Compact   35

(Allocation) Table 2.3 Supplementary information about fibres and resins in subsequent sections of Chapters 3 to 5

Section	Aspect	Subject of additional information
3.2.2.1	Anisotropy	Dependence of mechanical properties on fibre architecture (Fig. 3.1)
3.4	Viscoelasticity	Viscoelasticity of polymers (resins)
3.4.5.2		Creep rupture of glass, basalt, and carbon fibres (Fig. 3.26, right)
3.4.5.4		Relaxation of carbon fibre composite cables
3.5	Curing and physical ageing	Curing and physical ageing of polymers (resins)
3.7.1	Fracture	Fracture energies vs tensile strengths of glass, polyester, and epoxy (Table 3.4)
3.7.2.1		Fibre bridging caused by chopped strand mats (CSM)
3.7.3.3		Fibre bridging caused by continuous filament mats (CFM)
3.9.2.5/7	Fire	Elastic modulus and strength variations of glass fibres at elevated and high temperatures (Fig. 3.111)
4.2.2	Learnings from nature	Stress vs strain responses of glass and flax fibres (Fig. 4.2)
4.5.2.2	Function integration	Thermal conductivity of glass, basalt, carbon and aramid fibres, and polyester resin (Table 4.6)
5.2.2	Design	Material selection

Resins play a decisive role in ensuring the durability of composites, as they protect the fibres from environmental degradation, see Section 1.2; their intrinsic durability is thus of great importance. All three resins generally exhibit good resistance to moisture if appropriately cured. The best and lowest durability is normally exhibited by epoxy and unsaturated polyester, respectively, while the level of durability of vinylester lies in between. Concerning UV radiation, unsaturated polyester has good resistance, while vinylester and epoxy require a protective coating in most cases [1, 10].

More specific aspects of fibres and polymers are discussed in the sections listed in Allocation **Table 2.3**.

2.2.2 Sandwich core materials

2.2.2.1 Balsa wood

Balsa wood is a natural fibre-polymer composite. This hardwood exhibits a cellular structure, which is essentially composed of three types of cells: (i) tracheids (80–90% of the volume) and (ii) vessels, both in the longitudinal (or trunk axis or grain) direction, and (iii) rays, in the transverse (radial) direction, as shown in **Fig. 2.2** (for tracheids and rays). The tracheids exhibit multilayered cell walls composed of polymeric cellulose

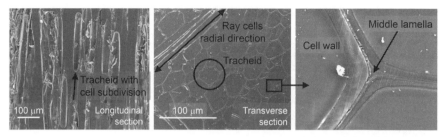

Fig. 2.2 Microstructure of balsa wood: longitudinal section (along grain) with tracheid cells and their subdivisions **(left)**; transverse section with tracheid cross sections and radial ray cells **(middle)**; detail showing tracheid cell walls with middle lamella **(right)**

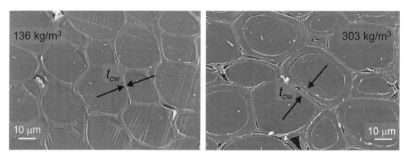

Fig. 2.3 Differences in cell wall thickness (t_{cw}) of low density (136 kg/m^3) **(left)** and high density (303 kg/m^3) balsa wood **(right)**

fibres (40–50% of the volume), embedded in a polymeric matrix of hemicellulose and lignin. Depending primarily on cell wall thickness, the density of the wood varies, see **Fig. 2.3**. The cellulose fibres are mostly crystalline and provide strength and stiffness to the material. Since they are mainly oriented in the grain direction, balsa wood exhibits orthotropic material properties. The cellulose fibres are almost insensitive to moisture, while the stiffness of the hemicellulose is strongly affected by moisture uptake and associated plasticisation. The latter reduces the composite action between the different materials and thus the overall strength and stiffness. More details about the microstructure of balsa wood and sensitivity to moisture are provided in [11].

In the context of engineering structures, balsa wood is normally used as a core material in composite sandwich panels, e.g., in the case of bridge decks or wind turbine blades. For this purpose, balsa trunks are cut into small blocks, which are then assembled and adhesively bonded together into end-grain panels, as shown in **Fig. 2.4**; the grain direction is thus normal to the face sheets. The pattern of the bond lines is usually continuous in one direction, while it is staggered in the other direction. In addition to the material orthotropy (mentioned above), such core panels are thus inhomogeneous, i.e., they are characterised by (i) a significant variation in block densities, (ii) the presence of numerous adhesive bond lines, and (iii) imperfections such as air gaps between blocks, caused by tolerances in the cutting and assembly process, see Fig. 2.4.

2. Composites – Compact

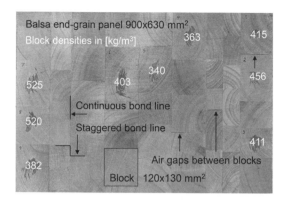

Fig. 2.4 Top view of balsa wood end-grain panel, nominal density according to manufacturer 285 kg/m^3, composed of blocks of varying densities and bonded together with polyurethane (PU) adhesive, continuous and staggered bond lines, air gaps between blocks, according to [12] (complemented)

More specific information about material properties and the application of balsa wood in sandwich panel cores can be found in the following sections, as indicated in Allocation **Table 2.4** (also see following Table 2.5).

(Allocation) Table 2.4 Supplementary information about balsa wood in subsequent sections of Chapters 3 to 5

Section	Aspect	Subject of additional information
3.2.2.2	Anisotropy	Reduction of material scatter by processing trunk material into laminated veneer lumber (LVL)
3.4.5.2	Viscoelasticity	Creep behaviour of balsa sandwich panel cores (Fig. 3.26, left)
3.7.3.3	Fracture	Mode-I static and fatigue debonding of composite face sheets from balsa sandwich panel cores
3.8.8	Fatigue	Fatigue-tailored detailing of balsa cores and advantages regarding fatigue of homogenous core vs web-core bridge decks
3.9.2	Fire	Temperature-dependent thermophysical and thermomechanical properties of balsa wood
3.9.2.5		Dynamic Mechanical Analysis (DMA) results of polyurethane adhesive used in balsa wood panels (Fig. 3.113)
3.9.3.3		Fire resistance behaviour of composite sandwich panels with balsa core
4.5.2.2	Function integration	Thermal conductivity values of balsa wood at ambient temperature (Table 4.6)
4.6.2	Hybrid construction	Complex core compositions to extend spans of hybrid bridges (Fig. 4.61)
4.6.3		In-plane composite action in composite balsa/timber sandwich bridge decks (Fig. 4.65)
5.2.2	Design	Material selection

2.2.2.2 Polymeric foams and comparisons to balsa

Polymeric structural foams are manufactured from (i) liquid polymeric components, as listed in **Table 2.5** for commonly used foams (PUR, PET, PVC), and (ii) blowing agents for the foaming process. All these foams have closed cells and exhibit anisotropic mechanical behaviour, which is primarily caused by the manufacturing process. In the case of PUR foam, for instance, the cells may stretch in the rise direction and thus create geometrical and mechanical anisotropy, see Section 3.2.2 [13]. Thermoset foams (PUR, PVC) can be shaped by machining, while thermoforming can be used for thermoplastic PET. The latter foams are also recyclable and waste parts resulting from manufacturing can be reused.

Table 2.5 Polymeric foams and balsa wood sandwich core materials, core types, application fields, and corresponding relevant properties (indicative values in rise, extrusion, or grain direction for selected densities), derived from [1, 14]

Property	PUR	PET	PVC	Balsa	
Composition[1]	Polyurethane from polyol and isocyanate	Polyethylene terephthalate	Cross-linked polyurea and polyvinyl chloride	Cellulose fibres, lignin, and hemicellulose matrix	
Polymer type	Thermoset	Thermoplastic	Thermoset	N/A (mixture)	
Density [kg/m^3]	100	140	250	100	250
Core type		Flexible			Rigid
Application		Buildings and pedestrian bridges			Vehicular bridges
Out-of-plane shear modulus [MPa]	9–11	29–30	81–97	200–220	270–350
Out-of-plane shear strength [MPa]	0.3–0.5	1.1–1.2	3.9–4.5	1.4–1.6	3.4–4.0
Out-of-plane compression modulus [MPa]	29–30	110–140	350–400	1 600–2 200	4 300–7 500
Out-of-plane compressive strength [MPa]	0.6–1.0	2.2–2.4	6.1–7.2	5.0–6.5	20–26
In-plane compressive strength [MPa]					220–230
In-plane tensile strength [MPa]		Irrelevant in most cases due to flexible core			0.7[2]–1.4[3]
Thermal conductivity [W/mK]	0.033	0.037	0.049	0.050	0.085

[1]   Excluding blowing agents
[2]   LVL balsa used in Avançon Bridge, characteristic value [15]
[3]   Mean value measured on panel from batch shown in Fig. 2.4 [16]

Depending on stiffness and geometrical properties, cores of sandwich panels exhibit different mechanical behaviours and are denominated accordingly [1]. Foam and lower density balsa cores normally only transmit shear forces and provide shear resistance; they are designated "flexible", see Table 2.5. On the other hand, "rigid" cores, such as high-density balsa cores in most cases, also take bending moments and contribute to the bending resistance. In this regard, confusion exists in the specification of PUR foams, where the terms "rigid" and "flexible" are used differently and thus have double significance. PUR foams are designated "rigid" if they exhibit closed cells and are highly cross-linked and appropriate for structural applications, while "flexible" PUR foams have open cells and inferior mechanical properties [13]. Structural PUR foams are thus normally "rigid" from their intrinsic constitution and "flexible" in their response as sandwich core.

Mechanical properties relevant for sandwich core applications of selected PUR, PET, and PVC foams are summarised and compared to end-grain balsa wood in Table 2.5; additional values for other properties and densities can be found in [1]. The values show that the mechanical properties generally increase with increasing density. The foam properties are normally inferior to balsa properties of the same density, except for the PVC out-of-plane shear strength, which is similar at the highest density. The most significant difference is in the out-of-plane compressive strength of high-density foams and balsa, which is significantly higher in balsa. These higher values enable the application of balsa in vehicular bridge decks, which need to withstand high tyre contact pressure from trucks. The applicability of foam and lower density balsa cores is thus typically limited to pedestrian bridges or building construction. On the other hand, the low in-plane (transverse-to-grain) tensile strength of high-density balsa, if acting as a rigid core and being thus subjected to bending, may represent a critical property in the core design (as was experienced in the design of the Avançon bridge, Appendix A.5). A corresponding potential formation of through-thickness core cracks may also lower the fire resistance, as discussed in Section 3.9.3.

Core material properties are normally more sensitive to temperature than composite face sheet materials. Foam cores, in particular, soften significantly below the range of glass transition temperatures (T_g) of composite face sheets, as shown in **Fig. 2.5** for the shear modulus. At service temperatures of 40–50°C, foam cores may already lose up to 30% of stiffness, while the reduction is much less for balsa wood. Such stiffness reductions can (among other things) affect the wrinkling resistance of the face sheets.

The mechanical properties of foam and balsa core materials are sensitive to moisture to varying degrees. Closed cell foam configurations usually hamper moisture ingress, while the ingress into balsa, in grain direction, is less hindered. However, since the face sheet-to-core interface strength may also be affected, the diffusion of moisture through the face sheets into the core should generally be prevented by adequate selection of the material and face sheet thickness, and appropriate constructive details [1].

Polymeric foams exhibit a very low thermal conductivity, which is also lower than balsa values, as shown in Table 2.5 (and Table 4.6). Foams are thus the preferred core materials in multifunctional applications such as building envelopes, where thermal insulation and structural functions can be merged, see Section 4.5.

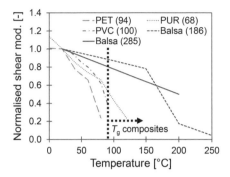

Fig. 2.5 Out-of-plane shear modulus (normalised by 20°C value) vs temperature for core materials: PET (94 kg/m^3) and PUR (68 kg/m^3) foam [17], PVC (100 kg/m^3) foam [1], balsa (186 kg/m^3) [18], balsa (285 kg/m^3) [16], and comparison to T_g range of composite laminates

(Allocation) Table 2.6 Supplementary information about polymeric foams in subsequent sections of Chapters 3 to 5

Section	Aspect	Subject of additional information
3.2.2.2	Anisotropy	Anisotropy of PUR, PET, PVC foams in mechanical properties and creep (PUR)
3.4.5.2	Viscoelasticity	Creep of PUR foam in web-core sandwich panel
3.9.3.2	Fire	Fire resistance data of composite-PUR core sandwich beams (Table 3.7)
4.3.2	Structural form	CNC milling of PUR foam blocks (Fig. 4.15)
4.4.4	Transparency	Composite skylights with thermally insulting foam (Fig. 4.44)
4.5.2.2	Function integration	Thermal conductivity values of PUR foam at ambient temperature (Table 4.6)
4.5.2.3		Thermal insulation performance of composite-PUR sandwich building envelopes
4.5.3.2		PUR foam in a composite hydronic radiant system (Fig. 4.54)
4.5.4.2		Solar cells encapsulated in composite-PUR core sandwich building envelopes
5.2.2	Design	Material selection

More specific information about polymeric foams and their use as cores of sandwich panels is provided in the following sections, as listed in Allocation **Table 2.6**.

2.2.3 Adhesives

Different types of thermoset adhesives exist which exhibit, at ambient temperature and moderate strain rate, responses that span from brittle (showing small deformations) to pseudo-ductile (at much larger deformations, see Section 3.6.2). As this categorisation

2. Composites – Compact   41

implies, the mechanical responses are strain rate- (i.e., time-) dependent, and furthermore they are temperature-dependent, as explained in Section 3.4.4. The physical and mechanical properties also depend on the curing degree and moisture content, see Section 3.5.3.

Typical stress vs strain responses at ambient temperature and moderate strain rate of a brittle epoxy-based (EP) adhesive and two pseudo-ductile adhesives, i.e., a polyurethane- (PU) and an acrylic-based (AC) adhesive, are compared in **Fig. 2.6**. The former two adhesives are thermosets, while the latter is a thermoset-elastomer. These adhesives have been used throughout the research work at the CCLab since 1999 (with few exceptions); references to adhesives in Chapters 3 and 4 generally refer to these adhesives, which were all provided by Sika, Switzerland. The adhesives are two-component adhesives, the EP adhesive is SikaDur-330, the PU adhesive is SikaForce-7851 (now discontinued), and the AC adhesive is SikaFast 5221 (and its successor since 2022, SikaFast-555L10).

True stress vs strain tension, compression and shear responses, and compression failure modes are shown in Fig. 2.6; selected values under tension are furthermore provided in **Table 2.7** (concerning the terms "true stress" and "true strain", see Section 3.4.4). The results were obtained according to EN ISO 527-2, ASTM D 695-96,

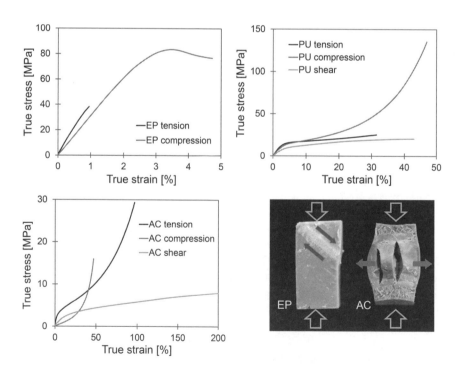

Fig. 2.6 Average true stress vs strain responses of fully cured epoxy (EP, **top left**); polyurethane (PU, **top right**); and acrylic (AC, **bottom left**) adhesives; and compression failure modes of quasi-brittle epoxy and pseudo-ductile acrylic adhesives (specimen sizes 12.7×12.7×25.4 mm) (**bottom right**), strain rate 5 mm/min in all cases, based on data in [19]

Table 2.7 Indicative values for true tension properties of different fully cured adhesives at ambient temperature and strain rate 5 mm/min, derived from bilinear modelling, average values from 5 specimens, based on [19]

Property	Epoxy	Polyurethane	Acrylic
Elastic modulus [MPa]	4 560	590	100
Stretching modulus [MPa]	N/A (brittle)	31	14
Strength [MPa]	39	25	29
Strain at failure [%]	1	32	97
Poisson's ratio [-]	0.37	0.42	0.40

and ISO 11003-1, respectively; the latter torsion method did not provide conclusive results for the EP adhesive due to small shear deformations. Further details and results can be found in [19].

Subjected to tension, the EP adhesive exhibits an almost linear response and brittle failure, while the PU and AC adhesives show a pronounced nonlinear behaviour, which can be classified as pseudo-ductile, according to Section 3.6.2. The elastic moduli differ greatly by a factor of almost 50 between the EP and AC adhesive. In the PU and AC cases, the viscous component of the viscoelastic response subsequently dominates, and the molecular chains start to be stretched; a corresponding stretching modulus can be derived from the slope of the curves, as also listed in Table 2.7. In the AC adhesive, this stretching develops much more, during an almost threefold strain up to failure compared to the PU, and a significant hardening is thus exhibited through the alignment of the molecular chains to the stress direction (see in-depth analysis in Section 3.4.2).

Under compression, the EP adhesive shows a quasi-brittle response, i.e., the failure mode exhibits the typical inclined shear failure plane of brittle materials, as shown in Fig. 2.6 (bottom right). The responses of the PU and AC adhesives are again pseudo-ductile due to a progressive crushing. The AC failure mode exhibits cracks developing parallel to the applied stress direction, caused by the crushing and associated transverse tension. PU specimens do not show these longitudinal cracks. The shear responses of the PU and AC adhesives are almost bilinear and again pseudo-ductile.

As already stated above, the adhesive responses are time- and temperature-dependent. The brittle response of EP described above, for instance, changes to pseudo-ductile at elevated temperatures, as demonstrated in Section 3.4.4 (Fig. 3.21, right). Further information, regarding fatigue and fire behaviours among others, can be found in Chapter 3, as indicated in Allocation **Table 2.8**.

2.3   COMPONENTS AND MEMBERS

2.3.1   Components

Following the classification of the European Technical Specification CEN/TS 19101 [1], composite components are constituents of members, such as (i) plies of laminates,

2. Composites – Compact   43

(Allocation) Table 2.8   Supplementary information about adhesives in subsequent sections of Chapters 3 to 5

Section	Aspect	Subject of additional information
3.4.2	Viscoelasticity	Strain hardening through molecular chain stretching in acrylic/elastomer adhesive
3.4.3		DMA results for epoxy/thermoset and acrylic/elastomer adhesives (Fig. 3.18)
3.4.4		Time dependence of acrylic and temperature dependence of epoxy adhesive stress vs strain responses (Fig. 3.21)
3.4.6		Effect of moisture on stress vs strain responses of epoxy adhesive (Fig. 3.29)
3.5.2	Curing and physical ageing	Differential Scanning Calorimetry (DSC) results of epoxy adhesive, curing degree, physical ageing, effect of moisture
3.5.3		Early-age development of physical and mechanical properties, effects of low curing temperature and moisture, for epoxy adhesives
3.5.4		Long-term development of physical and mechanical properties, epoxy adhesive
3.6.2	Ductility	Brittle and pseudo-ductile adhesive responses
3.8.2	Fatigue	Fatigue damage rates (slope coefficients) of acrylic and epoxy adhesives
3.8.6		Effects of stress (R-) ratio and temperature on fatigue life of epoxy adhesive (Fig. 3.94)
3.9.2.5	Fire	DMA results of polyurethane adhesive used in balsa wood panels (Fig. 3.113)
5.2.2	Design	Material selection

(ii) laminates as parts of profiles and face sheets of sandwich panels, and (iii) cores of sandwich panels.

A ply is a single layer (or lamina) of a laminate with a certain number of individual layers of fibres. Depending on the composition of these fibre layers, a ply can be uni-, bi-, or multidirectional, or quasi-isotropic. Micromechanics formulae for the determination of (i) ply stiffness properties, (ii) coefficients of linear thermal expansion, (iii) thermal conductivities, and (iv) failure criteria, as well as indicative values of such ply properties, can be found in [1, 10].

A laminate is a thin, flat, or curved composite component of members, with in-plane dimensions considerably larger than the through-thickness dimension. It is composed of individual plies, and – based on the ply composition and stacking sequence – balanced, symmetric, angle-ply and cross-ply laminates are ultimately differentiated. A balanced laminate is composed of pairs of rotated plies of the same composition, but opposite signs of the rotation angles, see **Fig. 2.7**. In a symmetric laminate, each ply is mirrored about the laminate's mid-plane. In symmetric balanced laminates under in-plane stresses, the axial, transverse, and in-plane shear strains are uniformly distributed through the thickness. Effects such as warping are thus prevented, and the laminate can be modelled as a single layer with orthotropic elastic

90	90	90	90
+45	+45	+45	+45
-45	+45	+45	-45
0	0	0	0
0	0	0	0
+45	+45	-45	-45
+45	+45	-45	+45
90	90	90	90
Unbalanced Asymmetric	Unbalanced Symmetric	Balanced Asymmetric	Balanced Symmetric

Fig. 2.7 Examples of un/balanced and a/symmetric laminates (numbers indicate angles of fibre direction) [14]

properties. It is generally recommended to conceive laminates with symmetric and balanced lay-up. Cross-ply laminates comprise plies stacked at $0°$ and $90°$, while other angles are used in angle-ply laminates. Cross-ply laminate responses are normally fibre-dominated while those of angle-ply laminates are matrix-dominated (see Section 3.2.3) [1].

The layered structure of laminates has the advantage that the fibre architecture can be optimised for specific in-plane stress cases. On the other hand, lacking through-thickness fibres make laminates very sensitive to out-of-plane tensile stresses, thus lowering the interlaminar shear strength considerably, which is the reason why delamination between plies often governs the design, see Section 3.7.1.

Stiffness and strength properties of laminates can be estimated based on the ply properties, the Classical Lamination Theory (CLT), and appropriate failure criteria, as described in [1, 10]. Indicative values of properties for symmetric balanced laminates, usable for preliminary design, are also provided in [1].

Cores of sandwich panels can either be composed of individual webs in web-core panels or be homogeneous and consist of polymeric foam or end-grain balsa materials, see the previous Section 2.2.2. Furthermore, sandwich cores can be conceived as "rigid" or "flexible", as also explained in Section 2.2.2. Cells between webs in web-core panels may also be filled. Such infills can have three functions: (i) to stabilise compressed face sheet and web laminates against buckling and provide higher resistances by enforcing a wrinkling failure mode, (ii) contribute to the shear and transverse compression resistance, and (iii) provide thermal insulation. Infills with structural function are nonetheless normally only justifiable in thicker sandwich panel members, where the web spacing is larger, as is the case in the Novartis Building roof, for instance, with a thickness up to 620 mm (Appendix A.4). Foam of sufficient density can fulfil both structural functions in such cases, as shown in Section 3.4.5, in addition to thermal insulation. In thinner panels, where the web spacing is narrower, infills to stabilise components and contribute to shear resistance are normally less appropriate, since the more numerous webs can easily be made slightly thicker, thus preventing the need for complicated manufacturing to include infills. A fourth more specific function of core infills is the anchorage of equipment elements in sandwich panels, such as inserts, or guardrails and crash barriers in pultruded bridge decks. In the latter cases, infills of higher strength materials are required, e.g., local concrete infills [1]. The anchoring of crash barrier posts in a composite-balsa sandwich bridge deck with homogeneous core is discussed in Section 2.6.2.

2.3.2 One-dimensional members

One-dimensional members used in composite structures are basically cables and profiles. Cables can be used as (i) hangers and main cables of suspended and cable-stayed structures, (ii) members of bracing structures, and (iii) for the anchorage of supports in the ground. Profiles can be implemented as (i) beams and columns, and both combined in frames, and (ii) bars in truss and grid-shell structures.

Cables are flexible and can therefore only transmit tension forces. Since they are often prestressed and thus subjected to significant permanent stress so they do not become slack, preferred materials are carbon or basalt fibres, and, to a much lesser extent, glass fibres, due to the high, medium, and low creep rupture resistance, respectively (see Fig. 3.26). Designing cable anchors can be a challenge, as their tension resistance should be as high as that of the cable cross section in order to be able to fully and thus economically use the material. The solution to this problem is more complex in the case of carbon fibre cables, due to the anisotropy of carbon fibres, as discussed in Section 3.2.3, and even more so if the cables are also subjected to fatigue [20].

Profiles are stiff and can transmit – more or less efficiently, depending on the shape of the cross section and fibre architecture – axial and shear forces, and bending and torsion moments. Closed (hollow) cross sections are often preferred over open sections with outstand flanges since closed sections (i) are less sensitive to local impact (see Sections 3.3.2 and 3.10.2), (ii) are less sensitive to local flange buckling, see **Fig. 2.8**, (iii) exhibit a higher fire resistance (Section 3.9.3), and (iv) have higher stiffness and strength under torsion.

In contrast to isotropic steel profiles, composite profiles are anisotropic, i.e., orthotropic in most cases (see Section 1.2, Fig. 1.1, left), and the transmission efficiency of internal forces thus depends on the fibre architecture. The transmission is efficient, i.e., deformations and sensitivity to time and temperature are relatively small, if the

Fig. 2.8 Outstand top flange of pultruded profile subjected to axial compression (through bending of whole profile): initial flange buckling, according to [21] **(left)**; subsequent delamination (at web-flange junction) caused by out-of-plane tensile stresses **(right)**

Fig. 2.9 Translucent glass fibre profile with complex cross section, 125 mm max. width **(left)**; large-scale hybrid glass and carbon fibre profile, 914 mm height **(right)**

behaviour is fibre-dominated (see Section 3.2.3). Pultruded profiles, which exhibit a (unidirectional) UD-dominated fibre architecture in most cases, are thus efficient in transmitting axial forces and bending. However, they show a matrix-dominated response under shear and torsion, which is associated with larger deformations and sensitivity to creep and elevated temperature. To exhibit a fibre-dominated response under shear and torsion would require a significant ±45° fibre contribution, which is, however, difficult to implement in pultrusion, see Section 2.5.2. A more appropriate process to produce a ±45° fibre architecture is filament-winding. However, the inclusion of a significant UD fibre volume fraction is difficult in that case. As this discussion shows, a close interdependence exists between structural efficiency, fibre architecture, and manufacturing process. The situation is further complicated if considering the joints between profiles. They ideally should exhibit an architecture different from that of the material between the joints, which is difficult to realise in continuous processes, also see Section 2.5.2.

Profiles from small to large scale and cross sections with complex shapes can be manufactured by pultrusion, the latter due to the flexibility of the die configuration, see Section 2.5.2. Furthermore, profiles can be conceived as translucent (see Section 4.4) and hybrid, i.e., carbon fibres can be added to glass fibres in the flanges to optimise the structural behaviour, see **Fig. 2.9**. Complex profiles can also be built up from two or more standardised profiles, preferably by adhesive bonding, as was done to achieve the built-up profiles of the Eyecatcher Building, see Section 3.10.2 and Appendix A.3.

Recently, curved pultruded profiles have also been appearing on the market. Profiles which are not yet fully cured when leaving the forming and curing die (see Section 2.5.2) are subsequently mechanically bent into a circular shape in a combined pulling and curving system [22]. To what extent the fibre-resin interfaces and the molecular structure of the resin are damaged by deforming the material in the last stage of cure is not, however, clear.

2.3.3 Two-dimensional members

Two-dimensional members used in composite structures are basically flat slabs and (curved) shells (if excluding membranes, see Section 1.3). Flat slabs can be conceived from web-core or homogeneous core sandwich panels, while only the latter are used for shells, since web-core panels cannot normally be curved (during manufacturing). Two-dimensional members can also be used as walls (subjected to axial compression), but such applications should be thoroughly evaluated, as discussed in Section 3.9.5.

Two basic types of web-core panels exist, whose differences mainly result from the manufacturing process, i.e., either pultrusion or vacuum infusion. In the first case, pultruded profiles are adhesively bonded together along their longitudinal edges to form the panel, as shown in **Fig. 2.10** (and Section 4.6.2, Fig. 4.63); the cells are thus empty. The cell shape of the pultruded profiles has a significant effect on the transverse load-bearing mechanism, stiffness, and ductility, as discussed in Section 3.6.3 (*Example-1*). In another configuration, closed profiles are arranged to build the core and are bonded to flat profiles, which form the face panels [23]. In the vacuum-infused case, the cells consist of foam blocks, around which the fibres are wound to form, after infusion, face panels and webs. The foam infill in the cells is thus mainly related to manufacturing and less to structural reasons.

While the structural response of pultruded panels is strongly anisotropic due to the different load-bearing mechanisms in longitudinal and transverse directions, the infusion process enables the implementation of transverse webs, which can partially counterbalance, but normally not eliminate, the anisotropy. The infusion process also allows optimisation of the fibre architecture of the critical web-face panel junctions to increase the resistance against delamination in the transverse panel direction, as shown in **Fig. 2.11**. The architecture of pultruded panels, at this location, is often dominated by imperfections, which may cause detrimental effects on strength and fatigue resistance, as discussed in Sections 2.5.2 (reasons for imperfections), 3.3.2 (effect on strength), and 3.8.8 (effect on fatigue resistance). Further differences concern the height of the panels, which is given by the die and thus constant in the pultruded case

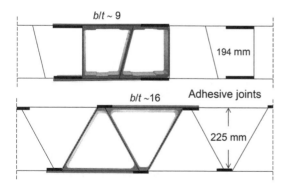

Fig. 2.10 Adhesively bonded pultruded profiles: with double-trapezoidal cell shape **(top)** and double-triangular cell shape **(bottom)**, both with outstand flanges for adhesive connections to form web-core sandwich panels, constant panel height, medium to high web spacing to face panel thickness ratio (b/t) (profiles at the same scale) [14]

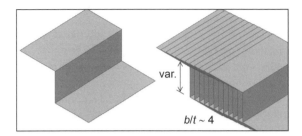

Fig. 2.11 Overlapping z-shaped fibre architecture of web-core panel made by vacuum infusion to increase resistance against web-face panel delamination, variable panel height, low web spacing to face panel thickness ratio (b/t)

(see Section 2.5); the span of underlying girders must be adapted accordingly. The height of infused panels can, however, be varied and thus adapted to the span. An important parameter regarding fatigue resistance, as further discussed in Section 3.8.8, is the ratio of the web spacing to the (upper) face panel thickness (b/t), which is higher in pultruded panels than in infused panels, where it is easier to reduce this ratio. A final difference concerns the fire resistance, which can be significant in panels with empty cells due to the insulation capacity of the air in the cells, as discussed in Section 3.9.3. On the other hand, the foam infill can result in high thermal gradients and cause face panel delamination, as emphasised in Section 2.6.2.

Sandwich panels with homogeneous core, consisting of foam or end-grain balsa in most cases, can be conceived fully bidirectional, if the fibre architecture of the face sheets is designed accordingly and the core material exhibits in-plane isotropy. Complex shapes can be realised if using a vacuum infusion process, as shown in Sections 2.5.3 and 2.7.4, and complex core assemblies can increase the span and composite action in hybrid girders, see Section 4.6.2 (Fig. 4.61). Depending on the thermal conductivity of the core material, a structural panel can also provide thermal insulation, as further elaborated in Section 4.5.2. The fracture and fatigue resistances of homogeneous sandwich panels with end-grain balsa core are characterised by small stress concentrations and a good face sheet-core interface strength, see Section 3.8.8. The fire resistance depends on the core material. Sandwich panels with balsa core typically exhibit better fire resistance than panels with foam core, since the wooden char has a significant insulation capacity. Such panels can reach fire resistances comparable to those of web-core panels with empty cells, see Section 3.9.3.

Further types of web-core or homogeneous core sandwich panels are shown in **Fig. 2.12**. In contrast to the previously discussed panels, which can also be applied for vehicular loads, these panels are conceived and limited to pedestrian loads. The web-core panels are composed of pultruded profiles and assembled to panels by adhesive bonding. In panels with the lowest load-bearing capacity, the bottom face panel is interrupted, i.e., replaced by narrow flanges below the webs. Also shown is a homogeneous core panel with a timber core and integrated water drainage (Fig. 2.12, bottom left); the panel is manufactured by vacuum infusion. The surface of the top face panels/sheets can be sanded for the direct use in pedestrian bridges, as shown in Fig. 2.12 (top and bottom left).

2. Composites – Compact

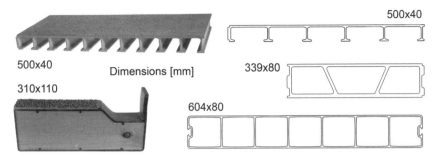

Fig. 2.12 Web-core and homogeneous (spruce) core panel components conceived for pedestrian loads, integrated sanding for non-slip surface **(top and bottom left)**, and additional dewatering channel **(bottom left)** (identical scale, dimensions in [mm])

Further information about components and members can be found in subsequent sections, according to Allocation **Table 2.9**.

(Allocation) Table 2.9 Supplementary information about components and members in subsequent sections of Chapters 3 to 5

Section	Aspect	Subject of additional information
3.2.2.1	Anisotropy	Dependence of mechanical laminate properties from fibre architecture (Fig. 3.1)
3.2.2.2		Orthotropy of balsa wood sandwich panel cores
3.2.3		Carbon fibre ground and rock anchors and flexible carbon fibre strap components
3.3.2	Imperfections and tolerances	Effect of defects on strain distribution in web-flange junctions (Fig. 3.11)
		Effect of imperfections on strength of laminates under axial compression (Fig. 3.12)
3.4.5.2	Viscoelasticity	Shear force and resistance redistribution due to creep in PUR-filled web-core sandwich roof (Fig. 3.27)
3.4.5.4		Relaxation in prestressed carbon fibre cables (Fig. 3.28)
3.6.3	Ductility	Effect of geometry and fibre architecture on in-plane shear behaviour and failure mode of pultruded/bonded web-core panels
		Effect of imperfections on failure mode in web-panel junctions
3.7.3	Fracture	Two-dimensional (2D) crack propagation in laminates and 2D face sheet debonding in sandwich panels
3.7.4		Multiple crack development in web-flange junctions

Section	Aspect	Subject of additional information
3.8.2	Fatigue	Tension-tension fatigue failure mode in pultruded laminates (Fig. 3.83)
3.8.4		Creep-fatigue interaction in angle-ply laminates
3.8.5		Effect of load interruptions on fatigue life of angle-ply laminates
3.8.8		Stress concentrations in web-core vs homogeneous core sandwich panels
		Mixed-mode static failure criterion and fatigue crack growth curves for debonding of glass fibre composite face sheet from balsa sandwich core (Fig. 3.100)
3.9.2	Fire	Thermophysical and thermomechanical properties of glass fibre composite laminates
3.9.3		Fire resistance behaviour of pultruded/bonded web-core and balsa-core sandwich panels, the latter also with steel inserts
3.10.2	Durability	Long-term durability of pultruded profiles subjected to different environments
4.3.3	Structural form and system	Anisotropic fibre architecture of pultruded profiles (Fig. 4.17)
4.4	Transparency	Transparency of glass fibre composite laminates
4.5.2.3	Function integration	Thermal transmittance of sandwich panel cores
4.5.2.5		Thermal break components
4.5.3.2		Thermally activated fibre-polymer composite building slab (Fig. 4.54)
4.5.4		Solar cell encapsulation into transparent glass fibre composite laminates
4.6.2	Hybrid construction	Complex core assemblies to increase span and composite action of sandwich slabs (Fig. 4.61)
		Web-core panels acting as top chord of hybrid bridge girders under static and fatigue loads (Fig. 4.63)
5.2.1	Design	Member composition and function integration

2.4 CONNECTIONS AND JOINTS

2.4.1 Overview

Bolted, adhesively bonded, and hybrid bonded/bolted connections and joints are distinguished in this section and discussed according to the aspects listed in **Table 2.10**. The difference between the terms "connection" and "joint" is that, in connections, forces are transmitted in only one plane, while joints are composed of several connections [1]. As an example, a double-lap shear "joint" has two transmission planes and thus is composed of two single-lap shear "connections", with one transmission plane each.

Table 2.10 Comparison of bolted, adhesively bonded, and hybrid bonded/bolted lap shear joints composed of pultruded glass or basalt fibre composite adherends [14]

Aspect	Bolted joint	Bonded joint	Hybrid joint
On-site application	Easy process if drilling and bolt hole sealing are done in factory and tolerances are appropriate	Demanding process due to dependence on quality of surface preparation and environmental conditions	
Deconstruction	Basically detachable, sometimes difficult due to bolt corrosion	Difficult, cutting required, e.g., in joint plane	
Durability	Ensured if bolt holes remain permanently sealed	Ensured if adhesive layer edges remain permanently sealed	
Stiffness [24]	Low due to bolt hole clearance	High with stiff and low with flexible adhesive	
Strength, i.e., joint efficiency [24] [1]	Low	High with flexible and low with stiff adhesive	
Ductility [24] [2]	Pseudo-ductile if pin-bearing failure, brittle for other failure modes	Pseudo-ductile with flexible and brittle with stiff adhesive	Pseudo-ductile with flexible adhesive or stiff adhesive and bolt pin-bearing failure[3], brittle in other cases
Creep [24] [2]	Insensitive	Sensitive with flexible and less sensitive with stiff adhesive	Insensitive due to bolt

[1] Joint efficiency is the ratio of adherend to joint strength.
[2] "Stiff" or "flexible" refer to typical nonlinear or uniform adhesive shear stress distributions along connection length, e.g., for brittle epoxy or pseudo-ductile acrylic adhesives, respectively, both under ambient conditions and moderate strain rate, see Section 2.4.3.
[3] Assuming that stiff adhesive connection resistance is equal or smaller than pin-bearing resistance.

Basically, there are two ways to join two adherends (members): either in a linear or in an orthogonal configuration, as shown in **Fig. 2.13** and **Table 2.11**. In the linear configuration, double-lap, double-strap, step-lap, scarf, or butt joints can be conceived. The first two cases can be adhesively bonded or bolted, while the remaining three cases can only be bonded. If bonded, the load transmission primarily occurs through shear stresses and only normal (axial) stresses in the first two cases and the last case, respectively, while combined shear-normal (axial) stresses are activated in the two intermediate cases. The first two cases furthermore exhibit significant eccentricities in the joint and associated through-thickness stresses. As will be discussed below (Section 2.4.3), minimising the eccentricity in joints is crucial to maintaining high joint resistance. Due to this fact, single-lap joints are not represented in Fig. 2.13, since they exhibit the highest eccentricity and lowest resistance, and their use should thus be avoided. The resistance of butt joints is also low, although there is no eccentricity, but mainly because the joint area is comparably small, and the normal stresses are thus

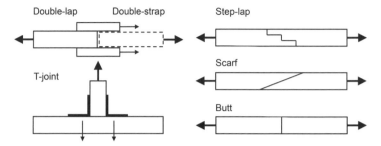

Fig. 2.13 Possible configurations to join two adherends (components or members), based on [25]

Table 2.11 Comparison of possible joint configurations (✓ = applicable, - = not applicable) [14]

Joint configuration	Bonded	Bolted	Adherend thickness [1]	Eccentricity	Joint strength
Scarf	✓	-	> 5 mm	Low	High
Step-lap	✓	-	> 5 mm	Low	High
Double-lap/strap	✓	✓	≤ 5 mm	High	Medium
Butt	✓	-	N/A	None	Low
T-joint	✓	✓	N/A	N/A	Low

high. Furthermore, the load transmission at the edges of the joint area is often reduced by manufacturing limitations and fracture initiation is facilitated at these locations. A further detail to be mentioned is that double-lap and double-strap joints exhibit the best use of material if they are "balanced", i.e., the two thinner outer laminates have the same resistance together as the thicker inner laminate.

In orthogonal configurations, T-joints are normally designed where the load transmission occurs through L-shaped cleats, which can be bonded and/or bolted to the adherends. Lap shear joints are normally preferred over T-joints as significant through-thickness tension in bonded T-joints or prying mechanisms in bolted T-joints may occur, which both reduce the joint resistance significantly. In this respect, T-joints can often be built-up from lap shear joints, if the adherends are multilayered, as is also done in timber construction, or in the multilayered angle joints discussed in Section 3.4.4 (Fig. 3.22). Such angle joints can also be conceived with larger, i.e., tube profiles, which can represent beams and columns of frame structures.

Consistent with the approach adopted throughout this chapter, information is kept compact and focuses on essentials. Further details are discussed in the following chapters, as indicated in Allocation **Table 2.12** at the end of this section. Further information on the design of bonded and hybrid connections and joints is provided in Section 5.3.

2.4.2 Bolted lap shear connections and joints

Bolted connections and joints are easy to implement on a construction site if the holes are pre-drilled in the factory and appropriate tolerances are taken into account, see Table 2.10. Furthermore, bolted joints are detachable and thus facilitate deconstruction at the end-of-life, provided they were appropriately maintained, i.e., steel bolts and nuts are not corroded.

A concern exists regarding the durability of bolted joints since the adherends need to be drilled. To prevent moisture ingress along the exposed fibre ends in the borehole walls, the latter should remain permanently sealed. Such sealing layers can, however, be destroyed during bolt installation, and moisture can ingress through undetected damage to the sealing, see Section 3.10.2 (Fig. 3.156). In general, the condition of a joint inside bolt holes is difficult to assess due to being covered by the washers.

Concerning structural aspects, the joint stiffness of bolted joints is comparatively low, which is mainly caused by the clearance, i.e., the gap between the borehole and bolt, see Section 3.6.3 (*Example-3*). Minimising the clearance is thus targeted, although tight boreholes increase the risk of damaging the wall sealing during bolt installation.

Strength and ductility of bolted joints primarily depend on the fibre architecture of the adherends and the associated bolt distances from the adherend edges and between bolts, both in the load direction. Accordingly, different failure modes may occur in lap shear joints, such as pin-bearing, shear-out, splitting, net-tension, or block-shear failure. Pin-bearing failure, i.e., progressive crushing of the adherends ahead of the bolt, is normally the targeted failure mode, since it can provide pseudo-ductility and damage tolerance, as demonstrated in Section 3.2.3 (Fig. 3.8). Geometrical conditions to produce this failure mode are provided in [1]. Nevertheless, bolted joint resistance is normally relatively low in the case of pultruded profiles, if expressed as joint efficiency, i.e., the ratio of adherend to joint resistance, see Section 3.6.3 (*Example-3*).

Over-tightening of bolts should be prevented to not damage the adherends, e.g., by the installation of spacer tubes, see Section 3.10.4 (Fig. 3.164). Applying pretightening torque can increase the joint resistance in the short term; the effect normally disappears, however, in the long term due to creep of the adherend matrix. Resistance increases due to pretightening should thus not be considered in the design [1].

2.4.3 Adhesively bonded lap shear connections and joints

The manufacturing of adhesive connections and joints is uncomplicated and reliable under controlled factory conditions, as long as protocols regarding surface preparation and adhesive application and curing are strictly followed. Adherend surfaces should be appropriately roughened and cleaned to ensure good adherence. Peel-ply (tear-off) fabrics, as shown in **Fig. 2.14**, which can be removed just before bonding, can provide an excellent surface texture and cleanliness in this respect (see also Section 4.7.3, Fig. 4.69, bottom left).

On-site execution of adhesive joints is, however, more demanding due to the dependence on environmental conditions, mainly related to temperature and humidity, similar to the case of on-site welding of steel. Furthermore, on-site curing at low

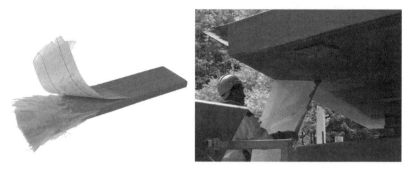

Fig. 2.14 Peel-ply fabric in pultruded profile (cross section 55×10 mm) **(left)**; and sandwich panel face sheet of Avançon Bridge (Appendix A.5) **(right)**

temperatures (during winter) is difficult, as discussed in Section 3.5.3. On the other hand, adhesive joints may compensate for member and installation tolerances during on-site application due to a certain flexibility in the adhesive layer thickness, as described in Section 3.3.3 (Fig. 3.13).

Bonded joints are not detachable, and members can only be separated by cutting, e.g., in the joint plane. However, bonded joints can be combined with detachable bolted joints to circumvent this problem during deconstruction, as done in the movable Eyecatcher Building (Appendix A.3).

Durability of adhesive joints is normally not a concern if joints are correctly manufactured and the joint edges sealed to prevent moisture ingress, see Section 3.10.4 (Fig. 3.163). Even in the case of the absence of a sealing, the penetration length of moisture into an epoxy adhesive connection layer is relatively small. After 25 years of exposure to continental subarctic and warm summer continental climates, epoxy adhesive joints did not exhibit any signs of damage, as shown in Section 3.10.2. Finally, since the design of adhesive joints should be fail-safe according to Section 5.3.1, an adhesive joint failure, due to insufficient durability, for instance, should not lead to the collapse of the entire structure.

To fully exploit the joint capacity, the failure mode should be either fibre-tear failure in the adherends (see Section 3.7.2, Fig. 3.66), or cohesive failure in the adhesive layer. Adhesive failure (i.e., failure in the adherend-adhesive interface) should be prevented by appropriate material selection and surface preparation [1], also see Section 5.3.3.

Stiffness, resistance, ductility, and creep behaviour of bonded joints depend on the adhesive type, adherend-to-adhesive stiffness ratio, and geometrical conditions. Under (i) service temperatures in the adhesive material, i.e., temperatures below T_g-20°C [1], (ii) moderate strain rates (see Section 3.4.4), and (iii) relatively low adherend-to-adhesive stiffness ratios (e.g., of approximately 2–6, as in the case of timber (spruce) or glass fibre composite adherends and an epoxy adhesive [19, 26]), the adhesive can be considered "stiff". The corresponding structural response, in the layer of fibre-tear failure, shows a typical highly nonlinear shear stress distribution along the lap shear length, with pronounced stress peaks at the transverse joint edges, and almost zero stress in the middle of the joint, see **Fig. 2.15** (left) [26, 27]. The

2. Composites – Compact

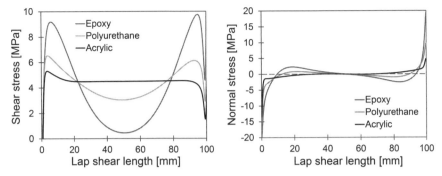

Fig. 2.15 Distributions of shear stresses **(left)** and through-thickness normal stresses **(right)** along lap shear length at fibre-tear location depth (0.5 mm inside inner adherend), in axially symmetric double-lap shear joints composed of glass fibre composite pultruded adherends of 100×5/10 mm cross section, 100 mm width and lap shear length, and epoxy, polyurethane, and acrylic adhesives, all stresses at 100 kN axial load (i.e., at ~25% joint efficiency), based on data in [27]

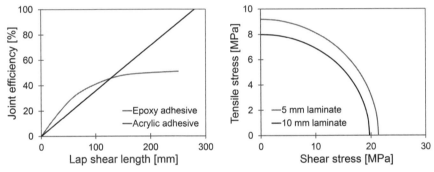

Fig. 2.16 Joint efficiency vs lap shear length of axially symmetric double-lap shear joints, composed of glass fibre composite pultruded adherends of 100×5/10 mm cross section and 100 mm width, with stiff epoxy and flexible acrylic adhesives, partially based on data in [14, 28] **(left)**; shear-tensile failure criteria for fibre-tear failure in glass fibre composite pultruded adherends of 100×5/10 mm cross section and flexible polyurethane adhesive, based on data in [29] **(right)**

joint stiffness is thus comparably high, while the joint resistance, in terms of joint efficiency, is comparably low, and an increase of resistance by increasing the lap shear length is limited, as shown in **Fig. 2.16** (left) [28]. The joint efficiency is defined as the ratio of joint to member resistance. Furthermore, the joint response is normally brittle (Section 3.6.3), but the joint sensitivity to creep is small [24].

For relatively high adherend-to-adhesive stiffness ratios (e.g., of approximately 100–300, as in the case of timber (spruce) or glass fibre composite adherends and an acrylic (elastomer-thermoset) adhesive [19, 26]), the adhesive can be designated as "flexible", since the shear stress distribution is almost uniform along the lap shear length, as also shown in Fig. 2.15 (left) [26, 27]. Accordingly, the joints stiffness is comparably low, but the joint resistance can be linearly increased by increasing the lap shear length, and 100% joint efficiency can thus be achieved, see Fig. 2.16 (left) and Section 3.6.3 (Fig. 3.56). Such joints can exhibit a high pseudo-ductility (Section 3.6.3), but sensitivity to creep deformations is significant [24]. Also shown in Fig. 2.15 is a polyurethane adhesive of intermediate adherend-to-adhesive stiffness ratio (approximately 80 [19]), which thus exhibits an intermediate shear stress response.

Eccentricities between adherend and axial load axes produce bending moments in lap shear joints, which lead to through-thickness (normal) tensile and compressive stresses. Through-thickness tensile stresses are particularly critical since fibres are missing in this direction, see Section 3.7.2. As for the shear stresses, the tensile through-thickness stresses exhibit significant peaks at the joint edges in the case of stiff adhesives while the peaks are reduced with increasing adhesive flexibility, see Fig. 2.15 (right) [27]. Since concurrent shear and tensile through-thickness stress peaks cause joint failure in most cases, it is advantageous to reduce joint eccentricities as much as possible and design lap shear joints with axial symmetry, e.g., as double-lap and not as single-lap shear joints, as already mentioned in Section 2.4.1. Failure criteria based on concurrent shear and through-thickness tensile normal stresses, for fibre-tear failure in pultruded adherends, were developed in [29], see Fig. 2.16 (right), and later complemented in [27], and now implemented in [1]. In sandwich panels, face sheets can be connected with scarf joints and eccentricities thus minimised, as shown in **Fig. 2.17**; the adhesive can be injected in such thin panel-to-panel connections (also see Sections 3.3.3, Fig. 3.13, and 3.8.8, Fig. 3.98).

Numerical studies often suggest that chamfering/tapering the adherends or adding adhesive fillets increases the joint resistance as stress peaks are reduced. However, experimental evidence could not be provided [30, 31]. Since these reduced stress peaks normally act on larger material volumes, the material strength is reduced due to a statistical (geometrical) size effect and thus prevents an increase in the joint resistance [32, 33].

The above-mentioned fail-safe condition can be met in very different ways, as CCLab projects demonstrate, see **Fig. 2.18**. In the Pontresina Bridge (Appendix A.2), the adhesive joints are compact (i.e., with small bonding areas) and were thus considered sensitive to manufacturing imperfections. Accordingly, back-up bolts were installed to absorb the adhesive forces in the event of adhesive joint failure. Furthermore, the structural concept, with multilayered crossed diagonals, is highly redundant

Fig. 2.17 Scarf joints in sandwich panel face sheets: tapered top face sheet **(left)**; scarf joints in top and bottom face sheets **(middle left)**; epoxy adhesive injection into panel-to-panel joint, inlet **(middle right)** and outlet **(right)** (Avançon Bridge, Appendix A.5)

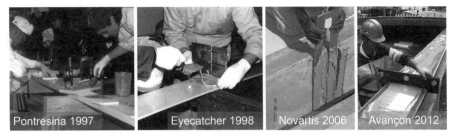

Fig. 2.18 Epoxy adhesive application in CCLab projects: small-area hybrid joints in Pontresina bridge **(left)**; large-area bonded connection along whole profile in Eyecatcher Building **(middle left)**; repetitive small-area bonded connection on webs between roof sandwich panel segments of Novartis Building **(middle right)**; large-area bonded connection of bridge deck on upper steel flange of Avançon Bridge **(right)**, projects see Appendices A2–5

and can withstand several joint failures. In the Eyecatcher Building (Appendix A.3), complex profiles are built up from standardised sections by adhesive bonding. Since the bonding areas are large and extend over the whole profile widths and lengths, the adhesive stresses are small and structural redundancy already exists at the member level. It was thus not found necessary to add back-up bolts in this case. The outer webs of the four large-scale sandwich panel segments of the Novartis Building roof (Appendix A.4) are bonded with small-area connections. The repetitive arrangement of the numerous connections along the webs provides redundancy, see also Section 4.7.2, Fig. 4.67 (bottom left). Even in the case of failure of all these connections, the roof would not collapse. In the Avançon Bridge (Appendix A.5), the bridge deck is bonded onto the top flanges of two steel girders. The large bonding areas again decrease the adhesive stresses significantly and a connection failure would not lead to a collapse of the bridge. However, the composite action between the deck slab and the girders would be lost. The steel girders were designed accordingly, i.e., to carry the full load, assuming, however, the connection failure as an accidental design situation, see Section 5.3.1. Failure of the transverse deck connections would not lead to the collapse of the structure either.

2.4.4 Hybrid bonded/bolted lap shear connections and joints

The aspect evaluation of hybrid-bonded/bolted joints is similar to that of bonded joints regarding on-site application, deconstruction, durability, and stiffness, see Table 2.10. Resistance, ductility, and creep behaviour significantly depend on whether the adhesive is stiff or flexible. In the case of stiff adhesives (e.g., epoxy), the bonded connection may fail before the bolted connection is activated due to the hole clearance. The bolted connection can thus only act as back-up to provide joint redundancy, if designed accordingly (see Section 5.3.3). The response is pseudo-ductile under pin-bearing failure conditions, otherwise, the behaviour is brittle. If the adhesive is sufficiently flexible to develop the bonded connection resistance during bolted connection pin-bearing failure, e.g., in the case of an acrylic adhesive, the two connection resistances can be summed and the joint resistance can thus significantly exceed that

(Allocation) Table 2.12 Supplementary information about bolted, bonded, and hybrid connections and joints in subsequent sections of Chapters 3 to 5

Section	Aspect	Subject of additional information
3.2.3	Anisotropy	Through-thickness anisotropy in adhesive connections
		Effect of fibre architecture on failure modes of bolted joints under static and cyclic action
3.3.2	Imperfections and tolerances	Effect of defects in adhesive layers on joint resistance
3.3.3		Compensation of member tolerances in adhesive joints (Fig. 3.13)
3.4.4	Viscoelasticity	Semi-rigid acrylic-hybrid angle joints, strain rate-dependent pseudo-ductility comparison with linear bonded double-lap joints
3.5.1	Curing	Temperature development in epoxy-bonded connection due to exothermic curing
3.6.3	Ductility	Summation of bonded and bolted connection resistances in hybrid joints with acrylic adhesive (Fig. 3.55)
		Acrylic-bonded and acrylic-hybrid joints with 100% joint efficiency (Fig. 3.56)
		Moment redistribution in composite two-span beams with pseudo-ductile acrylic-bonded connections (Fig. 3.58)
3.7.2.1	Fracture	Epoxy-bonded fracture mechanics joints and mixed-mode failure criterion for epoxy-bonded glass fibre composite joints (Fig. 3.62)
		Fibre-tear failure in adhesive joints (Fig. 3.66)
3.7.3.3		Fatigue crack growth rate of Mode-I epoxy-bonded glass fibre composite joints (Fig. 3.75 right)
3.8.2	Fatigue	Fatigue damage rates (slope coefficients) of epoxy-bonded joints
3.8.6		Effects of temperature and moisture on fatigue life of epoxy-bonded joints (Fig. 3.93)
3.9.3.2	Fire	Effect of adhesive connections on fire resistance of pultruded/bonded web-core slabs
3.9.3.3		Effect of adhesive joints in balsa core on fire resistance of composite-balsa core slabs
3.10.2.4	Durability	Durability of epoxy-bonded joints in Pontresina Bridge and Eyecatcher Building after 25 years of use
3.10.2.5		Rehabilitation of impact damage with epoxy-bonded laminates (Fig. 3.150)
3.10.2.7		Moisture ingress into bolted joints (Fig. 3.156)
3.10.3		Cracks along adhesive joints of sandwich roof (Fig. 3.159)
3.10.4		Penetration length of moisture into epoxy-adhesive connection layer (Fig. 3.163)
		Spacer tubes in bolted joints (Fig. 3.164)
3.10.5		Recommendations for durable adhesive joints

Section	Aspect	Subject of additional information
4.6.2	Hybrid construction	Adhesive bonding in hybrid composites-steel and composites-lightweight concrete members
4.6.3		Composite action and potential uplift in adhesive connections of hybrid members
4.7.2	Modular construction	Adhesive joints between large-scale sandwich panel segments (Fig. 4.67)
4.7.3		Non-detachable bridge deck-to-girder adhesive joints (Fig. 4.69) Detachable bridge deck-to-girder joint (Fig. 4.70)
5.2.4	Design	Concepts of joints and details
5.2.6		Preliminary dimensioning of adhesive joints
5.3		Detailed adhesive joint resistance verification

of bonded-only joints, see Section 3.6.3 (Fig. 3.55, right). The joint efficiency can be increased to 100%, as in bonded-only joints, but the geometry can be conceived more compactly, see Fig. 3.56. Excellent pseudo-ductility can be provided in such cases, and the bolted connection can limit creep deformations [24].

2.5 MANUFACTURING AND INSTALLATION

2.5.1 Hand lay-up

Hand lay-up (or wet lay-up) is a simple manufacturing method to produce prototypes or individual members where automated or semi-automated manufacturing is too expensive. A shape-giving mould is required on which, stepwise and alternatingly, the plies are positioned, and the resin is uniformly spread and applied with rollers to properly impregnate the fibres and remove trapped air. The laminate is subsequently cured, normally under ambient conditions. In this way, complex shapes can be quite easily manufactured. The achievable fibre volume fraction is comparatively low, however, and the variability of mechanical properties, number and magnitude of imperfections, and tolerances are higher than in the case of more advanced manufacturing methods. The following two examples demonstrate the nonetheless great potential of this manufacturing method in terms of complex shapes and scale.

- Example-1: Illuminated Flower Sculpture

The design of an illuminated Flower Sculpture, shown in Appendix A.6, was created by two artists. The sculpture was manufactured by (i) shaping the blooms and stems in foam (by the artists), (ii) laminating a glass fibre composite mould for the bloom manufacturing on the shaped foam, (iii) glass fibre-polymer hand lay-up of the blooms in the mould and wrapping of the stems, as shown in **Fig. 2.19**, and (iv) bonding blooms and stems together.

Fig. 2.19 Hand lay-up manufacturing of Flower Sculpture, Brig, 2002 (Appendix A.6): laminate wrapping on foam of stem **(left)**; mould for blooms **(middle)**; translucent bloom after curing **(right)**

Fig. 2.20 Assembly and hand lay-up of large-scale composite-PUR web-core sandwich panel segments of complex shape, for roof of Novartis Building (Appendix A.4): CNC-milled foam blocks **(left)**; wrapping of four-block modules **(middle left)**; assembly of four-block modules on flexible bed of spacers **(middle right)**; lamination of face sheets **(right)**

- *Example-2: Composite-PUR web-core sandwich roof of Novartis Campus Entrance Building*

The composite-PUR web-core sandwich roof of the Novartis Building (Appendix A.4) is composed of four large-scale segments, which were assembled on site, as shown in Section 4.7.2 (Fig. 4.67). The basis of the complex shape are 460 foam blocks of different geometry, see **Fig. 2.20**, which were manufactured by CNC milling, as shown in Section 4.3.2 (Fig. 4.15). In a first step, the foam blocks were wrapped with glass fibre fabrics by hand lay-up to form the grid of interior panel webs and the inner layers of the face sheets. Subsequently, four blocks were assembled and again manually wrapped with fabrics. In the third step, the four-block modules were positioned and adhesively bonded together on a flexible bed of spacers, which followed the bottom curvature of the segment. Lastly, the remaining layers of one face sheet were applied on the top surface of the segment. The segment was subsequently turned (see Fig. 4.67, top left) and the face sheet was completed on the other side [34].

2.5.2 Pultrusion

Pultrusion is a fully automated continuous process to produce composite structural profiles, which can be used as bars, columns, beams, girders, and in frames. Profiles

2. Composites – Compact

Fig. 2.21 Pultrusion process from left to right (**left**) and seen from the back (**right**)

can furthermore be assembled through adhesive bonding to form web-core sandwich panels (see Section 2.3.3). The pultrusion process is linear and driven by a pulling device, by which fibre rovings, fabrics, and mats are pulled through guides and a shape-giving die into a resin injection device, where the resin is injected and the profile subsequently cures under heat, see **Fig. 2.21**. Alternative processes use an open bath for the fibre impregnation, where, however, the escape of volatile solvents cannot be prevented. The length of the profiles is theoretically unlimited and can thus be easily customised by saw cutting. Pultrusion is overall a cost-effective process with high throughput and relatively low energy intensity and waste when compared to other processes, see Section 3.11.2 and [35].

Pultrusion also has its limitations, however. Since the profiles are straight, continuously curved structural shapes can only be approximated by polygons. Furthermore, cross section and fibre architecture are constant along the length and do not allow for an adaptation at the profile ends for joining. The joints thus determine the cross-sectional dimensions in most cases, and the material use between the joints is often comparably low, see Sections 2.4 (Table 2.10) and 3.10.2; the latter section shows that the profile capacity is used to only about 30% in the case of the Pontresina Bridge. This problem, in fact, is the same as in timber construction, see Section 1.2.

Since the process is based on pulling the fibres, the 0° (axial) fibre volume fraction needs to be high. To implement a significant amount of 45° or even 90° fibres is normally not possible and limited to some interspersed or external combined fabrics and mats, see Figs. 3.51 and 4.17. The anisotropy of pultruded profiles is thus high, and the shear and transverse properties are therefore low in comparison; typical values are provided in [1].

External and interspersed mat layers may, however, considerably increase the fibre-tear or interlaminar through-thickness fracture resistance and associated pseudo-ductility, in adhesively bonded joints and curved laminate regions, as demonstrated in Sections 3.7.2 and 3.6.3, respectively. External surface veils, furthermore, are very efficient in improving durability, see Section 3.10.2; their inclusion should be mandatory.

The pulling forces applied to the fibres generally provide a high fibre straightness and geometrical accuracy in the cross section, in the longitudinal direction. In the transverse direction, however, such stabilising forces are missing, and geometrical imperfections thus frequently occur, mainly in web-flange junctions, with associated detrimental effects on transverse strength and fatigue resistance, see Sections 3.3.2 and 3.8.8, respectively.

More detailed information about pultruded members, structured according to the respective aspects, can be found in the following chapters, as listed in Allocation **Table 2.13**, at the end of this section.

2.5.3 Vacuum infusion

Vacuum infusion is a closed mould process to produce one- or two-dimensional composite members, such as profiles, laminates, or sandwich panels, which can be used as beams, plates, slabs, or shells, see Sections 1.2, 2.6, and 2.7. Members from small to very large scale, with linear, flat, or complex curved shapes can be manufactured. The throughput is relatively low, compared to pultrusion, for instance, and a non-negligible amount of waste of material and by-products accumulates.

The mould consists of a bottom solid part and an upper flexible vacuum bag, as shown in **Fig. 2.22** and **Fig. 2.23** (left), which makes adaptations to complex shapes possible. The fibre layers and core are first laid up, as depicted in Fig. 4.69 (top left), in the case of the composite-balsa sandwich bridge deck of the Avançon Bridge. Vacuum bag and resin inlets and outlets are subsequently installed, and the resin is then driven by a vacuum, applied at the outlet, through the fibre layers, as also shown in Figs. 2.22 (left) and 2.23 (left). The resin flow mainly depends on the permeability of the laminate, the viscosity of the resin and the applied vacuum. Additional release agents and peel plies on the bottom mould and below the bag, respectively, facilitate demoulding after curing [36]. Accordingly, the bottom surface of an infused member is normally very smooth while the top surface may display a slight roughness. Imperfections may be produced by wrinkles in the vacuum bag, which can cause local wrinkles in the outermost fibre layers. Furthermore, it is sometimes difficult to control the laminate thickness, and slight deviations from nominal values are possible.

An advantage of vacuum infusion compared to pultrusion is that the fibre architecture and core composition can be tailored and adapted to the subsequently applied stresses. Sandwich core materials can be varied or combined, and inserts installed, as shown in Sections 4.6.2 (Fig. 4.61) and 3.9.3 (Fig. 3.127), respectively. Highly

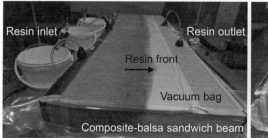 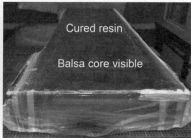

Fig. 2.22 Vacuum infusion of glass fibre-balsa sandwich beam (1.5×0.3×0.12 m), with moving resin front during infusion **(left)**; cured translucent face sheet with visible end-grain balsa core composition, vacuum bag not yet removed **(right)**

2. Composites – Compact

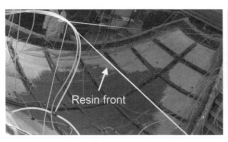 

Fig. 2.23 Element manufacturing of domes of Orthodox Trinity Cathedral, Paris (Fig. 2.47): vacuum infusion of double-curved glass fibre-PET sandwich segment (approx. 7.5×4×0.05 m), resin front visible **(left)**; demoulded element with inside flanges for bolted connections **(right)**

complex shapes can be realised, even in sandwich construction, and flanges for connections can be integrated, as shown in **Figs. 2.23–24**. Complex members can also be built up from different components, as depicted in **Fig. 2.25**. The girders of the Rhyl/Foryd Bridge (shown in Section 2.6, Fig. 2.33, right) are composed of different components bonded together.

Fig. 2.24 Prototype of basic carbon fibre-PET sandwich half-module of 30 and 40 mm thickness **(left)** and assembly of two half-modules **(right)**, TSCB Bridge project (Appendix B.4)

Fig. 2.25 Combined assembly of web-core cantilever girder from Rhyl/Foryd Bridge: vacuum-infused bottom sandwich panel of complex shape and bonded longitudinal and transverse web laminates, longitudinal half of one girder **(left)**; a finished web-core girder with top sandwich panel and surfacing **(right)**

(Allocation) Table 2.13 Supplementary information regarding manufacturing and installation in subsequent sections of Chapters 3 to 5

Section	Aspect	Subject of additional information
3.2.3	Anisotropy	Effect of pultruded profile fibre architecture on bolted joint resistance, ductility, and failure mode (Fig. 3.8)
3.3.2	Imperfections and tolerances	Imperfections in web-flange junctions of pultruded profiles (Fig. 3.11)
		Residual stresses and distortion of members due to manufacturing
4.4.5	Transparency	Infusion instead of hand lay-up to reduce defects and increase transparency
4.6.2	Hybrid construction	Infused composite-balsa sandwich slab with core-integrated composite arch (Fig. 4.61)
		Pultruded face panels with T-upstands which also serve as formwork (Fig. 4.62)
4.7.2	Modular construction	Transport and installation of hand lay-up manufactured sandwich roof of Novartis Building (Fig. 4.67)
4.7.3		Fibre and balsa core lay-up for vacuum infusion of Avançon bridge deck, and transport and installation (Fig. 4.69)
5.2.5	Design	Manufacturing and installation
5.4		Execution

More details about vacuum infusion and members produced by this process are provided in the following chapters, as listed in Allocation Table 2.13.

2.5.4   Prepreg moulding

Prepregs are pre-impregnated fibres or fabrics, which are laid up in a mould and subsequently consolidated in a vacuum bag and then heat cured in an oven or autoclave. The separation of the fibre impregnation from the member manufacturing process provides high laminate quality and makes high fibre volume fraction possible. In building construction, this process is used to manufacture carbon fibre composite roofs, as shown in Section 2.7.4.

2.5.5   Installation

Thanks to their light weight, composite structures can be prefabricated in large modules, easily transported to the construction site, and rapidly installed, e.g., using small cranes, see **Fig. 2.26** and Fig. 4.66. The factory manufacturing using semi- or fully automated processes ensures a high quality and normally low tolerances. The on-site joining of the modules is quite easy if bolting is used, and more demanding in the case of adhesive bonding, see Section 2.4.

Fig. 2.26 Rapid installation by helicopter or crane of large-scale lightweight composite members and structures: Pontresina bridge, span 2×11.5 m, weight 1.6 t/span **(left)**; Eyecatcher Building, frame height 14.6 m **(middle left)**; Novartis Building, segment length 18.5 m, average weight 7 t/panel [34] **(middle right)**; Avançon Bridge, length 12.0 m **(right)**, for projects see Appendices A.2–5

2.6 COMPOSITES IN BRIDGE CONSTRUCTION

2.6.1 Introduction

My work on composite vehicular bridges began in 1999 with the commissioning of a state-of-the-art report by the Swiss Federal Roads Authority (FEDRO), on the use of composites in bridge construction; the report was later published by IABSE [5]. Supported by the previous experiences with the Pontresina Bridge (1997, Appendix A.2) and Eyecatcher Building (1998, A.3), an outcome of this work was the recommendation to develop fully adhesively bonded connections between composite bridge decks and main girders of vehicular bridges (which are subjected to fatigue). These connections were executed, at that time, by mechanical fastening or hybrid techniques [5]. A subsequent research project to develop such fully bonded connections was again funded by FEDRO and supported by Martin Marietta Composites (USA) and Fiberline (Denmark), by providing both their pultruded web-core bridge deck systems for the experimental work. The outcome of this project laid the foundation for the execution of several vehicular and pedestrian bridges with fully bonded deck-to-girder connections in Europe (e.g., the bridge shown in Section 1.2, Fig. 1.6, right). In 2010, I became involved in the development of a novel bridge deck system ("Colevo", funded by 3A Composites, Switzerland), consisting of vacuum-infused sandwich panels, composed of composite face sheets and a homogeneous core of end-grain balsa wood. The Avançon (vehicular) Bridge was built with this system in 2012 (Appendix A.5), and several pedestrian bridges were subsequently constructed in Switzerland (see Section 2.6.3, Fig. 2.34).

Over some 25 years of involvement in composite bridge construction, I have seen many proposals for novel composite bridge deck and girder designs appear – and subsequently disappear, even though they were labelled "innovative" and were often accompanied by a significant experimental effort. My conclusions from these experiences, and market observations in addition, are twofold: (i) that composite bridge decks for vehicular bridges should be simple to manufacture and install in order to be economical, and robust in design to resist fatigue loads, and (ii) composite main girders cannot compete economically with steel or concrete girders, since the durability of composite girders is not an advantage if the steel or concrete girders are protected from critical environmental impact by the bridge deck.

The following Section 2.6.2 on vehicular bridges thus focuses on bridge decks, which, in the past and also today, have received a certain acceptance from bridge authorities and were, or are, subjected to real traffic conditions. The subsequent Section 2.6.3 on pedestrian bridges is, however, not limited to decks and covers whole bridge superstructures. It must be noted that concrete bridge decks with composite rebars are not within the scope of this monograph. Furthermore, the bridge decks discussed above and below use exclusively glass fibres, which will thus no longer be specified.

Research on composite bridge decks for vehicular bridges began in the early 1980s [37], and the first decks were installed in the mid-1990s [5]. Looking back over the period since the appearance of such composite decks reveals two main phases of development: (i) a first phase of rapid growth but subsequent equally rapid slowdown in the mid-2000s, almost exclusively in the USA, and (ii) a subsequent (current) second phase of moderate but continuous application, again mainly in the USA, and to a lesser extent in Europe and other regions. These two phases of vehicular bridge deck applications are addressed below.

2.6.2 Vehicular bridge decks

2.6.2.1 First phase of vehicular bridge deck application

In the USA, in the mid-1990s, about 40% of the bridges were considered structurally deficient or functionally obsolete. Based on this knowledge, a ten-year research and development programme, CONMAT (CONstruction MATerials), was launched in 1995 targeting infrastructure renewal and emphasising the use of high-performance construction materials and systems. From the planned investment of US$2 billion, more than 40% was allocated to composites for the "creation of a new generation of bridge, marine, and utility structures with the benefits of reduced life-cycle costs and construction time" [38].

In line with this programme, composite bridge decks were developed and installed from 1996 onwards, and a bright future was predicted for composite bridges (also see survey in Preface). Four proprietary deck systems emerged and dominated the deck installations, i.e., the decks developed by the companies Hardcore Composites, Kansas Structural Composites, Martin Marietta Composites (MMC), and Creative Pultrusions (CP) [5, 39]. The Hardcore deck was a vacuum-infused web-core sandwich deck with variable deck thickness, where an orthogonal web system was formed in the core by assembling fibre-wrapped foam blocks. The web-spacing was about 80% of the web height, in the example shown in [3]. The Kansas deck was a sandwich deck with a honeycomb core also of variable thickness, manufactured by hand lay-up techniques. The remaining two deck systems were pultruded/bonded web-core decks of constant thickness, called "DuraSpan" (MMC), see Section 2.3.3 (Fig. 2.10, top) and "Superdeck" (CP). Two bridges using the MMC deck system, an all- and a hybrid-composite bridge, are shown in **Figs. 2.27–28**, respectively, in addition to different types of crash barrier fixations. The former bridge, named "Tech 21" (Materials Technology for the 21st Century), was developed based on transferring

Fig. 2.27 Smith Road (Tech 21) Bridge, Butler County, Ohio, 10.1×7.3 m, 1997 (deck manufacturer Martin Marietta Composites) **(left)**; two of three U-shaped hand-laminated glass fibre composite box girders **(middle)**; connections of crash barriers to deck **(right)** (07 June 2001)

Fig. 2.28 Darke County Bridge, State Route 47, Ohio, 15.2×14.0 m, 1999 (deck manufacturer Martin Marietta Composites) **(left)**; steel longitudinal and transverse girders below deck panels **(middle)**; connections of crash barriers to steel girders below deck **(right)** (07 June 2001)

military technologies, according to a plate fixed to one abutment. After only four years of use, however, the bridge exhibited significant corrosion in the abutment walls (see Fig. 2.27), which was caused by leaking expansion joints. In most of these bridges, the deck surfacing was a thin (10–20 mm) layer of polymer concrete.

Until 2003, about 120 composite vehicular bridge decks or whole composite vehicular superstructures were installed in the USA [40, 41]. After 2003, however, the installations abruptly stopped [42]. This stop is likely related to the appearance of damage in some of the installed decks, after only short periods of use in some cases [3, 43, 44]. Delamination of the top face panels from the cores and leakages in the joints occurred in sandwich decks, and the thin polymer surfacing frequently cracked above field joints of all deck types. These types of damage were mainly attributed to temperature variations and through-thickness gradients, insufficient manufacturing quality, and too large tolerances. Through-thickness temperature gradients were particularly significant in the Hardcore deck due to the insulation effect caused by the foam core. The deck installations demonstrated, however, that in comparison to reinforced concrete decks, installation times could be significantly reduced, and lane closures avoided due to prefabrication and the light weight of the composite panels [3]. The four deck types mentioned above, nevertheless, subsequently disappeared from the market.

In Europe, only one relevant composite deck system was developed for vehicular bridges: a pultruded/bonded web-core deck produced by Fiberline, Denmark, named "ASSET" (see Section 2.3.3, Fig. 2.10, bottom). The deck was installed, from

2002–2014, on several vehicular bridges [45], an example is shown in Section 1.2, Fig. 1.6 (right). The production of the deck was subsequently stopped, however.

Based on the analysis performed in Section 3.8.8, the debonding of the face panels in the Hardcore deck may also be traced back to the relatively large web spacing and thus uneven support of the face panels against tyre loading, in addition to the above-mentioned temperature gradients. Some reasons for the disappearance of the pultruded/bonded web-core decks might have been (i) too little flexibility in application due to geometrical limitations and orthotropy, and (ii) limited fatigue resistance, as described in Section 3.8.8.

2.6.2.2 Second phase of vehicular bridge deck application

A second phase of vehicular bridge deck applications began around 2010, mainly with the appearance of three companies on the construction market: Composite Advantage in the USA, and FiberCore and 3A Composites in Europe. These companies developed proprietary sandwich deck systems based on vacuum infusion, namely the "FiberSpan" system in the USA, the "FiberCore/InfraCore" system in the Netherlands, and the "Colevo" system in Switzerland.

The FiberSpan and FiberCore systems are composed of web-core panels with narrowly spaced webs integrated in a foam core, as shown in Section 2.3.3 (Fig. 2.11) for the European system. The narrowly spaced webs provide a much more continuous support of the top face panels and thus increase fatigue resistance compared to the first phase, as discussed in Section 3.8.8. The fibre architecture of the critical web-face panel junction is furthermore optimised in the European system to better resist delamination (Fig. 2.11). The infusion moulding also allows more versatility compared to pultruded decks. Features such as camber, slopes, kerbs, drain scuppers, and panel joint details can be integrated into the deck during manufacturing; furthermore, complex shapes can be conceived, as demonstrated in **Fig. 2.29**. The Colevo deck has a homogeneous end-grain balsa core and thus provides a fully continuous support to the top face sheet, see Section 3.8.8; simple kerb and dewatering details can be conceived, as shown in Section 3.10.4 (Fig. 3.162).

Four application fields emerged in the USA during this second phase, in which composite decks were seen to offer economic benefits. These fields are (i) movable

Fig. 2.29 Kruisvaartbrug Bridge, Utrecht, 13.4×11.8 m, 2014 (deck manufacturer FiberCore) **(left)**; lightweight large-scale composite sandwich bridge deck of complex shape during installation **(right)**

Fig. 2.30 De Locht Bridge, Kerkrade, 53 m span, 2021 (deck manufacturer FiberCore) **(left)**; concrete deck replacement by four composite sandwich panels **(right)**

Fig. 2.31 Sunset Lake Floating Bridge (limited to 12 t vehicles), Brookfield, Vermont (USA), 76.5×6.9 m, 2015 (manufacturer Composite Advantage) **(left)**; five sandwich panels during installation **(right)**

bridges, (ii) historic steel truss bridges, (iii) bridges with steel gratings, and (iv) lateral pedestrian deck additions to existing vehicular bridges. The light weight of composite decks is the main advantage in all these applications [46]. In Europe, most of the installed decks also involve deck replacements and movable bridges. Bridges can be upgraded by replacing heavy concrete decks with lightweight composite decks, as shown in **Fig. 2.30**. The replacement of existing concrete superstructures with lightweight composite decks on steel girders even allows for the widening of existing bridges, if the additional traffic weight on new lanes can be compensated by the weight reductions from changing the superstructure, see the Avançon Bridge (Appendix A.5). A further interesting application of lightweight sandwich panels are floating bridges, as shown in **Fig. 2.31**. It should also be noted that composite bridge decks do not require a waterproofing layer, as concrete decks do, which simplifies the kerb detailing and associated maintenance significantly, see Fig. 3.162 (Section 3.10.4).

2.6.2.3 Design aspects of vehicular bridge decks

Composite vehicular bridges can be designed based on the European Technical Specification CEN/TS 19101 [1] and the associated Commentary [10]. The required design verifications are provided, together with generic examples of details, such as bearings, expansion joints, kerbs, adhesive deck-girder connections, and crash barrier fixations. Critical design aspects are also discussed in Chapters 3 to 5, according to Allocation Table 2.14. They primarily concern fatigue resistance and durability, which both significantly depend on appropriate detailing. Details such as web-face panel junctions, deck panel-to-panel, and deck-to-girder connections should be designed in such a way as to prevent significant stress concentrations and not create locations of moisture ingress in the latter two cases; see Sections 3.8.8 and 3.10.4, respectively.

Another relevant aspect concerning fatigue is fatigue testing. Depending on the (European) traffic categories, 10 to 800 million axle loads may pass over a bridge during a 100-year design service life, the former number for local traffic on Category-4 bridges (on local roads) and the latter for long-distance traffic on Category-1 bridges (motorways) [10, 47]. Fatigue testing is, however, often performed at only two million fatigue cycles to limit the test duration. This relatively low cycle number may, nevertheless, be acceptable if the experimental set-up and fatigue test load are conceived accordingly, i.e., in such a way that the same cumulative damage is produced during testing as occurs on the real bridge during the design service life; see Section 3.8.9. Considering the still remaining uncertainties regarding the fatigue resistance of composite bridge decks (see Section 3.8.8), it seems wise, however, to currently limit their application to Category-3 and -4 bridges (on main and local roads), until more long-term data are available, to prevent the negative experiences of the first phase of composite bridge deck application.

A further aspect concerns concrete deck replacement, as described in the previous section, which is often cited as the most promising application of composite decks. In many cases, however, the replacement is not so simple, as the longitudinal in-plane stiffness of composite decks may not be sufficient to provide the same composite action as concrete decks; this problem is further discussed in Sections 3.1 and 4.6.2.

Crash barrier fixations are another critical aspect of composite bridge decks. Energy dissipation during impact should occur in the posts and fixation parts of the crash barriers, which are replaceable, and thus not damage the deck and corresponding irreplaceable deck inserts, as shown in **Fig. 2.32** for the case of the Avançon Bridge (Appendix A.5 and Fig. 3.162). Different concepts of crash barrier fixations are shown in Figs. 2.27–28, and generic examples are depicted in [1].

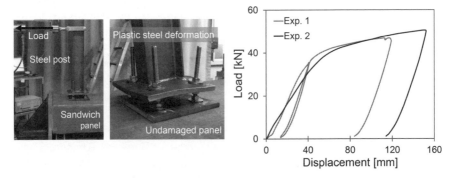

Fig. 2.32 Experimental verification of crash barrier post fixation in composite-balsa sandwich bridge deck: experimental set-up (**left**); plastic deformation in replaceable steel connections and undamaged sandwich panel (**middle**); corresponding load vs horizontal displacement responses showing ductile energy dissipation (**right**)

A last aspect to be mentioned concerns the surfacing. It is understandable that a lightweight thin layer of polymer concrete applied on the lightweight decks was the preferred solution in the first phase. As mentioned above, the durability of this surfacing type may be limited, however, and it may even contribute to a reduction in the fatigue resistance of the deck, see Section 3.8.8. A more robust solution was therefore preferred for the Avançon Bridge (Appendix A.5), i.e., a 60 mm thick medium-temperature asphalt layer, although its weight is an additional 75% of the deck weight [15]. However, the thicker surfacing layer better distributes concentrated tyre loads and thus reduces stress concentrations. Since the layer is applied after deck installation, the additional weight does not compromise the advantages of lightweight construction and, compared to live loads, still represents a small fraction.

Linked to the surfacing is a matter of public safety. Since lightweight composite decks have a much lower thermal mass than the roadway on either side of the bridge, a drop in temperature during winter may cause ice to form on the bridge surface much earlier than on the road to either side [44].

2.6.3 Pedestrian bridges

2.6.3.1 Design concepts

Composite pedestrian bridges have been built since the early 1980s. The first bridges, known as "Techtonics Bridges" and built in the USA, were conceived as truss bridges composed of standardised pultruded profiles [5], thus mimicking steel or timber truss bridges; the decks were made of timber in most cases. A more material-tailored bridge concept was developed for the Aberfeldy Bridge, inaugurated in 1992, see **Fig. 2.33** (left) [48]. The bridge girders and deck are composed of pultruded multicellular panels, three-way and toggle connectors, a system developed by Maunsell Structural Plastics, and today commercialised by Strongwell, see Fig. 2.12 (bottom right). The cable-stayed bridge represents a lightweight form-active system and thus requires stabilisation measures, see Section 4.3.4. In this case, additional weight was added by filling deck panel cells with concrete at mid-span to reduce oscillations and uplift caused by wind.

Fig. 2.33 Aberfeldy (pedestrian) Bridge, UK, 113 m length, 1992, tailored pultruded profiles (manufacturer GEC Reinforced Plastics) **(left)**; Pontresina (pedestrian) Bridge, Switzerland, 2×12.5 m span, 1997, standardised pultruded profiles (manufacturer Fiberline) **(middle)**; Rhyl/Foryd (cycle/pedestrian bascule) Bridge, UK, 2×32 m span, 2013, complex shape through vacuum infusion, additional carbon fibre reinforcement (manufacturer AM Structures) **(right)**

Since these first applications of composites for pedestrian bridges, numerous other bridges have been built. They can be classified into all-composite and hybrid-composite bridges. In the former case, two basic design concepts can be identified: material-substitution and material-tailored concepts.

All-composite material-substitution designs are mostly composed of standardised pultruded profiles, as the first bridges were. The profiles are used in typical beam, truss, arch, or cable-stayed systems. Examples include the Pontresina Bridge with two lateral trusses (Appendix A.2), as shown in Fig. 2.33 (middle), or the Kolding Bridge with a cable-stayed system (Fig. 4. 26, right). When looking at these bridges, the novel material can normally not be recognised, as they have the appearance of steel bridges. The required material quantity of such bridges is comparably high in most cases, since the cross-sectional dimensions are determined by the compact joints, which normally exhibit a low joint efficiency, see Section 2.4.

The required material quantity can be significantly reduced by developing material-tailored systems, where the shape and fibre architecture are adaptable to the internal forces and stress fields, respectively, and compact joints can be prevented or their number minimised. Such bridges are often composed of large-scale prefabricated sandwich modules, produced by vacuum infusion, which also allows the manufacturing of complex shapes. An example is the Rhyl/Foryd Bascule Bridge [49], shown in Fig. 2.33 (right), where carbon fibres were able to be easily added in critical sections using the infusion method.

In hybrid bridges, the main load-bearing system is normally made of conventional materials and only the decks are composed of composites, using commercial deck systems for pedestrian loads, as shown in Section 2.3.3 (Fig. 2.12). Examples built in Switzerland with steel and timber primary structures and lightweight balsa-composite sandwich decks (Colevo system) are shown in **Fig. 2.34**.

Fig. 2.34 Hybrid pedestrian bridges with steel or timber primary structures and balsa-composite sandwich bridge decks, built in Switzerland (deck manufacturer 3A Composites): Boncourt cable-stayed bridge, 33.0×2.2 m, 2013 **(top)**; La Tène arch bridge, 66.0×5.5 m plus ramp of 89.0×3.3 m, 2015 **(middle)**; Bremgarten three-span girder bridge, 35.7×3.5 m, 2015 **(bottom)**

2.6.3.2 Design aspects

As mentioned above for vehicular bridges, composite pedestrian bridges can also be designed based on the European Technical Specification CEN/TS 19101 [1] and the associated Commentary [10]. Design aspects such as durability and system stabilisation of form-active systems are further addressed in Chapters 3 and 4, according to Allocation **Table 2.14**.

A critical design aspect of pedestrian bridges are vibrations; excessive vibrations should be prevented so as not to cause discomfort to the users. To further prevent resonance effects, the natural frequencies of the structure, or of parts thereof, should be higher than the excitation frequencies exerted by pedestrians or wind. In this respect, natural frequencies should be higher than 3 Hz in the vertical mode (derived from [47]). My own design experiences have also shown that if stiffness is sufficient to meet deflection limits, vibration verifications are also met – this observation cannot, however, be taken as a rule. Furthermore, non-load bearing guardrails can have significant effects on the natural frequency and may therefore be considered in vibration analyses.

(Allocation) Table 2.14 Supplementary information regarding bridge construction in subsequent sections of Chapters 3 to 5

Section	Aspect	Subject of additional information
3.3.3	Imperfections and tolerances	Execution of adhesively bonded bridge deck-to-girder connections (Fig. 3.13)
3.4.5.2	Viscoelasticity	Creep in lightweight composite bridges
3.4.5.3		Reversible creep-fatigue effects in composite bridges due to interrupted fatigue
3.4.5.4		Stress relaxation in pretensioned carbon cables of Verdasio Bridge (Appendix A.1) (Fig. 3.28)
3.8.8	Fatigue	Comparison of fatigue behaviour of web-core and homogeneous core sandwich bridge decks
3.8.9		Fatigue load model for composite bridges
3.9.1	Fire	Standard fire curve for bridges (Fig. 3.105)
3.9.3.1		Requirements regarding fire resistance of bridges
3.9.5		Assessment of effects of fire below bridges
3.10.2	Durability	Durability behaviour of Pontresina Bridge (Appendix A.2) over 25 years of use, including adhesive joints
3.10.3		Durability behaviour of Avançon Bridge (Appendix A.5) over 11 years of use, including adhesive joints
3.10.4		Detailing of composite bridge deck kerb without waterproofing layer (Fig. 3.162)
		Detailing of composite bridge deck-to-steel girder adhesive connection (Fig. 3.163)
3.10.5		Conclusions regarding detailing to ensure durability

Section	Aspect	Subject of additional information
4.2.3	Learnings from nature	Discussion about usefulness of a 100-year design service life for bridges
4.3.4.2	Structural form and system	Stabilisation of bridges with form-active structural system (Fig. 4.24)
4.6.2	Hybrid construction	Examples of hybrid composite-balsa/timber sandwich slab bridges and bridge decks with improved composite action (Fig. 4.61)
		Example of hybrid composite-concrete sandwich bridge decks (Fig. 4.62)
		Examples of hybrid composite web-core deck-steel girder bridges (Fig. 4.63)
4.6.3		Composite action in composite bridge decks adhesively bonded to steel girders
4.7.3	Modular construction	Modular bridge composed of prefabricated composite-balsa sandwich deck modules, adhesively bonded onto two steel main girders (Fig. 4.69)
		Modular bridge concept with composite deck panels detachable from steel girders (Fig. 4.70)
5.2	Design	Conceptual design
5.3		Detailed fatigue resistance verification for bridges
5.4		Execution
5.5		Maintenance and rehabilitation

2.7 COMPOSITES IN ARCHITECTURE

2.7.1 Introduction

Thanks to their physical properties and free formability, composites can fulfil, in addition to structural functions, building physics functions normally attributed to the non-load bearing building envelope (such as thermal insulation), and architectural functions (such as complex shapes and transparency) [50]. The opportunities that may result from this possible function integration in composite members in architecture, and concerning sustainability and economy, are discussed in Section 4.5.

This section focuses on composite applications in architecture, where composite members have a significant structural role. Composite building envelopes composed of small-scale cladding elements, attached to a closed-mesh underlying structure made of steel and/or concrete, are not further considered, although a majority of composite building applications fall into this category. In the context of integrating texture in composite members, an example of composite cladding is shown, however, in Fig. 4.59 (right), and further examples can be found in [4, 51], for instance.

In accordance with Section 2.3, applications of composite laminates, profiles, and sandwich panels will be discussed, see **Table 2.15**. As will be shown, these components and members are almost exclusively used in the building envelope (including

Table 2.15 Integration of structural, building physics, and architectural functions into composite building envelopes (✓= applicable, (✓) = limited applicability, - = not applicable) [14]

Building envelope constituents	Functions			
	Primary structural	Building physics	Architectural	
		Thermal insulation	Shape	Transparency
Laminates	(✓)	-	✓	✓
Profiles	✓	(✓)	(✓)	(✓)
Sandwich panels	✓	✓	✓	-

roofs); the interior building structure is rarely composed of primary composite members. An exception is the Eyecatcher Building (see Appendix A.3). The reason for this specific application may reside in the fire reaction behaviour and fire resistance, the latter being particularly critical for members under axial compression, such as interior columns or walls, see Sections 2.7.5 and 3.9.5.

Laminates, profiles, and sandwich panels make various combinations of function integration possible, as summarised in Table 2.15. A primary structural function can be provided by profiles and sandwich panels, while the structural role of laminates is normally secondary, i.e., an underlying primary structure is necessary (or an integration as face sheet into sandwich panels). Considering building physics functions, thermal insulation in particular, only sandwich panels, preferably with homogenous foam core, can fully meet such requirements. Envelope-penetrating glass fibre composite profiles can form thermal bridges; these may be acceptable, however, due to the low thermal conductivity when compared to steel profiles. Glass fibre composite laminates cannot provide a sufficient thermal insulation due to their low thickness, although the thermal conductivity is low compared to steel or concrete. Using basalt instead of glass fibres would improve the thermal resistance of profiles and laminates considerably, see Section 4.5.2 (Table 4.6). Regarding architecture, complex (double-curved) shapes can be manufactured for laminates and sandwich panels, while profile structures are limited to polygonal shapes (Section 2.5.2). A high degree of transparency can be reached with laminates, while profiles can offer some translucency, see Fig. 2.9. The local removal of the core and merging of both face sheets can combine the advantages of laminates and sandwich panels, however, and provide full function integration, as shown in Fig. 4.44.

The following discussion first analyses the initiation of composites in architecture in the Space Age and emerging building concepts. Subsequently, drawing on several examples, a discussion will follow on how composite laminates, profiles, and sandwich panels can be used in today's building applications and how function integration can be implemented. As will be shown, these applications can basically be categorised into (i) space units of smaller scale, (ii) lateral building envelopes (not to be confused with the above-mentioned cladding), (iii) primary vertical structures, and (iv) roof constructions. Finally, critical design aspects will be raised, such as fire resistance, sound absorption, and indoor climate.

2.7.2 Initiation phase of composites in architecture

The appearance of composites in building construction was closely linked to the beginning of the Space Age, i.e., the era of space exploration which began with the launch of the first Sputnik satellite in 1957. A spirit of optimism and hope for the future emerged after World War II, and design was influenced by the futuristic shapes of satellites and space technology. Streamlined modern architecture, however, had already emerged earlier, as shown in **Fig. 2.35**, and also influenced the first generation of "plastic architects" [52]. Composites were poised to respond to the spirit of the time, as their characteristics enabled complex shapes to be manufactured and materials to be made transparent and coloured.

Further driven by the chemical industry, such as Monsanto in the USA or BASF in Europe (both of which were keen to enter the construction market), numerous designs of housing units emerged, such as (i) the first all-composite house in 1955, influenced by the growth of a snail's shell and exhibiting a moulded-in interior (Fig. 4.6), (ii) the famous Monsanto "House of the Future" in 1957, designed in collaboration with MIT (Fig. 4.66), or a bit later, (iii) the "Tube House" in **Fig. 2.36**, and (iv) the "Futuro House" (Fig. 4.16); details of these houses are provided in **Table 2.16a** and in

Fig. 2.35 Frank Lloyd Wright: Johnson Wax Building, 1936–39, reinforced concrete: exterior view **(left)** and interior view **(right)**

Fig. 2.36 "Tube House", St. Gallen, Switzerland, 1969 (architect Franz Ulrich Dutler, engineer Heinz Isler): function-integrated and space-forming circular foam core sandwich structure **(left)**; interior view showing coherent formal language of windows and "plastic" furniture **(middle)**; manufacturing on rotating steel mandrel, coating in progress **(right)** [53]

2. Composites – Compact 77

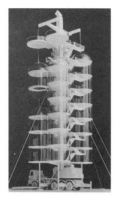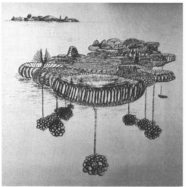

Fig. 2.37 Composite utopias: "Hotel Mobile", 1972 (architect Gernot Nalbach, 80-bedroom hotel erected in one day by mobile crane) **(left)** and "Plastics Floating City", late 1950s (architect William Katavolos) **(right)** [52]

[52–56]. In parallel, utopias were developed, such as towers with attached composite units or floating cities, as depicted in **Fig. 2.37** [52].

All these housing units shared some common characteristics in that they (i) had a modular design and were targeted for serial production, (ii) were built in glass fibre composite sandwich construction, (iii) had curved shapes mainly to provide geometrical stiffness, and (iv) integrated structural, building physics, and architectural functions into the sandwich components, in most cases. Composite laminates and members could, furthermore, already bridge significant spans and be conceived with a certain transparency, as demonstrated by the three examples in **Figs. 2.38–40**. In the third case, the roof is composed of lozenge-shaped translucent laminates, which comprise a central crease for stiffening and inside-turned flanges for assembly, and these

Fig. 2.38 Service Station, Thun, Switzerland, 1960 (design Heinz Isler): integration of structural and architectural functions in web-core sandwich roof composed of translucent glass fibre composite laminates **(left)**; manufacturing by assembling building blocks **(top right)**; lifting of roof **(bottom right)** [53, 56]

Table 2.16a Details of composite projects in architecture discussed in this section – until 1970 [14]

Project	Figure	Application (Composition)	Member type	Member shape	Dimensions Thickness	Laminate Core	Finish outside Inside (fire)	Manufacturing Connections
Snail Shell House, Paris, 1955 [56]	4.6	Space unit (Panels)	Sandwich panels, radial beams	Flat Straight	ø 4 m (core) 30 mm	Glass-polyester Paper honey-comb	Gel coat None	Hand lay-up Bolted
Monsanto House, California,1957 [56]	4.66	Space unit (8 half shells, concrete core)	Sandwich panels	Double-curved	4.8 m (cantilever) 100 mm	Glass-polyester PUR foam (sprayed)	Gel coat None	Hand lay-up Bolted & bonded
Tube House, St. Gallen, 1969 [53, 56]	2.36	Space unit (Tube)	Sandwich panel	Circular	14.9 m (length) ø 4.8 m 58 mm	Glass-polyester PVC foam	Gel coat None	Hand lay-up on rotating steel mandrel
Futuro House, various locations, 1968 [56]	4.16	Space unit (16 segments)	Sandwich panels	Ellipsoid	ø 7.8 m 48 mm	Glass-polyester PUR foam	Gel coat None	Hand lay-up Bolted
Service Station, Thun, 1960 [53]	2.38	Roof (Jointless)	Sandwich panel	Flat	22×14 m 500 mm	Glass-polyester Web-core	None None	Hand lay-up None
Pavillon Expo'64, Lausanne [56, 57]	2.39	Roof (24 umbrellas)	Laminates in steel frame	Hyperbolic paraboloid	18×18 m 3 mm	Glass-polyester None	None None	Hand lay-up Bonded
Sulphur Factory, Rome, 1966 [56]	2.40	Roof (Segments)	Laminates folded	Rhombic	24×10×5 m 2.72×1.20 m (segm.) 1–2 mm	Glass-polyester 30% transparent None	None None	Press method Bolted

Fig. 2.39 Pavillon "Les échanges" Expo'64, Lausanne, 1964 (engineer Heinz Hossdorf) **(left)**; installation of one translucent umbrella composed of glass fibre composite laminates bonded to steel frame [57]

Fig. 2.40 Mobile roof of Sulphur Factory, Rome, 1966 (architect Renzo Piano) **(left)**; installation of translucent panels with inside-turned flanges for assembly, forming crossed arches **(right)**

flanges then generate a form-active system of crossed arches, see Section 4.3.4. In conclusion, most of the design options listed and discussed in Chapter 4, suggested as opportunities for composites to be explored today, already existed in this first phase of composite application in building construction.

This phase of optimism and economic growth ended abruptly, however, with the first and second oil crises during the Arab-Israeli and Iran-Iraq wars in 1973 and 1979, respectively. Composite housing development also stopped, the reasons for which were, in addition to the economic crisis, (i) a low acceptance due to the impossibility of adaptation to individual needs, associated with serial production, (ii) the impossibility of building more than one storey, due to limitations in material properties and connection technology, (iii) as yet insufficiently developed automated processes, and (iv) deficiencies in detailing. According to Heinz Isler [58], in Switzerland, the cost doubled due to the oil crises, the quality of manufacturing was low, maintenance was missing, and architects did not show any interest due to the negative connotation of composites as a poor material. Furthermore, the houses generally did not respond to building acoustic and fire behaviour requirements defined in the standards.

After the decline of composite applications in building construction at the end of the 1970s, it took more than a decade before a revival of composites in architecture occurred. In the meantime, the material properties and manufacturing methods, such as vacuum infusion and pultrusion, significantly improved. Sandwich panels and pultruded profiles became available, which exhibited mechanical properties and shape sizes that even made possible the construction of vehicular bridges (see Section 2.6), multi-storey buildings, and large-scale roofs, as will be discussed below.

2.7.3 Laminated and profile structures

Composite laminates can be used as outer skins of building envelopes. Due to their low thickness and bending stiffness, they cannot, however, contribute to vertical load transmission, but mainly take the horizontal actions from wind. To do so, again depending on their stiffness, they require a narrow underlying supporting structure. The material stiffness can be increased, however, by applying curvature to the laminates, i.e., geometrical stiffness, to activate a shell mechanism with associated membrane stresses, see Section 4.3.4.

Fig. 2.41 Sheraton Hotel Milan Malpensa Airport, Italy, 2010 (Roselli Architetti Associati, manufacturer 3B-the fibreglass company) **(left)**; pultruded glass fibre composite sheets curved on site on subframe of pultruded composite profiles and steel arches **(right)**

Fig. 2.42 House of Dior, Seoul, 2015 (architect Christian de Portzamparc, manufacturer Kolon Composites Seoul) **(left)**; installation of panel on temporary framework **(right)**

Fig. 2.43 Eyecatcher Building, Basel, Switzerland, 1998 (architect Felix Knobel, profile manufacturer Fiberline) [62] **(left)**; three pre-assembled frames composed of pultruded glass fibre composite profiles, installation of last frame **(right)**

Two examples are shown in **Figs. 2.41–42**; related details of the buildings are given in **Table 2.16b**. In the first case, 20 m long pultruded sheets of 1 m width, which are mechanically interlocked along the longer edges, form the vertical envelope of one side of the three-storey building. The sheets are positioned vertically, exhibit a single curvature, and are supported by a narrow subframe of horizontal pultruded composite profiles and vertical steel arches. The thermal insulation layer is not curved and is detached from the laminate structure. In the second example, eleven large-scale, partially overlapping composite panels of complex shape envelop the building interior and characterise the architectural expression, which symbolises the drapery of fabrics. The panels were manufactured by vacuum infusion and are reinforced, on the inner side, by a grid of ribs. Thermal insulation panels are positioned on the inner side.

Pultruded composite profiles can serve as the primary load-bearing structure for buildings, as is demonstrated, for instance, in the five-storey Eyecatcher Building, see **Fig. 2.43** and Appendix A.3. The vertical load transmission structure consists of three parallel frames, which are integrated into the two lateral side envelopes and located in the building centre. The frames are three-layered and built up from adhesively bonded pultruded profiles, see Fig. 3.147; the slabs are made of timber. The joints between horizontal (central) and vertical/inclined (lateral) profiles are bolted to enable the dismantling of the building, as shown in Fig. 4.23 (right). An active fire protection is provided by a sprinkler system (Fig. 3.106, left). The lateral envelopes consist of 50 mm thick translucent, corrugated glass fibre composite sheets filled with translucent aerogel insulation; the inclined envelope is made of glass. Timber staircases are integrated into a timber box on the back side, to which an elevator is also attached [50].

2.7.4 Sandwich structures

A first type of application of sandwich structures concerns space units where panels are assembled to directly conceive smaller living spaces or installed on an underlying structure to form larger spaces. A structural foam core normally also serves as

Table 2.16b Details of composite projects in architecture discussed in this section – from 1990 onwards [14]

Project	Figure	Application (Composition)	Member type	Member shape	Dimensions Thickness	Laminate Core	Finish outside Inside (fire)	Manufacturing Connections
Sheraton Hotel, Milan, 2010 [59]	2.41	Envelope (Skin on sub-frame)	Laminates	Single-curved	20.0×1.0 m 5 mm	Glass-polyester None	Polyurethane varnish None	Pultruded Mechanical interlock
House of Dior, Seoul, 2015 [59]	2.42	Envelope (11 panels)	Ribbed laminates	Double-curved	20×6 m (max.) 12 mm (laminate)	Glass fibres None	Primer plus top coat None	Vacuum-infused None
Eyecatcher, Basel, 1998 [50]	2.43	Primary vertical structure	Profiles	Linear	15.9 m (max. length) 18 mm (max.)	Glass-polyester None	None None (Sprinkler)	Pultruded Bonded & bolted
Spacebox, Utrecht, 2008 [59]	2.44	Space unit (5 panels)	Sandwich panels	Flat	7.5×3.0×2.8 m (unit) 88–110 mm (panels)	Glass-polyester PIR foam	Dyed polyester Spray plaster	Vacuum-infused Bonded
Halley VI, Antarctic, 2013 [60]	2.45	Space units (Panels on steel frame)	Sandwich panels U=0.113 W/m^{2}K	Flat Polygonal nose cones	~10.0×3.9×3.9 m (nose cone) ~195 mm	Glass-polyester PIR foam	Gel coat & UV paint Intumescent paint	Vacuum-infused Bolted on site Nose cones factory bonded
Bus Station, Hoofddorp, 2003 [59]	2.46	Space unit	Sandwich construction	Double-curved	50×10×5 m Variable	Glass-polyester EPS foam full unit thickness	Paint N/A	Laminates sprayed on site on foam
Trinity Cathedral, Paris, 2016 [59]	2.47	Roof (5 domes, large dome composed of 14 segments)	Sandwich panels	Double-curved	Large dome: ø 12 m 7.5 m (segment) 56 mm (constant)	Glass-epoxy PET foam	86 000 gold foil leaves None (no fire requirements)	Vacuum-infused Bolted on inside flanges

Project	Figure	Application (Composition)	Member type	Member shape	Dimensions Thickness	Laminate Core	Finish outside Inside (fire)	Manufacturing Connections
Yitzhak Rabin Center, Tel Aviv, 2005 [61]	2.48	Roof (5 shells composed of 75 segments)	Sandwich panels	Double-curved	31 m (max. shell) 9 m (cantilever) 200–300 mm (constant)	Glass-polyester Web grid inside PIR foam	Paint Sprayed plaster	Vacuum-infused Bonded
Novartis Building, Basel, 2006 [34]	2.49	Roof (4 segments, 70 kg/m^2)	Sandwich panels	Double-curved	21.6×18.5 m 12.5 m span 70–620 mm (variable)	Glass-polyester Web grid inside PUR foam	Gel coat Gel coat (Structural redundancy)	Foam CNC milled, hand lay-up Segments bonded
Steve Jobs Theatre, Cupertino, 2017 [59]	2.50	Roof (44 circle sectors connected to central hub, 42 kg/m^2)	Sandwich beams	Inverted U-shape, straight flanges	ø 47.2 m ø 41.1 m (support) 50–1380 mm (edge-hub, variable)	Carbon-epoxy SAN polymer foam	PU-based coating (Blankets with aluminium foil, sprinkler and rock wool in U-cells)	Prepreg moulding Bolted, silicon sealings

Fig. 2.44 Spacebox modular housing system, University of Utrecht campus, over 300 units, 2003–2008 (architect Mart de Jong, manufacturer Holland Composites) **(left)**; installation of boxes **(right)**

Fig. 2.45 Halley mobile research station VI, Brunt Ice Shelf in Antarctic, 2013 (Hugh Broughton Architects, manufacturer Millfield Composites)

thermal insulation. The housing units of the initiation phase described in the previous Section 2.7.3 belong to this type. More recent examples, assembled from flat panels, are shown in **Figs. 2.44–45**, and Table 2.16b summarises related basic information. The living quality inside such space units strongly depends on the design targets. While compromises were made in student housing concerning sound insulation and indoor climate (i.e., overheating during summer), as shown in the first example, a more robust construction, as in the second example, can provide comfortable living conditions even at temperatures as low as -56°C and wind speeds up to 160 km/h [60]. Flat panel assemblies, however, do not yet demonstrate the impressive potential of sandwich construction in respect to architecture, as is demonstrated in the "building sculpture" in **Fig. 2.46**. Freeform shapes were achieved by directly cutting the foam core accordingly. Since the member thickness was not limited in this case, i.e., the stresses were small due to compact members, laminates could be applied on site (under a tent) by simply spraying impregnated short fibres on the pre-shaped core, rendering a complex mould unnecessary.

Sandwich panel members of a complex shape are also used for roof constructions, as shown in Figs. 2.47–49, see also Table 2.16b. Curved shapes can originate from structural or purely architectural intentions. In the first case, the target is usually to activate a shell mechanism, i.e., membrane stresses, and thus reduce material consumption.

2. Composites – Compact

Fig. 2.46 Bus Station, Hoofddorp, Netherlands, 2003 (NIO Architects) **(top)**; on-site assembly of pre-shaped foam core under tent **(bottom)**

Based on specific boundary conditions, the shape then results from an optimised flow of forces and thus cannot be freely selected, see Section 4.3.4. In the second case, if the shape is freely selected, the consequence is that the structural behaviour is that of a curved plate (in bending) instead of that of a shell, and the material consumption is thus elevated. In the ideal case, however, both intentions are merged, which requires a close collaboration between architect and structural engineer.

In the case of the Trinity Cathedral, shown in **Fig. 2.47**, the onion shape of the dome is an imposed feature, characteristic of Russian orthodox churches and therefore not structurally optimised. Compared to the conventional cladded steel frame construction, the material substitution by composites, nevertheless, made possible a significant reduction of construction time, weight, maintenance and thus overall cost. Furthermore, the shape is smooth and free of sharp edges. The five domes were

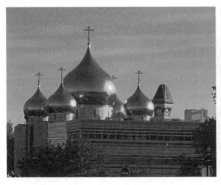 

Fig. 2.47 Orthodox Trinity Cathedral, Paris, 2016 (architects Wilmotte & Associés, manufacturer Multiplast) **(left)**; on-site assembly of central dome **(right)**

Fig. 2.48 Yitzhak Rabin Center, Tel Aviv, Israel, 2005 (architect Moshe Safdie, manufacturer Holland Composites) **(left)**; on-site assembly of two sandwich shells from smaller prefabricated segments **(right)**

manufactured from sandwich panel segments of constant thickness, which were bolted together on site, see Fig. 2.47 (right). No requirements regarding thermal insulation or fire had to be met; the foam is purely structural.

The five sandwich roofs of the Yitzhak Rabin Center, **Fig. 2.48**, also exhibit double curvature and constant thickness; their shape is associated with the petals of a flower. A shell mechanism seems to be activated in the central parts but not in the large overhangs. The main structural action is caused, however, by the sun's radiation and associated through-thickness temperature gradient due to the thermal insulation provided by the foam. This load case made the insertion of webs into the foam necessary to connect both face sheets [61]. For transportation reasons, the five shells were manufactured in smaller segments, which were adhesively bonded on site, see Fig. 2.48 (right). The foam core serves as thermal insulation, and fire protection is provided by sprayed plaster on the inside.

The intention of the architect of the Novartis Building, **Fig. 2.49** (Appendix A.4) was to have a fully transparent building without any visible load-bearing structure, i.e., only a glass envelope, and a wing-shaped roof with large cantilevers. The required light weight and complex form could only be realised with a composite sandwich structure. In contrast to the previous examples, the mid-plane of the roof is flat, but both face sheets are double-curved to form the wing shape. The roof thus works in pure bending and the thickness is adapted accordingly with the highest values in the centre. The shear force transmission in the core is shared by an orthogonal grid of webs and a foam core, whose density is adapted to the shear forces. The roof furthermore stabilises the glass walls and participates in the transmission of the horizontal wind actions. To avoid the cost of a complex mould, 460 foam blocks were CNC-milled into shape and four large-scale roof segments were then manufactured through hand lay-up, and adhesively bonded together on site, see Figs. 2.20, 2.49 (right), and 4.67. The foam also provides thermal insulation, and acoustic panels are integrated into recesses of the sandwich structure. In the event of a fire, a 2×2 m^2 burn-through can occur at any location without causing a collapse of the building – the fire resistance is thus provided by a redundant structural concept [34].

The Steve Jobs Theatre in Cupertino, USA, consists of a circular carbon fibre composite roof, which is supported by a likewise circular glass wall, see **Fig. 2.50**; dimensions and further details concerning materials are listed in Table 2.16b. Carbon

2. Composites – Compact

Fig. 2.49 Novartis Campus Entrance Building, Basel, Switzerland, 2006 (architect Marco Serra, manufacturer Scobalit) **(left)**; installation of glass fibre composite sandwich panel segment **(right)**

Fig. 2.50 Steve Jobs Theatre, Cupertino, USA, 2017 (architect Foster + Partners, manufacturer Premier Composite Technologies, Dubai)

fibres were mainly selected to minimise the roof weight in view of (i) the pure glass supporting structure with limited load-bearing capacity, (ii) the high-risk seismic zone, and (iii) the long-distance transportation from the manufacturing location in Dubai. The roof is divided into 44 narrow circle sectors, which are conceived as radial beams with inverted U-shaped sandwich cross sections of variable width and height and are connected to a central circular hub. The connections between beam webs and beam and hub webs are bolted. Similar to the previous example, the roof transmits forces in bending and shear. Active and passive fire protection systems are installed in the U-shaped cells; the rock wool also serves as thermal insulation.

It is interesting to compare the conceptionally related lens- or wing-shaped carbon and glass fibre composite roofs of the Steve Jobs Theatre and Novartis Building (Fig. 2.49), respectively. Although significant differences exist in the roof plan shapes (circular vs rectangular), spans (ø 41.1 vs 12.5 m unidirectional), and cantilevers (3.1 vs 5.0 m, respectively), basic quantities may be compared, such as (i) slenderness, i.e., span/mid-height ratio, and (ii) unit surface weight. Based on the values in Table 2.16b, the slendernesses of the carbon and glass fibre roofs are 29.8 vs 20.2 and the average unit weights are 42 vs 70 kg/m^2, respectively. The higher slenderness and lighter weight in the former case can be traced back to the use of carbon instead of glass fibres. Furthermore, the carbon fibre roof is function-separated, i.e., the thermal insulation is added and not load-bearing, while the glass fibre roof is function-integrated, i.e., the enclosed foam is thermally insulating and has a structural function.

2.7.5 Design aspects in architecture

The following technical design aspects, specific to architecture, are briefly addressed: fire resistance, function integration, sound absorption, and indoor climate.

As has been shown in the previous sections, profiles and sandwich panels can be used for vertical load transmission. However, if profile components or sandwich panel face sheets (and webs in case of web-core panels) are subjected to axial or in-plane compression, their responses become matrix-dominated and thus sensitive to creep and elevated temperature, see Section 3.2.3. The response to temperature is particularly critical in the event of a fire, since the failure of one member, e.g., a critical column, may lead to the collapse of the whole building. The vertical load-bearing structure should thus be protected by active and/or passive fire protection systems. As discussed in Section 3.9.1, the decomposition of the composite material under fire is usually not the problem, however; the real problem arises much earlier, i.e., when the temperature approaches the glass transition temperature, and the matrix softens. The stabilising effect of the matrix is then lost, and face sheets can wrinkle or webs buckle. The fire protection should thus be able to keep the temperatures below the glass transition temperature, i.e., below 80–120°C. Furthermore, the structural concept for the vertical load transmission should include a certain redundancy to prevent a building collapse in the event of the failure of individual members.

The situation may be less critical for floor slabs or roofs if the fibres are well anchored in zones not subjected to fire and thus remain load-bearing in tension during the fire. Such slabs or roofs should not collapse since they are retained by an active fibre network; they may, however, experience considerable decomposition and deformation. It is thus recommended to also protect such members with an effective fire protection system.

The advantages of function integration into building members have already been discussed. To realise such potential, however, requires a close collaboration between architect, structural engineer, building physicist, and manufacturer.

Since composite structures are lightweight, sound insulation also needs to be considered in the building design [63]. Structure-borne sound can be interrupted by an appropriate arrangement of damping elements or layers. Design solutions can be found in timber construction, where the problems are similar. If the material surfaces are smooth, they are reflective and do not absorb airborne sound. An appropriate interior design can usually solve this problem.

Finally, the indoor climate concerning temperature and humidity should be addressed with care. The thermal insulation performance should be designed accordingly, and it should be possible to absorb potential excessive indoor humidity through appropriate wall or ceiling coatings, interior design, or remove it by natural or mechanical air conditioning.

Additional information about the use of composites in architecture is provided in the following chapters, according to Allocation **Table 2.17**.

2. Composites – Compact

(Allocation) Table 2.17 Supplementary information regarding composites in architecture in subsequent sections of Chapters 3 to 5

Section	Aspect	Subject of additional information
3.9.1	Fire	ISO 834 (indoor) standard fire curve (Fig. 3.105)
		Sprinkler system in Eyecatcher Building (Fig. 3.106) (Appendix A.3)
3.9.3		Fire resistance behaviour of pultruded and sandwich panel members
3.9.4		Fire dynamic simulation of indoor fires
3.10.2	Durability	Durability of Eyecatcher Building (Appendix A.3) over 25 years of use, including adhesive joints
3.10.3		Durability of sandwich roof of Novartis Building (Appendix A.4) over 17 years of use, including adhesive joints
3.10.5		Conclusions regarding detailing to ensure durability
4.2.3	Learnings from nature	Floor plan of Dock Tower, derived from plant stem cross section (Fig. 4.4) (Appendix B.1)
		Discussion on usefulness of a 50-year design service life for buildings
4.2.4.2		First composite house, 1955/56, inspired by growth of a snail shell (Fig. 4.6)
		Kinetic composite envelope of Thematic Pavilion, Expo 2012, Yeosu, South Korea (Fig. 4.7)
4.3.1	Structural form and system	Lightweight pavilions with identical space programmes in timber, aluminium, and composite construction (Fig. 4.12)
4.3.2		Material-tailored support detail of sandwich roof of Novartis Building (Fig. 4.14) (Appendix A.4)
4.3.3		Material-tailored structural form: Futuro House, Matti Suuronen, 1968 (Fig. 4.16)
4.4.4	Transparency	Integration of skylights and opaque solar cells into composite sandwich roof structure (Fig. 4.44)
		Integration of dye-transparent solar cells into composite sandwich roof structure (Fig. 4.45)
4.5.1	Function integration	Merging of load-bearing structure and insulating envelope and effect on architectural expression (Fig. 4.46)
4.5.2.3		Example of envelope thickness reduction by function integration (Fig. 4.49)
4.5.2.4		Zero-emission buildings
4.5.3.2		Composite hydronic radiant system (Fig. 4.54)
4.5.4		Solar cell encapsulation into glass fibre composite laminates
4.5.5		Coloured translucent glass fibre composite laminates with white backlights, Zurich Main Station, 2003 (Fig 4.57)
		Composite enclosure of HR Giger Bar, Gruyères, 2003 (Fig. 4.59, left)
		Composite cladding texture of San Francisco Museum of Modern Art expansion, 2015 (Fig. 4.59, right)
		Texture of sandwich panels of House of Dior, Seoul, 2015 (Fig. 4.60)

Section	Aspect	Subject of additional information
4.7.2	Modular construction	Monsanto "House of the Future", 1957 (Fig. 4.66)
		Installation of sandwich roof of Novartis Building (Fig. 4.67) (Appendix A.4)
4.7.3		Densification of cities by adding new storeys to existing buildings (Fig. 4.68)
5.2	Design	Conceptual design

2.8   COMPOSITES IN SCULPTURE

Composite materials have always been appreciated by artists because of their free formability, ease of application in hand lay-up technique, light weight, availability in any colour, and ease in applying paint. Due to the strength and stiffness of the materials, and their durability, they are also preferred materials for large-scale outdoor sculptures. The most important factor in this respect is that artists can produce the sculptures themselves, in contrast to large-scale metal sculptures, where a dependence on manufacturers exists, and full control over the final product is thus limited. Furthermore, sculptures can be developed on a small-scale and then scaled up using the same materials and techniques [64, 65].

To retroactively trace back the precise nature of the materials used in such sculptures is not always simple. Terms like "epoxy", which do not address the fibre component, or "fabric drenched in glue", without specifying the nature of the materials, can be found in the literature. If the fibre component is addressed, the term "fibreglass" is used in most cases, which permits better identification of the material [66–68]. There are essentially two methods of composite manufacturing of large-scale sculptures: hand lay-up of impregnated fabrics (i) on pre-shaped supports, such as foams or wire-meshes, and (ii) in moulds made of foam, or more rarely of wood, to create thin shells [64].

A more detailed discussion follows on the application of composites in sculpture by renowned artists Jean Dubuffet (1901–1985), Niki de Saint Phalle (1930–2002), Angel Duarte (1930–2007), Yayoi Kusama (1929), Miguel Berrocal Ortiz (1933–2006), and Jonathan Borofsky (1942). Large-scale (mostly outdoor) sculptures are shown to establish a link to architecture in terms of (i) further highlighting the formal potential of composites and (ii) demonstrating how the ease of experimenting with these materials can support form finding. Further composite sculptures are discussed in other chapters, as specified in Allocation **Table 2.18** at the end of this section.

Jean Dubuffet was a French painter and sculptor and representative of the "art brut" (Outsider art) style [69]. His large-scale outdoor sculptural work is characterised by the use of "fibreglass" and phenoxy resin. After finding that painting applied directly to expanded polystyrene (EPS) foam had insufficient durability and mechanical resistance, Dubuffet conducted a thorough material search and began to use fibreglass due to its improved durability, free formability, and robustness [70]. The materials were hand laid up in expanded polystyrene moulds to manufacture thin shells, which were then attached on both sides of grids of metallic or polymer profiles to form thicker sandwich panels of complex form. The polystyrene moulds lent a characteristic rough texture to the fiberglass surfaces, on which the typical painting patterns and colours were then applied [66].

Two of Dubuffet's works are pictured in **Figs. 2.51–53**: the *Jardin d'émail* (1974) [67] and the *Tour aux figures* (1988) [65] (for details see figure captions). The first work represents a large rectangular one-storey building, see Fig. 2.51 (left), on which a tall tree and two smaller sculptures are fixed (Fig. 2.52, left). The two leaves of the tree were manufactured in one piece in the above-described sandwich panel technique, and subsequently painted (Fig. 2.52, right and middle). Furthermore, the concrete roof slab of the building, modelled on top with a complex relief (Fig. 2.51, right), was coated with a fiberglass layer and the characteristic white painting with black lines was then applied. The top part of the tree leaves demonstrated good durability over more than forty years, while still exhibiting the original painting, until a renovation of the whole building sculpture took place in 2016–20. A 2×3 m architectural model, also made of fibreglass, served as a basis for the realisation of the work in 1974 [71].

Fig. 2.51 *Jardin d'émail* building sculpture, 30×20×10 m, Kröller-Müller Museum, Otterlo, NL, 1974 (artiste Jean Dubuffet) **(left)**; artist on concrete relief before applying "fibreglass" coating, 2 April 1974 **(right)**

Fig. 2.52 Tree sculpture in *Jardin d'émail* **(left)**; painting of tree leaves, March 1974 **(middle)**; one of two "fibreglass" shells of tree leaves, moulded in polystyrene mould

Fig. 2.53 *Tour aux figures* building, 24 m tall, Île Saint-Germain, Issy-les-Moulineaux, France, 1988 (artiste Jean Dubuffet) **(left)**; painted "fibreglass" shells with steel back-frames, waiting on installation on steel skeleton **(middle)**; hand lay-up of "fibreglass" shells in expanded polystyrene moulds, finished unpainted shell element with steel back-frame (in foreground) and installation of steel back-frame on shell (in background) **(right)**

The sculpture *Tour aux figures* is a 24 m tall multi-storey building, composed of an outer steel skeleton, concrete slabs, and masonry walls, see Fig. 2.53 (left and middle). The envelope elements are conceived as fibreglass shells with steel back-frames to be fixed on the skeleton, as shown in Fig. 2.53 during manufacturing (right) and installation (middle). Similar to the case of the *Jardin d'émail*, Dubuffet developed the project on the basis of various small-scale polymer and polystyrene models [65].

Niki de Saint Phalle was a French-Swiss painter and sculptor, who became known for her Nana sculptures, which she first started manufacturing in 1964, using fibreglass and polyester resin on chicken wire mesh, among other materials [64, 72]. One of her largest works is *She – A Cathedral*, a giant female (temporary indoor) sculpture, which she created together with Jean Tinguely and Per-Olof Ultvedt for the *Hon-en historia* exhibition in the Moderna Museet, Stockholm, in 1966, see **Fig. 2.54** (left) [68, 73]; a model of the sculpture was developed simultaneously, as shown in Fig. 2.54 (middle) and **Fig. 2.55** (right, in the background). The animated walk-through sculpture has two storeys (Fig. 2.54, right) and is composed of an inner steel structure enveloped with chicken wire (Fig. 2.55, left), on which resin-impregnated fabrics were applied (Fig. 2.55, middle). Subsequently, the sculpture and the model were simultaneously painted by Niki de Saint Phalle (Fig. 2.55, right).

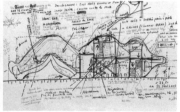

Fig. 2.54 *She – A Cathedral* sculpture, 24.6×9×6 m, Moderna Museet, Stockholm, Sweden, June–September 1966 (artists Niki de Saint Phalle, Jean Tinguely, Per-Olof Ultvedt) **(left)**; model of sculpture, 133 cm long, paper mache on chicken wire, 1966 **(middle)**; sculpture animated with bar, exhibition space, cinema, slide, aquarium **(right)**

2. Composites – Compact

Fig. 2.55 Two-storey steel frame of *She – A Cathedral* sculpture during installation **(left)**; "glue-drenched fabric" application on chicken wire envelope **(middle)**; Niki de Saint Phalle applying paint with model of sculpture in background **(right)**

Fig. 2.56 Climbing sculpture, 5×5×5 m, open-air bath in Tramelan, 1967–70 (artist Angel Duarte, engineering work by Heinz Isler): view on sculpture **(left)**; people climbing on sculpture **(middle)**; module and pull-out-test of module connectors **(right)**

Angel Duarte was a Spanish painter and sculptor interested, among other things, in geometric abstraction and the interactivity of plastic space, using the hyperbolic paraboloid as a module. Based on the model *Module blanc 13* (90×90×73.5 cm, polymer injection moulding), he designed the large-scale climbing sculpture, shown in **Fig. 2.56**, which was engineered and materialised by Heinz Isler. The more than 100 hyperbolic modules are assembled by invisible connectors.

Yayoi Kusama is a contemporary Japanese artist, who became known for her polka dots applied on canvas and sculptures. Particularly well known are her pumpkin sculptures, some of which were also manufactured using fibreglass, see **Fig. 2.57** [74].

Fig. 2.57 *Pumpkin* sculpture, 2 m tall, Benesse Art Site Naoshima, Japan, 1994, "fibreglass" shell (hollow inside) **(left)**; artist Yayoi Kusama together with sculpture **(right)**

Fig. 2.58 *Citius Altius Fortius*, 3.0 m tall, kinetic torso, Olympic Museum Lausanne, Switzerland, 1991/92 (artist Miguel Berrocal Ortiz): sculpture closed **(left)**; lateral elements translated and rotated **(right)**

Miguel Berrocal Ortiz was a Spanish sculptor who is known for his puzzle sculptures, which can be disassembled into abstract elements. The sculpture shown in **Fig. 2.58** is composed of six upright elements of irregular shape and named *Citius Altius Fortius*, which refers to the Olympic motto (faster, higher, stronger). "The mobile artwork illustrates the athletes' effort and physical and mental concentration" (description near the base). The sculpture has a polystyrene foam core and carbon and aramid fibre composite face sheets. Composites have been selected for their light weight, which facilitates the translating and rotating movements of the elements.

Jonathan Borofsky is a contemporary American sculptor and printmaker. His most famous works are the monumental two-dimensional *Hammering Man* sculptures with motorised arms, made of steel. In another large-scale sculpture, the *Walking Man*, see **Figs. 2.59–60**, only the interior structure is made of steel, while the exterior skin is made of fibreglass. The sculpture was scaled up from a model, which is composed of a steel armature wrapped with CelluClay (a form of fibre mache), see Fig. 2.60 (left). Nine elements were built and pre-assembled in Los Angeles (Fig. 2.60) and then shipped to Munich and re-assembled there (Fig. 2.59). The artist selected fibreglass for the shell because of its durability (when protected properly), the possibility of modelling and shaping by hand, and the ability to conceal the seams between the elements after assembly on site [75].

Fig. 2.59 *Walking Man* sculpture, 17 m tall, Munich Re building, Germany, 1995 (artist Jonathan Borofsky) **(left and right)**; on-site assembly of sculpture consisting of nine elements **(middle left and right)**

Fig. 2.60 Jonathan Borofsky with model in front of unpainted foot of sculpture **(left)**; application of "fibreglass" skin on inner steel structure, La Paloma factory, Los Angeles **(middle left and right)**; storage and pre-assembly of nine elements **(right)**

(Allocation) Table 2.18 Supplementary information regarding composites in sculpture in subsequent sections of Chapters 3 to 5

Section	Aspect	Subject of additional information
4.2.4.1	Learnings from nature	TCy Tensegrity Sculpture, Geneva, 2022 (Fig. 4.5, right) (Appendix A.7)
4.5.5.4	Function integration	D-Tower with LED backlight of different colours, NOX Architecture, Doetinchem, Netherlands, 2004 (Fig. 4.58)

2.9 REFERENCES

[1] *CEN/TS 19101: Design of fibre-polymer composite structures*. European Committee for Standardization, Brussels, 2022.

[2] *EN 1990: Basis of structural and geotechnical design*. European Committee for Standardization, CEN/TC 250, Brussels, 2002.

[3] R. M. W. Reising, B. M. Shahrooz, V. J. Hunt, A. R. Neumann, A. J. Helmicki, and M. Hastak, "Close look at construction issues and performance of four fiber-reinforced polymer composite bridge decks", *J. Compos. Constr.*, vol. 8, pp. 33–42, 2004, doi: 10.1061/(asce)1090-0268(2004)8:1(33).

[4] J. Knippers, J. Cremers, M. Gabler, and J. Lienhard, *Construction manual for polymers + membranes*. Birkhäuser, Basel, 2011.

[5] T. Keller, *Use of fibre reinforced polymers in bridge construction*. Structural Engineering Documents, SED 7, International Association for Bridge and Structural Engineering, IABSE, Zurich, 2003.

[6] J. Thomason, L. Yang, and F. Gentles, "Characterisation of the anisotropic thermoelastic properties of natural fibres for composite reinforcement", *Fibers*, vol. 5, 2017, doi: 10.3390/fib5040036.

[7] M. L. Minus and S. Kumar, "The processing, properties, and structure of carbon fibers", *Jom*, vol. 57, pp. 52–58, 2005, doi: 10.1007/s11837-005-0217-8.

[8] K. L. Pickering, M. G. A. Efendy, and T. M. Le, "A review of recent developments in natural fibre composites and their mechanical performance", *Compos. Part A Appl. Sci. Manuf.*, vol. 83, pp. 98–112, 2016, doi: 10.1016/j.compositesa.2015.08.038.

[9] L. Yan, N. Chouw, and K. Jayaraman, "Flax fibre and its composites: A review", *Compos. Part B Eng.*, vol. 56, pp. 296–317, 2014, doi: 10.1016/j.compositesb.2013.08.014.

[10] J. R. Correia, T. Keller, J. Knippers, J. T. Mottram, C. Paulotto, J. Sena-Cruz, and L. Ascione, *Design of fibre-polymer composite structures: Commentary to European Technical Specification CEN/TS 19101:2022*. CRC Press, London, 2025.

[11] N. Vahedi, "Fire performance evaluation of GFRP-balsa sandwich bridge decks", PhD thesis, EPFL, Lausanne, 2022. [Online]. Available: 10.5075/epfl-thesis-9364.

[12] N. Vahedi, C. Tiago, A. P. Vassilopoulos, J. R. Correia, and T. Keller, "Thermophysical properties of balsa wood used as core of sandwich composite bridge decks exposed to external fire", *Constr. Build. Mater.*, vol. 329, 2022, doi: 10.1016/j.conbuildmat.2022.127164.

[13] S. Yanes-Armas, J. de Castro, and T. Keller, "Long-term design of FRP-PUR web-core sandwich structures in building construction", *Compos. Struct.*, vol. 181, pp. 214–228, 2017, doi: 10.1016/j.compstruct.2017.08.089.

[14] Original contribution.

[15] T. Keller, J. Rothe, J. De Castro, and M. Osei-Antwi, "GFRP-balsa sandwich bridge deck: Concept, design, and experimental validation", *J. Compos. Constr.*, vol. 18, 2014, doi: 10.1061/(ASCE) CC.1943-5614.0000423.

[16] N. Vahedi, J. R. Correia, A. P. Vassilopoulos, and T. Keller, "Effects of core air gaps and steel inserts on thermomechanical response of GFRP-balsa sandwich panels subjected to fire", *Compos. Struct.*, vol. 313, 2023, doi: 10.1016/j.compstruct.2023.116924.

[17] M. Garrido, J. R. Correia, and T. Keller, "Effects of elevated temperature on the shear response of PET and PUR foams used in composite sandwich panels", *Constr. Build. Mater.*, vol. 76, pp. 150–157, 2015, doi: 10.1016/j.conbuildmat.2014.11.053.

[18] N. Vahedi, C. Wu, A. P. Vassilopoulos, and T. Keller, "Thermomechanical characterization of a balsa-wood-veneer structural sandwich core material at elevated temperatures", *Constr. Build. Mater.*, vol. 230, 2020, doi: 10.1016/j.conbuildmat.2019.117037.

[19] J. De Castro and T. Keller, "Ductile double-lap joints from brittle GFRP laminates and ductile adhesives, Part I: Experimental investigation", *Compos. Part B Eng.*, vol. 39, pp. 271–281, 2008, doi: 10.1016/j.compositesb.2007.02.015.

[20] U. O. Meier, A. U. Winistörfer, and L. Haspel, "World's first large bridge fully relying on carbon fiber reinforced polymer hangers", in *Sampe Europe Conference Amsterdam*, 2020.

[21] T. Keller and M. Schollmayer, "Plate bending behavior of a pultruded GFRP bridge deck system", *Compos. Struct.*, vol. 64, pp. 285–295, 2004, doi: 10.1016/j.compstruct.2003.08.011.

[22] T. Q. Liu, P. Feng, Y. Wu, S. Liao, and X. Meng, "Developing an innovative curved-pultruded large-scale GFRP arch beam", *Compos. Struct.*, vol. 256, 2021, doi: 10.1016/j.compstruct.2020.113111.

[23] S. Satasivam, Y. Bai, Y. Yang, L. Zhu, and X. L. Zhao, "Mechanical performance of two-way modular FRP sandwich slabs", *Compos. Struct.*, vol. 184, pp. 904–916, 2018, doi: 10.1016/j.compstruct.2017.10.026.

[24] L. Liu, X. Wang, Z. Wu, and T. Keller, "Resistance and ductility of FRP composite hybrid joints", *Compos. Struct.*, vol. 255, 2021, doi: 10.1016/j.compstruct.2020.113001.

[25] J. L. Clarke (Ed.), *EUROCOMP design code and handbook – Structural design of polymer composites*. E&FN Spon, 1996.

[26] M. Angelidi, A. P. Vassilopoulos, and T. Keller, "Ductile adhesively-bonded timber joints – Part 1: Experimental investigation", *Constr. Build. Mater.*, vol. 179, pp. 692–703, 2018, doi: 10.1016/j.conbuildmat.2018.05.214.

[27] J. De Castro and T. Keller, "Ductile double-lap joints from brittle GFRP laminates and ductile adhesives, Part II: Numerical investigation and joint strength prediction", *Compos. Part B Eng.*, vol. 39, pp. 282–291, 2008, doi: 10.1016/j.compositesb.2007.02.016.

[28]   T. Keller and T. Vallée, "Adhesively bonded lap joints from pultruded GFRP profiles. Part I: Stress-strain analysis and failure modes", *Compos. Part B Eng.*, vol. 36, pp. 331–340, 2005, doi: 10.1016/j.compositesb.2004.11.001.

[29]   T. Keller and T. Vallée, "Adhesively bonded lap joints from pultruded GFRP profiles. Part II: Joint strength prediction", *Compos. Part B Eng.*, vol. 36, pp. 341–350, 2005, doi: 10.1016/j.compositesb.2004.11.002.

[30]   T. Vallée and T. Keller, "Adhesively bonded lap joints from pultruded GFRP profiles. Part III: Effects of chamfers", *Compos. Part B Eng.*, vol. 37, pp. 328–336, 2006, doi: 10.1016/j.compositesb.2005.11.002.

[31]   T. Vallée, J. R. Correia, and T. Keller, "Probabilistic strength prediction for double lap joints composed of pultruded GFRP profiles part I: Experimental and numerical investigations", *Compos. Sci. Technol.*, vol. 66, pp. 1903–1914, 2006, doi: 10.1016/j.compscitech.2006.04.007.

[32]   T. Vallée, J. R. Correia, and T. Keller, "Probabilistic strength prediction for double lap joints composed of pultruded GFRP profiles – Part II: Strength prediction", *Compos. Sci. Technol.*, vol. 66, pp. 1915–1930, 2006, doi: 10.1016/j.compscitech.2006.04.001.

[33]   T. Vallée, T. Keller, G. Fourestey, B. Fournier, and J. R. Correia, "Adhesively bonded joints composed of pultruded adherends: Considerations at the upper tail of the material strength statistical distribution", *Probabilistic Eng. Mech.*, vol. 24, p. 358–366, 2009, doi: 10.1016/j.probengmech.2008.10.001.

[34]   T. Keller, C. Haas, and T. Vallée, "Structural concept, design, and experimental verification of a glass fiber-reinforced polymer sandwich roof structure", *J. Compos. Constr.*, vol. 12, pp. 454–468, 2008, doi: 10.1061/(ASCE)1090-0268(2008)12:4(454).

[35]   M. Volk, O. Yuksel, I. Baran, J. H. Hattel, J. Spangenberg, and M. Sandberg, "Cost-efficient, automated, and sustainable composite profile manufacture: A review of the state of the art, innovations, and future of pultrusion technologies", *Compos. Part B Eng.*, vol. 246, 2022, doi: 10.1016/j.compositesb.2022.110135.

[36]   A. Hindersmann, "Confusion about infusion: An overview of infusion processes", *Compos. Part A Appl. Sci. Manuf.*, vol. 126, 2019, doi: 10.1016/j.compositesa.2019.105583.

[37]   A. Zureick, "Fiber-reinforced polymeric bridge decks", Structural engineering, mechanics and materials research report No. 97-1, prepared for the National Seminar on Advanced Composite Material Bridges, 1997.

[38]   H. M. Bernstein and H. R. Bosch, "The CONMAT initiative: Charting an innovative path to the next century", *Public Roads, U.S. Dep. Transp. Fed. Highw. Adm.*, vol. 59, 1996.

[39]   J. P. Busel, J. D. Lockwood, and K. D. Walshon, "Product selection guide: FRP composite products for bridge applications", Market Development Alliance of the FRP Composites Industry, Harrison, NY, 2000.

[40]   "Global FRP use for bridge applications: Vehicular", Market Development Alliance of the FRP Composites Industry, Harrison, NY, 2003.

[41]   L. N. Triandafilou and J. S. O'Connor, "FRP composites for bridge decks and superstructures: State of the practice in the U.S.", in *Proceedings of International Conference on Fiber Reinforced Polymer (FRP) Composites for Infrastructure Applications*, 2009.

[42]   P. B. Potyrała, "Use of fibre reinforced polymer composites in bridge construction – State of the art in hybrid and all-composite structures", Polytechnic University of Catalonia, Barcelona, Spain, 2011.

[43]   T. Hong and M. Hastak, "Construction, inspection, and maintenance of FRP deck panels", *J. Compos. Constr.*, vol. 10, pp. 561–572, 2006, doi: 10.1061/(asce)1090-0268(2006)10:6(561).

[44] L. N. Triandafilou and J. S. O'Connor, "Field issues associated with the use of fiber-reinforced polymer composite bridge decks and superstructures in harsh environments", *Struct. Eng. Int. J. Int. Assoc. Bridg. Struct. Eng.*, vol. 20, pp. 409–413, 2010, doi: 10.2749/101686610793557663.

[45] "Fibre-reinforced polymer composite bridges in Europe", University of Plymouth; UK, 2023. https://ecm-academics.plymouth.ac.uk/jsummerscales/composites/bridges.htm (accessed Aug. 10, 2023).

[46] S. Reeve, "FRP bridge decking – 14 years and counting", *Reinf. Plast.*, 2010, [Online]. Available: https://www.reinforcedplastics.com/content/features/frp-bridge-decking-14-years-and-counting/

[47] *prEN 1991-2: Actions on structures – Part 2: Traffic loads on bridges*. European Committee for Standardization, Brussels, 2021.

[48] C. J. Burgoyne and P. R. Head, "Aberfeldy Bridge – an advanced textile reinforced footbridge", in *TechTextil Symposium*, Frankfurt, 1993.

[49] "Rhyl Harbour Bridge, Wales" https://www.gurit.com/rhyl-harbour-bridge-rhyl-wales-uk/ (accessed Sep. 01, 2023).

[50] T. Keller, "Fibre reinforced polymer materials in building construction", in *IABSE Symposium Towards a Better Built Environment*, Melbourne, 2002.

[51] Q. Truong, *Composite architecture*. Birkhäuser, Basel, 2020.

[52] A. Quarmby, *The plastics architect*. Pall Mall Press, London, 1974.

[53] H. Isler, "Tragende Bauteile aus Kunststoff: Anwendungsbeispiele", *Schweizerische Bauzeitung*, vol. 95, pp. 13–20, 1977.

[54] A. Schwabe and H. Saechtling, *Bauen mit Kunststoffen*. Ullstein AG, Berlin, 1959.

[55] R. Doernach, *Bausysteme mit Kunststoff*. Deutsche Verlags-Anstalt, Stuttgart, 1974.

[56] P. Voigt, *Die Pionierphase des Bauens mit glasfaserverstärkten Kunststoffen (GFK) 1942 bis 1980*. PhD thesis, Bauhaus-Universität Weimar, 2007. [Online]. Available: https://e-pub.uni-weimar.de/opus4/frontdoor/index/index/docId/821

[57] P. Dietz and J. A. Torroja, *Heinz Hossdorf – Das Erlebnis ein Ingenieur zu sein*. Birkhäuser, Basel, 2003.

[58] H. Isler, personal communication on 22 January 1998.

[59] Data obtained from manufacturer or designer.

[60] R. Slavid, *Ice Station: The creation of Halley VI Britain's pioneering Antarctic research station*. Park Books AG, Zurich, 2015.

[61] M. Eekhout and S. Wichers, *Lord of the wings: The making of free form architecture*, vol. 12. IOS Press, Amsterdam, 2015.

[62] T. Keller, J. De Castro, and M. Schollmayer, "Adhesively bonded and translucent glass fiber reinforced polymer sandwich girders", *J. Compos. Constr.*, vol. 8, pp. 461–470, 2004, doi: 10.1061/(ASCE)1090-0268(2004)8:5(461).

[63] M. Proença, A. Neves e Sousa, M. Garrido, and J. R. Correia, "Acoustic performance of composite sandwich panels for building floors: Experimental tests and numerical-analytical simulation", *J. Build. Eng.*, vol. 32, 2020, doi: 10.1016/j.jobe.2020.101751.

[64] L. Beerkens and F. Breder, "Temporary art? The production and conservation of outdoor sculptures in fiberglass-reinforced polyester", *Conserv. Perspect. GCI Newsl.*, 2012.

[65] M. Loreau, *Catalogue des traveaux de Jean Dubuffet: Tour aux figures – amoncellements, cabinet logologique*. Weber Editeur, Lausanne, 1973.

2. Composites – Compact

[66] E. E. Nagy, A. Lippincott, D. Lippincott, J. Chasse, and R. Prosser, "The treatment of a large painted outdoor sculpture: Kiosque l'évidé (1970) 1984, by Jean Dubuffet", *J. Am. Inst. Conserv.*, vol. 62, pp. 28–57, 2023, doi: 10.1080/01971360.2022.2031456.

[67] *Dubuffet Jardin d'émail*. Rijksmuseum Kröller-Müller, Otterlo Nederland, 1974.

[68] *Hon-en historia*. Moderna Museet, Stockholm, 1967.

[69] "Fondation Dubuffet", https://www.dubuffetfondation.com/ (accessed Oct. 22, 2023).

[70] E. Pratt, P. Houlihan, and E. Ordonez, "An investigation into a transferred paint/cast resin sculpture by Jean Dubuffet", in *From marble to chocolate – The conservation of modern sculpture. Tate Gallery Conference*, 1995.

[71] "Jardin d'émail," Kröller-Müller Museum Otterlo, NL. https://krollermuller.nl/en/jean-dubuffet-jardin-d-email (accessed Oct. 23, 2023).

[72] "Niki de Saint Phalle", https://nikidesaintphalle.org/ (accessed Oct. 23, 2023).

[73] "Remembering *She – A Cathedral*", Moderna Museet Stockholm. https://www.modernamuseet.se/stockholm/en/exhibitions/remembering-she-a-cathedral/ (accessed Oct. 23, 2023).

[74] "Yayoi Kusama", http://yayoi-kusama.jp/e/information/ (accessed Oct. 23, 2023).

[75] J. Borofsky, personal communication on 3 November 2023.

CHAPTER 3

COMPOSITES – OVERCOMING LIMITATIONS

3.1   INTRODUCTION

As introduced in the Preface and Chapter 1, composites may offer unique properties in structural engineering and architecture when compared to conventional structural materials. However, composites may also exhibit some limitations in properties, design, or use, not in general but under certain circumstances, when compared to conventional materials such as reinforced concrete or steel. Such potential limitations and associated circumstances are addressed in this chapter, and solutions or measures of how limitations can be prevented or overcome are discussed.

The chapter is structured in sections according to aspects of composites that may cause limitations, as listed in **Table 3.1** (Sections 3.2–11). The selection of these aspects is based on my experience in research and design of composite structures, as emphasised in the Preface. For each aspect, typical potential limitations are summarised as keywords in Table 3.1; the list is not complete, however, and more details are given in the corresponding sections. Similarly, measures to overcome these limitations are listed; they will also be discussed in detail throughout the chapter.

An aspect not contained in Table 3.1 is stiffness, i.e., the comparatively low elastic modulus of glass fibre composites, which in most cases governs the design, if deflection limits are critical. A typical glass fibre composite laminate may exhibit an elastic modulus of 30 GPa, which is, on the one hand, seven times lower than that of steel (210 GPa), but on the other hand, of the same magnitude as the values for reinforced concrete, or approximately three times higher than the elastic moduli of timber (8–12 GPa). The low stiffness of glass fibre composites can often be overcome quite easily by simply increasing the depths of the cross section, which does not in fact increase the material consumption significantly. Although the elastic modulus of a composite bridge deck is similar to a reinforced concrete deck, the required depth of the composite deck is approximately 30–50% larger, since the cross section is thin-walled and not filled with material, as in the case of the concrete deck. The replacement of existing concrete by composite bridge decks may thus cause problems since the depth cannot often be changed. The situation is similar if composite beams are used in place of steel beams, which would require a significant increase in depth of the former. A feasible solution in such cases may be the partial use of carbon fibres, which exhibit a much higher elastic modulus than glass fibres (but also are much more expensive). Due to these obvious measures that can be taken concerning stiffness, no separate section is dedicated to this aspect.

Table 3.1 Aspects causing potential limitations of composites and measures to overcome them

Aspects of composites	Potential limitations	Measures to overcome
3.2 Anisotropy	Time- and temperature-dependent matrix-dominated response	Appropriate fibre architecture to trigger fibre-dominated response
3.3 Imperfections and tolerances	Unexpected reduction in load-bearing capacity	Appropriate testing Nonlinear analysis Appropriate detailing
3.4 Viscoelasticity	Time and temperature dependence of mechanical properties	Fibre-dominated response Post curing
3.5 Curing and physical ageing	Low glass transition temperature Curing at low temperature	Post curing Fibre-dominated response
3.6 Ductility	Brittle material or member response	Structural redundancy to provide system ductility Hybrid members and systems
3.7 Fracture	Low fracture resistance	Appropriate fibre architecture to activate fibre bridging
3.8 Fatigue	High damage rate	Appropriate detailing to limit stress concentrations Prevention of creep-fatigue interaction
3.9 Fire	Low fire resistance of material	Fibre-dominated response Cellular section members Redundant structural systems Fire protection systems
3.10 Durability	Irreversible damage caused by environmental conditions	Appropriate material selection Appropriate detailing Protective coatings Inspection and maintenance
3.11 Sustainability	High environmental cost	Consideration of whole life cycle Appropriate application

As can be seen in Table 3.1, some aspects are related since they may cause similar limitations, e.g., anisotropy and viscoelasticity. Anisotropy may result from the fibre architecture, which can trigger, depending on the loading direction, either fibre- or matrix-dominated structural responses. In the latter case, the viscous component of the generally viscoelastic behaviour of composites is activated, which results in time and temperature dependence of the responses and may thus cause limitations in the design. Imposing a more fibre-dominated response, through a correspondingly tailored fibre architecture, can overcome such limitations. An appropriate fibre architecture can also prevent limitations occurring from other aspects, e.g., curing and physical ageing, fracture, and fire.

This discussion across the aspects will be continued in Section 6.1 of Chapter 6, where the measures derived from the individual aspects in this Chapter 3 will be regrouped into five groups of measures to overcome limitations of composites, and final conclusions will be drawn.

3.2 ANISOTROPY

3.2.1 Introduction

The term "anisotropic" designates the direction dependence of material properties, in contrast to "isotropic", which characterises direction-independent properties. A third recurrent term is "orthotropic", which represents a special case of "anisotropic", where the properties differ along three orthogonal axes. Typical isotropic and orthotropic conventional materials are steel and timber, respectively, while reinforced concrete, in most cases, is also orthotropic – based on the typical orthogonal layout of the steel reinforcement, which is embedded in the intrinsically isotropic concrete (if noncracked). Composite pultruded profiles, on the other hand, are strongly orthotropic, as already pointed out in Section 1.2 (Fig. 1.1, left).

Composites can thus essentially be compared with reinforced concrete, since they also comprise linear components – fibres instead of rebars – which are embedded in an isotropic matrix – a polymer instead of a concrete matrix, see Section 1.2. The situation is, however, more complicated than in concrete because (i) the fibres can be anisotropic themselves (e.g., carbon or natural fibres) and (ii) the fibre architecture is normally more complex than a simple grid of steel rebars. Combinations of different materials at the member level, e.g., orthotropic face sheets and core materials in sandwich panels, can further complicate the mechanical responses of composite structures. In the following, anisotropy is first discussed at the material level, and the possible effects on the structural responses at the member or joint levels are addressed subsequently, before concluding on how anisotropy can improve (rather than compromise) the structural performance.

3.2.2 Anisotropy in composite and core materials

3.2.2.1 Fibres and laminates

While glass and basalt fibres are isotropic materials, carbon and natural fibres are strongly anisotropic, as discussed in Section 2.2.1. At the laminate level, the degree of anisotropy varies depending on the fibre architecture. A laminate comprising unidirectional fibres is orthotropic, while a quasi-isotropic laminate can be conceived by a multidirectional architecture. Fibres can also be mixed, as shown in **Fig. 3.1** (left), where glass fibres (white colour) and carbon fibres (black) are arranged in a bidirectional/orthotropic woven twill fabric.

The possible effects of anisotropy on the mechanical laminate properties are depicted in Fig. 3.1 (right). The variation of the elastic modulus is shown for different fibre architectures, as a function of the direction of the applied stress, the latter varying from 0° to 90° related to the fibre direction. The laminates have the same number of layers and fibre volume fraction. If the laminate is unidirectional (UD) and the stress is applied in the same 0° (axial) direction, the elastic modulus is a maximum, while

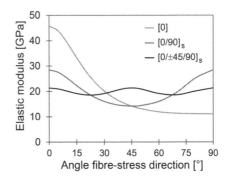

Fig. 3.1 Bidirectional/orthotropic woven twill fabric composed of glass (white) and carbon (black) fibres (left); dependence of elastic modulus on direction of applied stress, for composite laminate with different fibre architectures, fibre volume fraction of 0.6, fibre/matrix elastic moduli of 74/3.5 GPa (right)

the modulus values decrease with increasing angle of stress, exhibiting a minimum at 90°, which approaches the elastic modulus of the matrix. In the case of a symmetric bidirectional/orthotropic [0/90]$_s$ laminate, the 0° and 90° values are identical. Since only half of the fibres are in the 0° direction, the modulus in the 0° direction is much lower than in the UD case. The lowest value is found, however, in the 45° direction, which is without fibres. A quasi-isotropic laminate, with an elastic modulus almost independent of the stress angle, can be achieved by adding fibres in further directions, e.g., in a symmetric [0/±45/90]$_s$ laminate. The 0° value of the modulus, however, further decreases compared to the bidirectional laminate, while the 45° value increases.

The fibre architecture of a laminate can thus be tailored to the actual stresses resulting from the applied forces. If the direction of a force does not change, all the fibres can be arranged in that direction and their capacity can be fully used. However, if a force and its direction are variable, a more isotropic architecture is appropriate, with the consequence that the directional properties decrease (compared to the UD case) and only a proportion of the fibres can be used to their full capacity.

The anisotropy of composite laminates also affects their fracture behaviour, as explained in Section 3.7.1, where interlaminar, intralaminar, and translaminar fracture or crack planes are differentiated (Fig. 3.60).

3.2.2.2 Sandwich core materials

Commonly used core materials in composite sandwich panels are foams, such as polyurethane (PUR) and polyethylene terephthalate (PET) foams, and end-grain balsa wood, see Section 2.2.2. These all exhibit an overall orthotropic behaviour, while they can be assumed isotropic perpendicular (transverse) to their rise (PUR), extrusion (PET), or growth (balsa) direction. In the case of PET, weld-lines parallel to the extrusion direction may, however, slightly affect the transverse isotropy. Out-of-plane and in-plane properties should thus be basically differentiated in sandwich core applications [1].

In the case of PUR foam, as described in Section 2.2.2.2, a geometrical anisotropy in the cell morphology may occur, which depends on the manufacturing process

3. Composites – Overcoming Limitations

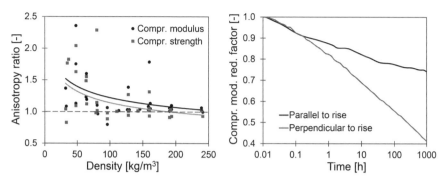

Fig. 3.2 Anisotropy of polyurethane foam (PUR): compressive strength and compression modulus anisotropy ratio vs density **(left)**; compression modulus reduction factor due to creep vs time, parallel and perpendicular to rise direction, density 60 kg/m^3, linear viscoelasticity at 20–24°C **(right)**, based on data in [2]

and density. At lower densities, the cell shape may be more elongated in the rise direction than at higher densities. This geometrical anisotropy also causes a mechanical anisotropy, which can be characterised by the strength or modulus anisotropy ratio between values in rise and perpendicular-to-rise directions, as shown in **Fig. 3.2** (left) for compression. As can be seen, the anisotropy ratio is significant for lower densities up to 80–100 kg/m^3 and is more pronounced for modulus than for strength. Similarly, a pronounced anisotropy may occur under creep conditions, as depicted in Fig. 3.2 (right), where creep effects are considered by a reduction of the compression modulus over time. The creep deformations are significantly higher in the perpendicular than in the rise direction, which can be explained by the higher stiffness in the rise direction (Fig. 3.2 , left) [2].

Sandwich cores made of balsa wood exhibit orthotropic behaviours at two levels, i.e., the material and balsa panel levels, as described in Section 2.2.2.1. At the material level, anisotropy may cause different responses in the three principal shear planes, i.e., the shear plane (i) parallel to end-grain (Eg), (ii) parallel to flat-grain (Fg/P), and (iii) transverse to flat-grain (Fg/T), as shown in **Fig. 3.3** (left). The corresponding shear stress vs strain relationships are depicted in Fig. 3.3 (right). Pseudo-ductile responses can be recognised in all cases (according to Fig. 3.48), while the highest resistance and pseudo-ductility are exhibited by the end-grain configuration, where a tendon-action of the cellulose fibres is activated, similar to the case of LVL balsa materials, shown below (Fig. 3.5, top left). The flat-grain parallel configuration has medium strength and lowest pseudo-ductility [3].

At the panel level, the anisotropy may cause different fracture or crack planes, depending on (i) the selected trunk cut-outs for the balsa panel, (ii) their in-plane orientations in the panel, and (iii) their densities. Fracture may thus develop in (i) radial-longitudinal planes (RL), (ii) transverse-longitudinal planes (TL), and (iii) adhesive interface planes between balsa blocks (IF), as shown in **Fig. 3.4**. Fracture primarily occurs in the RL or interface planes of the lowest density blocks and subsequently propagates to the next low-density blocks through TL planes [4].

Since balsa wood is a natural material, the mechanical properties exhibit a significant scatter, in line with the density variation. One way of reducing this scatter is

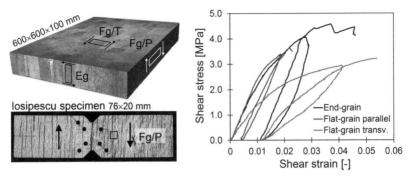

Fig. 3.3 Definition of shear planes in end-grain balsa panel, average density 291 kg/m^3 (**top left**); Iosipescu shear specimen with failure parallel to flat-grain (Fg/P) (and video-extensometer marks, **bottom left**); corresponding shear stress vs strain relationships in three shear planes, all specimens of similar density and exhibiting pseudo-ductility (**right**), based on data in [3]

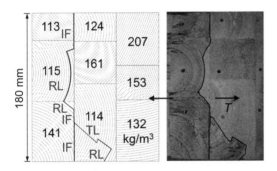

Fig. 3.4 Top view on crack pattern under in-plane tension (T) in end-grain balsa wood panel composed of blocks of different trunk segments, in-plane orientations, and densities (RL = radial-longitudinal crack/fracture plane, TL = transverse-longitudinal plane, IF = adhesive interface plane between blocks), based on [4]

to compose panels with blocks of similar densities, as shown in Fig. 3.4. However, as can be seen in the figure, the density variation is still significant. Another approach to make mechanical properties more uniform is to recompose the trunk material into a laminated veneer lumber (LVL) material. The balsa core of the Avançon Bridge deck slab, for instance, is composed of laminated end-grain veneer layers of 250 kg/m^3 average density [5] (see Appendix A.5). A further advantage of this approach is that the orientations of the layers can be varied, thus attenuating the effects of the basic material anisotropy. Results of shear experiments performed on a 0/90° LVL balsa material are shown in **Fig. 3.5**. In an in-plane layer arrangement, the through thickness shear strength is significantly increased by the activation of tension in the 0° cellulose fibres, which also provides a pseudo-ductile response ("shear tendon" case in Fig. 3.5, right). The shear strengths in the other planes cannot, however, benefit from the material recomposition and thus remain comparably small. Compared to Fig. 3.3 (right), shear modulus and strength are significantly lower, since the average density is much lower. Responses from such standardised small-scale experiments, such as tendon

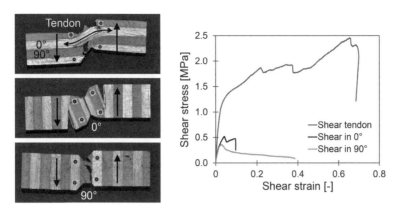

Fig. 3.5 Shear failure modes of LVL 0/90° balsa Iosipescu specimens of different configurations (layer thickness 6 mm, average density 186 kg/m^3) **(left)**; shear stress vs strain responses **(right)**, based on data in [6]

activation, must, however, be carefully checked to ensure that they correspond to the real material behaviour in the full-scale member, and whether values obtained from small-scale experiments can thus be used for the design.

3.2.3 Effects of anisotropy on structural response

The interactions between (i) the anisotropy, i.e., fibre architecture, (ii) the viscoelastic nature of composites, (iii) the range of stress directions, and (iv) the stress type (tension, compression, or shear) may produce very different responses of composite members, which generally can be grouped into two categories, i.e., fibre-dominated and matrix-dominated responses. In fibre-dominated cases, the elastic component of the viscoelastic material always dominates the responses, while in matrix-dominated cases, the viscous component is dominant. If the elastic component is dominant, the extent of the effects of elevated temperature, moisture, and creep and relaxation on the structural responses is quite low, while the extent may be significant if the viscous component is dominant, as further discussed in Section 3.4.1. To be more specific, applying tension to a laminate in the fibre direction, for instance, causes a fibre-dominated response. However, the same tension transverse to the fibres induces a matrix-dominated response. Compression and shear almost always produce a matrix-dominated response. A fibre-dominated response under shear is possible if the fibres act in the principal stress direction, as can be the case in the web of a beam with ±45° fibre architecture, for instance. Furthermore, a typical matrix-dominated response is progressive fibre-tear failure in adhesive connections, caused by combined through-thickness shear and tension, see Section 3.7.2.

As has been discussed above, anisotropic carbon fibres have excellent tension properties while the transverse properties are low. Together with the fact that they also exhibit a high creep rupture strength (see Section 3.4.5), carbon fibres are the preferred materials for pre- or post-tensioned tendons and cables. However, the anchoring

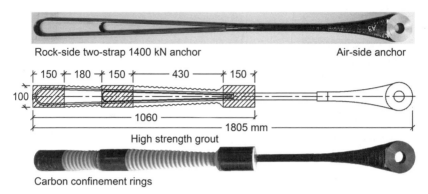

Fig. 3.6 Carbon fibre ground or rock anchor of 1400 kN capacity, comprising rock-side two-strap anchor, air-side one-strap anchor, high-strength grout anchor body, and carbon fibre confinement rings at rock-side strap ends and deviation point [8, 11, 13]

of these tension members is difficult due to the low interlaminar shear and transverse compressive strength. Cone-type anchors of varying degrees of sophistication were therefore and are still being developed, which are filled with different types of casting materials to generate transverse compression and associated friction and also to prevent stress concentrations at the entrance of the anchor. However, the basic concept remains as it was when introduced in 1996 [7] and has also been implemented in the strengthening of the Verdasio Bridge with carbon fibre cables, see Appendix A.1. A gradient filling material was used in that case to reduce stress concentrations.

A more material-tailored anchor for ground and rock applications was developed at the CCLab [8, 9]. The anchor is based on laminated pin-loaded carbon fibre straps, as conceived in [10]. The carbon fibres are continuous inside the anchor in this approach, and the strap radius is designed so that the radial deviation stresses do not exceed the transverse compression resistance. The rock-side anchor can be composed of multiple straps, which are embedded in a prefabricated high-strength grout anchor body. Further carbon fibre confinement rings are arranged at the strap ends and other deviation points, as shown in **Fig. 3.6**. In a three-strap configuration, an anchor capacity of 2500 kN can be reached [11, 12].

Further concepts based on prestressed carbon fibre straps were developed at the CCLab to strengthen existing reinforced concrete slabs against punching shear failure, as shown in **Fig. 3.7**. Four crossed non-laminated and thus flexible straps are arranged around the column, i.e., they are installed through inclined boreholes and subsequently prestressed. In addition to increasing the live load, the prestressing also allows the partial unloading of the slab from the self-weight, and thus further increases the punching resistance [14–16].

The anisotropy of composite laminates also has a significant effect on the load-bearing capacity of joints and of bolted joints in particular. Depending on the fibre architecture, a bolted joint response can vary from brittle with low resistance to pseudo-ductile with high resistance, as shown in **Fig. 3.8**. In the former case, a unidirectional architecture of the adherends causes a brittle splitting failure at low

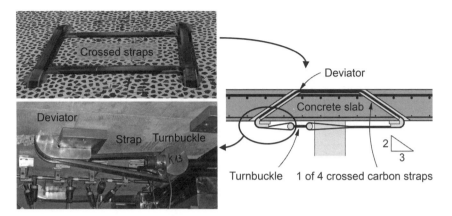

Fig. 3.7 Prestressed flexible carbon fibre strap system to strengthen existing reinforced concrete slabs against punching failure, top and bottom view of experimental set-up with four crossed carbon straps (**left**); section drawing with one installed flexible strap (**right**) [13–15]

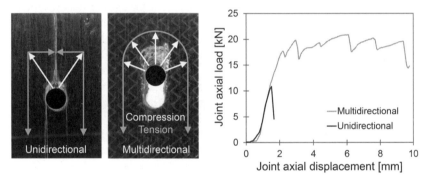

Fig. 3.8 Splitting and pin-bearing (crushing) failure modes (**left**); brittle vs pseudo-ductile response (**right**) of bolted double-lap joints with pultruded adherends of uni- and multidirectional basalt fibre architecture, respectively, both adherend types with similar total fibre volume fraction, based on [17]

resistance since only the matrix resists the transverse tension stress induced by the load introduction. In a multidirectional architecture, however, fibres in the transverse direction prevent a splitting failure and result in a pseudo-ductile pin-bearing (crushing) failure at a much higher resistance.

A more complex failure is shown in **Fig. 3.9**, where the adherends are pultruded and thus exhibit the typical architecture of this manufacturing process with (unidirectional) rovings in the centre and combined mats and fabrics on the outer sides. The bolted double-lap joint is subjected to reversed cyclic loading. In the first phase, the outer mats and fabrics can maintain a double-sided pseudo-ductile pin-bearing failure. However, when the distance to the free edge becomes too short, a comparably brittle shear-out failure of the material between bolt and edge occurs, which is mainly caused by the low shear resistance of the matrix in the inner roving core. Such a shear-out

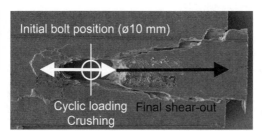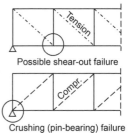

Fig. 3.9 Pultruded glass fibre composite adherend of bolted double-lap joint exhibiting complex fibre architecture (rovings in centre and combined mats/fabrics) **(left)**; combined initial crushing (pin-bearing) and final shear-out failure mode under reversed cycling loading **(middle)**; direction of diagonals in bolted truss system **(right)**

failure, which limits the resistance, cannot occur in the other direction, away from the edge. Pin-bearing failure would continue there under monotonic loading, and the resistance could be increased or at least maintained during the pseudo-ductile failure. Consequently, when considering the structural level and taking the example of trusses, it may be advantageous to arrange the diagonals in the compression direction to prevent shear-out failure, as shown in Fig. 3.9 (right), as is occasionally implemented in timber trusses with bolted joints.

The resistance of adhesively bonded joints is also affected by anisotropy. Not so much by anisotropy in the laminate plane, as by anisotropy through the laminate thickness, because the stresses in the adhesive layer must always first pass through the outer resin layer of the adherend until they can be transmitted by the underlying fibres. Since adhesive bonding areas are large (compared to the contact areas of bolts, for instance), the stresses are relatively small and can thus be borne by the resin layer of the adherend.

3.2.4 Conclusions

If one excludes anisotropic carbon fibres and core materials, anisotropy in composite members is mainly related to the fibre architecture and can have positive and negative effects on the structural response and performance. If anisotropy triggers a fibre-dominated behaviour, structural stiffness and resistance usually increase and the sensitivity to environmental conditions (elevated temperature, moisture) and time (creep, relaxation) decreases. On the other hand, the structural response may tend to be brittle, at the material or member level. If anisotropy causes a matrix-dominated response, the effects on the structural responses are opposed, i.e., stiffness and resistance may be reduced and sensitivity to environmental conditions and time increased; the response may tend, however, to pseudo-ductile behaviour, again at the material and member level. A reasonable design strategy and conception of the fibre architecture for composite structures may thus target (i) a fibre-dominated response, to reach optimum stiffness and resistance, and (ii) a redundant structural system, to provide a pseudo-ductile response at the system level, as discussed in Section 3.6.

Similar opposing requirements may exist in the design of members and their joints. A high degree of unidirectional fibres may be desired to increase the member stiffness, for instance, while in the joint area, a multidirectional architecture is preferred to increase the resistance of bolted or hybrid joints. Members and joints can thus not always be optimised individually, and an overall compromise solution should be found. This particularly applies for pultruded members, where the fibre architecture cannot be changed in the joint area.

3.3 IMPERFECTIONS AND TOLERANCES

3.3.1 Introduction

Imperfections in structural engineering primarily concern deviations from the theoretical or prescribed geometry of members or structures, which arise during manufacturing of the former or erection of the latter. Tolerances, on the other hand, are allowable geometrical deviations from the theoretical geometry and can thus be designated as allowable geometrical imperfections. They are specified by manufacturers for their products or defined in standards.

Geometrical imperfections may negatively or can positively affect structural performance. In thin-walled or slender members or structures subjected to compression, they can have destabilising effects, as shown in the example of **Fig. 3.10**. The axis of the slender column has an initial lateral geometrical imperfection, which is w_0 at mid-height. Subjected to axial compression, the bending moment and lateral deflection disproportionally increase until the column fails, which normally occurs below the elastic bifurcation buckling load. Such slender members generally fail under predominant nonlinear (or second-order) bending by reaching the material's bending strength, and not through "buckling", which, strictly speaking (and against common practice), is not a failure mode, since the elastic (Euler) buckling load depends only on the elastic modulus and is not limited by material strength. The same applies for snap-through "buckling", as demonstrated in Section 4.3.4 for prestressed arches (or bending-active

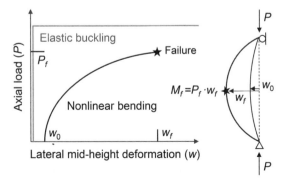

Fig. 3.10 Nonlinear bending of a column subjected to axial compression (P) caused by geometrical imperfection (w_0) and comparison to elastic bifurcation buckling (w_f, P_f, M_f are the lateral deformation, load, and bending moment at failure, respectively)

structures), which also fail when they reach the bending strength. Geometrical imperfections can, however, also have positive effects if stabilising tension forces or tensile stresses are activated, and the resistance is increased accordingly.

Current design codes developed for conventional materials such as steel, concrete, and timber consider geometrical imperfections. According to [18], geometrical imperfections are defined as deviations of geometrical data and should be considered if the design of a structure is sensitive to such data. Imperfections can be considered (i) on the loading side of the resistance verification, (ii) in the output of the resistance model (R), and (iii) in the partial factor (γ_{Rd}) accounting for the uncertainty of the resistance model and geometric deviations if these are not modelled explicitly. Furthermore, it is specified (i) that tests to derive partial factors for material properties should take into account imperfections if they can influence the results, and (ii) deviations of geometrical properties that are within tolerances are assumed to be catered for by the partial factors for actions, material properties, and resistance models.

The problem of imperfections and their effects on the structural performance becomes more complicated for composites, since the material compositions and manufacturing methods are more complex (compared to steel, for instance); consequently, imperfections are not limited to geometrical imperfections. In the following, the different types of imperfections in composite materials, members and structures, and their possible consequences on the structural performance are discussed first. Tolerances are then addressed before concluding with the ways in which imperfections and tolerances can be addressed to least affect the structural performance.

3.3.2 Types and consequences of imperfections in composites

Reviewing the literature covering imperfections in composites reveals that many types of imperfections are discussed and characterised. Nevertheless, available information about the quantification of imperfections is rather limited, and a consistent classification of imperfections is missing. Concerning the latter point, as an example, "residual curing stresses" are subsumed under "geometrical imperfections" in [1], while they are classified under "manufacturing defects" in [19]. This example, however, highlights what can be considered as three "umbrella terms" for a classification of imperfections in composites. The terms can be related to the composite (i) material, (ii) member and joint, and (iii) structure levels, as summarised in **Table 3.2**. Geometrical imperfections are attributed to the member and joint, and structural levels. The further distinction used in [1], i.e., local and global geometrical imperfections, is in line with this attribution. Residual stresses arise during curing procedures and thus concern the member and joint level. Lastly, manufacturing defects are related to the material level and concern fibres, matrix, and interfaces.

Global geometrical imperfections mainly comprise sway imperfections of frames or imperfections of arches and shells, while local geometrical imperfections are related to members, such as profiles or sandwich panels, and bonded, bolted, or hybrid joints. They may comprise, amongst other things, bow imperfections (out-of-straightness), unevenness, eccentricities in load application and inside joints, and non-parallel face sheets and face sheet waviness in sandwich panels [1, 20]. As mentioned above, such

3. Composites – Overcoming Limitations

Table 3.2 Classification of imperfections [13]

Type of imperfections	Level		
	Material	Member, joint	Structure
Geometrical imperfections	-	Local	Global
Residual stresses	-	Curing	-
Manufacturing defects	Fibres, matrix, interfaces	-	-

geometrical imperfections can reduce the load-bearing capacity, in particular of members that are subjected to axial compression.

Residual stresses mainly arise during curing and can be caused by (i) temperature and curing degree differences inside members, (ii) cure shrinkage of the matrix, and (iii) manufacturing tool-member interaction (e.g., through different thermal expansions or locking mechanisms) [19, 21, 22]. They can induce distortion of members or reduction of the load-bearing capacity, since such residual stresses must be added to the stresses from external loading and cannot be relieved by yielding, as is the case in steel, for instance.

Manufacturing defects found in fibres can include misalignments, waviness and wrinkles, and non-uniform distribution. Matrix defects comprise cracks, voids, porosity, and resin-rich regions. Interface defects include deficient adhesion and delamination, either between fibres and matrix or in adhesive joint interfaces [19, 20, 23–27]. Manufacturing defects increase the uncertainty in the estimates of material properties, which is expressed in increased partial factors for materials, if the latter are appropriately derived from tests (see above). In CEN/TS 19101 [1], for instance, an additional strength correction factor is introduced if sub-laminate or ply testing occurs, since the uncertainty in the derived laminate properties increases due to, amongst other things, the influence of imperfections.

Critical locations in respect to manufacturing defects are web-flange junctions (WFJs) of pultruded profiles, which often exhibit multiple defects, such as mat wrinkles, discontinuities in mat layers, and resin pockets, as shown in two cases in **Fig. 3.11** (left) and Fig. 3.52 (web-face panel junctions, WPJ). In the former case, a mat wrinkle makes the response matrix-dominated and causes increased principal tensile stresses, and subsequent failure occurs at this location (Figs. 3.11, middle and right). In the latter case, as discussed in *Example-2* of Section 3.6.3, depending on the type of imperfection, the WPJs exhibit either a matrix-dominated brittle or ductile failure mode. A third case concerns outstand flanges of open-section profiles, where transverse impact can cause cracks in the WFJs, as discussed in Section 3.10.2 (Fig. 3.149). It is difficult to prevent imperfections in WFJs/WPJs since they occur transverse to the pultrusion direction where no tension stretches the fibres and holds them in position. Since these imperfections also make WFJs/WPJs sensitive to fatigue, as discussed in Section 3.8.8, the use of pultruded profiles may be critical if they are subjected to

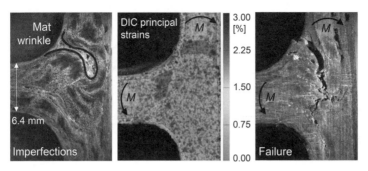

Fig. 3.11 Mat wrinkle defect in web-flange junction of pultruded I-profile **(left)**; effect on principal strain distribution close to failure if subjected to transverse bending (M) **(middle)**; resulting failure mode at wrinkle location **(right)**, according to [28]

fatigue in the transverse direction, as is the case in bridge decks composed of pultruded profiles.

In another case, it is shown how non-uniform distributions of roving fibres cause eccentricities, which may increase the nonlinear bending of pultruded laminates under axial compression, and thus reduce the failure load as a function of the non-dimensional slenderness, as shown in **Fig. 3.12**. Defects in the adhesive layer can also reduce the static and fatigue performance of adhesive joints, particularly if they are located at the overlap ends where high stress concentrations already occur without imperfections [29, 30]. Furthermore, as shown in Section 3.10.2 (Fig. 3.151), misalignments of surface veils can lead to early fibre blooming and may thus reduce the durability of composite members.

3.3.3 Tolerances

Tolerances, i.e., allowable imperfections (see above), can be specified at all three levels shown in Table 3.2 – either by manufacturers in agreement with designers or in standards. Tolerances for dimensions, straightness of edges, squareness and flatness of composite members and pultruded profiles are provided in [20] and [23], respectively.

Adhesive joints are particularly sensitive to tolerances, since they often already incorporate unfavourable eccentricities (e.g., in lap shear joints), and additional eccentricities due to geometrical imperfections of the adherend members worsen the problem. On the other hand, adhesive joints also allow compensation for tolerances of the adherend members by adapting the layer thickness. An example is shown in **Fig. 3.13**, where tolerances in composite bridge deck panel thicknesses are compensated for in panel-girder adhesive connections. Spacers on top of the steel girder flanges assure a minimum adhesive layer thickness, while the adhesive is distributed using a camber-shaped application device which has a larger thickness in the middle. Unnecessary adhesive is laterally pressed out through this geometry by the weight of the deck panels, and full contact in the entire adhesive layer interfaces is assured.

3. Composites – Overcoming Limitations

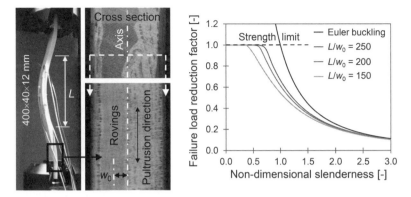

Fig. 3.12 Pultruded laminate under axial compression **(left)**; centroid eccentricity (w_0) of roving layer **(middle)**; imperfection-dependent reduction factor of failure load (first eigenmode shape assumed with maximum imperfection = L/w_0) **(right)**, derived from data in [25]

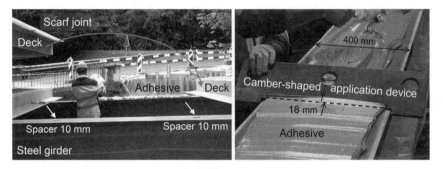

Fig. 3.13 Handling of tolerances in bridge deck-to-girder adhesive connections, with spacers **(left)** and camber-shaped application device **(right)** (Avançon Bridge, see Appendix A.5)

3.3.4 Conclusions

Imperfections in composite construction may occur during manufacturing at the material and member or joint levels and during execution at the structural level where members are assembled. If imperfections exceed the tolerances that were assumed during the design (i.e., allowable imperfections), they may reduce the theoretical load-bearing capacity and thus compromise structural safety. Critical cases in this respect are members or structures subjected to axial compression, where geometrical imperfections can cause nonlinear increases of bending and subsequent reaching of material strength. Web-flange junctions of pultruded profiles are another critical example, where manufacturing defects at the material level can affect (i) the static resistance to transverse loads and lead to premature crippling failure, (ii) the transverse impact resistance of outstand flanges in open sections, and (iii) the fatigue resistances of bridge decks.

Allowable imperfections (tolerances) should be specified in agreement with the manufacturer and considered in the design of members, joints, and whole structures. In critical cases, as mentioned above, nonlinear (second-order) analysis should be performed to cover the effects of geometrical imperfections. Material characterisation tests to derive partial safety factors should include the effects of potential residual stresses and manufacturing defects, i.e., coupon test specimens should be representative of the manufactured members regarding imperfections. Testing at the member or joint level allows one to take all types of imperfections into account, as is specified in [1] for fatigue testing, for instance.

Measures to compensate for geometrical imperfections can be implemented in the design of details and joints. Structures that are highly sensitive to imperfections can furthermore be conceived in a redundant or fail-safe way, where local failure does not induce the collapse of the whole structure. Finally, existing imperfections should be controlled during execution and compared with the assumptions made in the design.

3.4   VISCOELASTICITY

3.4.1   Introduction

Composites with a thermoset matrix show a pronounced viscoelastic load-bearing behaviour, i.e., they exhibit the synergistic features of elastic solids and viscous fluids. In elastic solids, external forces are directly added to those between atoms and molecules; these materials store and completely release elastic energy during loading and unloading. Viscous fluids, however, tend to flow away under applied forces through large changes in shape, whereby energy is continuously and totally dissipated. Accordingly, viscoelastic materials do both, i.e., store and release, and dissipate energy during loading/unloading cycles [31, 32].

The viscous component of the material, in particular, makes the mechanical responses time-, temperature-, and moisture-dependent, and may strongly affect the serviceability and structural safety of composite structures, especially if (i) elevated temperatures, significant permanent loads and moisture contents are involved, (ii) the behaviour is matrix-dominated (see Section 3.2.3), and (iii) the polymer matrix is not fully cured. Under such conditions, composites may exhibit viscoelastic phenomena such as softening, creep, recovery, and relaxation. In this respect, the state of a composite material is of importance, e.g., if the material is in the glassy or rubbery state. The viscous component significantly increases during glass transition, i.e., during the transition from the glassy to the rubbery state.

Viscoelastic phenomena may be linear or nonlinear. If the material state does not change and the material remains undamaged, the material behaviour is linearly viscoelastic. In this case, stresses are proportional to strains at a given time and the linear Boltzmann superposition principle is applicable. If the material state changes and damage initiates, as may occur in the secondary stage of creep (when secondary bonds are progressively lost and molecular chains are aligned to the stress direction, accompanied by the appearance of microcracks, see below), the material behaviour changes to nonlinearly viscoelastic and then depends, among other things, on the stress level [1, 31, 33].

This Section 3.4 is written from an engineering perspective and focuses on aspects related to viscoelasticity that are relevant for the design of composite structures. Certain complex behaviours of viscoelastic composites are thus simplified and others, which may be important from the perspective of material scientists, are not addressed. Composites and adhesive joints based on thermosets and elastomers are considered, since these are the most widely used polymers in composite structures. Engineering and phenomenological models are presented first to explain the background of the viscoelastic behaviour. These models are then used to discuss (i) glass transition behaviour, (ii) time and temperature sensitivity of material properties, (iii) creep, recovery, and relaxation behaviours, and (iv) the reversible effects of moisture on material properties.

3.4.2 Engineering and phenomenological models

The configuration and response to stress of the molecular structure of a thermoset or elastomer polymer can be described by the engineering model shown in **Fig. 3.14**, which differentiates an elastic from a viscous state. In the elastic state, i.e., if subjected to low stress levels, the molecular chain network is stabilised by the secondary bonds and can be seen as a triangulated, almost rigid truss, which mainly acts as an elastic solid, see Fig. 3.14 (left); the stored energy is almost completely released after unloading.

If the stress level increases, the chains start to align with the stress direction and the secondary bonds diminish and then almost disappear because the chain distances increase in the stress direction; the network flexibility increases and finally transforms into a mechanism (in the structural design sense), with its behaviour approaching that of a viscous fluid. The increasing deformation – e.g., stretching of the chains in the case of applied tension, as shown in Fig. 3.14 (right) – is accompanied by the formation of microcracks (damage) in the matrix and friction between the crack flanks and moving chains. Damage causes strain softening, while, on the other hand, stretching of the molecular chains results in strain hardening in the loading direction, see **Fig. 3.15**

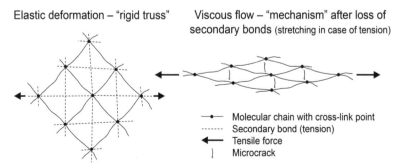

Fig. 3.14 Engineering model for molecular network structure of a thermoset polymer, elastic state **(left)**; viscous (stretched) state **(right)** [13]

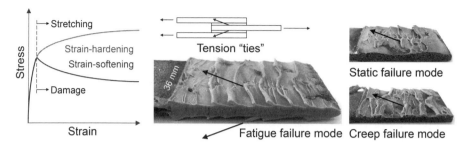

Fig. 3.15 Strain-hardening and strain-softening definitions **(left)**; stretching and alignment with principal stress direction of pseudo-ductile acrylic (elastomer) adhesive in composite double-lap joints, similar cohesive failure modes under monotonic/static, creep, and fatigue loading (at stress ratio $R = 0.1$), according to [34] **(middle and right)**

(left) and Section 2.2.3 [32]. All these processes dissipate energy, as further discussed in Section 3.6. If damage occurs, irreversible permanent deformation will remain in the viscoelastic material after unloading. This irreversible viscoelastic deformation should, however, not be confused with the irreversible viscoplastic deformation in thermoplastic materials, which is not caused by damage or fracture.

The alignment of the molecular chains to the stress direction is also illustrated by the experimental results shown in Fig. 3.15. Composite double-lap joints bonded with the acrylic elastomer adhesive, introduced in Section 2.2.3, are subjected to axial tension, under monotonic/static loading at ambient and elevated temperatures, creep, and fatigue loading. The failure modes are similar in all three cases, i.e., a cohesive failure occurs in the adhesive, where the initially homogenously distributed material separates into almost parallel ties (or bands), which are aligned with the tension principal stresses. In cases where the molecules are fully stretched at failure, i.e., at elevated temperatures, creep, and medium or high cycle fatigue, cohesive failure always occurs at the same strain, independent of the loading type [34], which can thus be associated with the failure strain of the primary bonds.

To simulate the viscoelastic response of polymers and composites, phenomenological models are used in many cases. Based on the theory of viscoelasticity and depending on the material behaviour, different combinations of springs and dashpots are used, the former representing the elastic and the latter the viscous behaviour. These combinations and constants, i.e., spring stiffnesses (S_i) and dashpot viscosity coefficients (η_i) are normally fitted to experimental results. An example is shown in **Fig. 3.16**, where the behaviour of the same acrylic elastomer adhesive (as above), applied in composite double-lap joints subjected to axial monotonic tension and reversed cyclic tension-compression, is modelled. The adherends are pultruded laminates of 100×10 mm cross section and the joint area is 100×100 mm. Two parallel Maxwell units (consisting of one spring and one dashpot in series in each unit) are used to model the pseudo-ductile adhesive responses. A nonlinear spring is introduced in the second unit to capture the strain hardening due to molecular stretching, while the softening due to damage is simulated by the dashpot. The load vs joint displacement curves in the monotonic case demonstrate how the modelling response is composed:

3. Composites – Overcoming Limitations   119

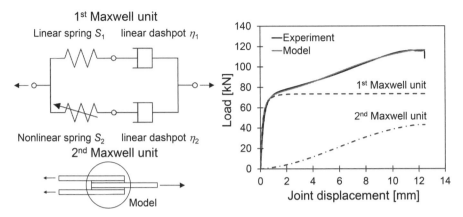

Fig. 3.16 Phenomenological model for pseudo-ductile acrylic (elastomer) adhesive in composite double-lap joints subjected to axial monotonic tension and reversed cyclic tension-compression **(left)**; load vs joint displacement responses in monotonic case, experimental and from modelling, including both Maxwell unit contributions **(right)**, based on data in [35]

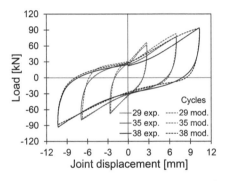

Fig. 3.17 Load vs joint displacement responses in reversed cyclic tension-compression case, experimental (exp.) and from modelling (mod.), based on data in [35]

the first unit captures the elastic and subsequent nonlinear response, while the second unit adds the nonlinear stretching (hardening) and subsequent softening (damage) contribution, see Fig. 3.16 (right). The sum of both matches the experimental result well [35]. Similarly, the experimental and modelling responses of the reversed cyclic tension-compression loading agree well. The hardening at smaller joint displacements and softening towards the maximum displacements are well simulated by the model, as shown in **Fig. 3.17**.

3.4.3   Glass transition temperature

An experimental method to quantify the elastic and viscous responses of polymers and composites is Dynamic Mechanical Analysis (DMA), where a small specimen is subjected to an oscillating stress or strain of a certain amplitude and frequency

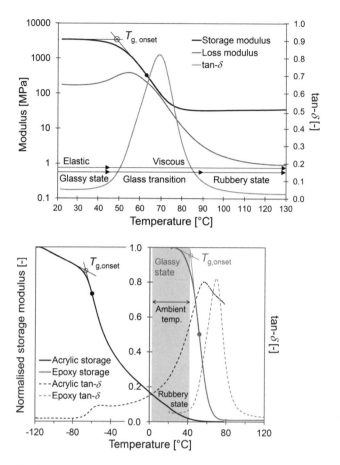

Fig. 3.18 DMA temperature-dependent responses of a cold curing epoxy adhesive, definition of glass transition temperature **(top)**; comparison of cold curing epoxy (thermoset) and cold curing acrylic (elastomer) adhesive, values adopted from [34] **(bottom)**

during an increasing temperature; the support conditions of the specimen can also be varied. Resulting metrics are the storage modulus (E'), loss modulus (E''), and loss factor (tan-δ), as shown in **Fig. 3.18** (top). The storage modulus represents the elastic stiffness and is proportional to the stored energy during a loading cycle, while the loss modulus is proportional to the dissipated energy during the loading cycle. The loss factor is the ratio of loss to storage modulus, i.e., tan-$\delta = E''/E'$, and thus corresponds to the ratio of the viscous to the elastic response [36–38].

Two material states and a material transition can basically be discerned in Fig. 3.18 (top), i.e., the glassy state, glass transition, and rubbery state [39]. In the glassy state, the response is mainly elastic, and the storage modulus is high and the loss modulus low. During glass transition, the secondary bonds are progressively lost; the flexibility of the molecular structure and the viscous response thus increase, and

3. Composites – Overcoming Limitations

more energy is dissipated. Accordingly, the storage modulus sharply decreases, while the loss modulus and loss factor reach a peak. In the rubbery state, storage modulus, loss modulus, and loss factor stabilise at a low level. The response during glass transition is normally reversible, i.e., the secondary bonds can re-establish during cooling, while thermal damage may occur if entering further into the rubbery state.

DMA also allows the differences between two cold curing adhesives, i.e., an epoxy thermoset and an acrylic elastomer (both discussed in Section 2.2.3), to be understood from the structural point of view, as shown in Fig. 3.18 (bottom). At ambient temperature, i.e., at about 0–40°C, the epoxy adhesive is in the glassy state, while the acrylic adhesive is in the rubbery state. The response of the former is thus almost purely elastic without any energy dissipation, while the viscous component and associated energy dissipation are significant in the latter, at this temperature range. Accordingly, the epoxy adhesive exhibits an almost brittle and the acrylic adhesive a pseudo-ductile stress vs strain response at ambient temperature, as further discussed in Section 3.6.2.

An important metric from the structural design point of view is the glass transition temperature (T_g) of a polymer or composite. Based on DMA results, several definitions exist for this value, i.e., it can be assumed as (i) the inflexion point of the storage modulus curve, (ii) the peak of the loss modulus or loss factor curve, or (iii) the onset value of the storage modulus decrease ($T_{g,onset}$) defined in [1, 37, 38], as indicated in Fig. 3.18 (top). A further definition of T_g is derived from Differential Scanning Calorimetry (DSC), see Section 3.5.2. This value is based on changes in the specific heat capacity and thus not related to mechanical behaviour, and normally it is also higher than $T_{g,onset}$, see **Fig. 3.19** (left). As can be seen in Figs. 3.18 (top) and 3.19 (left), $T_{g,onset}$ is the lowest of these values and the structural stiffness, which is in line with the storage modulus, may already have decreased significantly at the other higher values. In structural design, $T_{g,onset}$ obtained from DMA should therefore be used as the relevant value [1, 40].

The curves shown in Fig. 3.19 (left) also demonstrate that T_g increases with curing time, which is caused by the continuation of curing, as also shown in Fig. 3.19

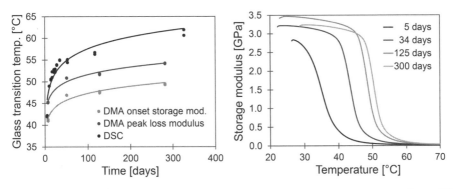

Fig. 3.19 Comparison of glass transition temperature of a cold curing epoxy adhesive according to different definitions **(left)**; DMA storage modulus curves **(right)**, both with development over curing time at 21°C, based on data in [43]

(right), where the storage modulus curves are shifted to higher temperatures with increasing curing time. Moisture uptake, on the other hand, may have an effect inverse to that caused by the curing time and decrease the T_g-values compared to dry conditions, see below (Section 3.4.6). It should also be noted that DMA results depend on the selected heating rate, and this rate must be limited so as not to overestimate the T_g-values [41, 42] (see also below, Section 3.4.4).

Storage and loss moduli and glass transition temperature of a composite material normally deviate from the values obtained for the polymer matrix of the composite due to the influence of the fibres. If the composite behaviour is matrix-dominated, the differences between composite and polymer values may remain small. If the behaviour is fibre-dominated, however, the storage modulus curve is shifted to higher temperatures, and the glass transition temperature is thus also higher than in the case of the pure polymer.

3.4.4 Time and temperature dependence

Time dependence applies for physical and mechanical material properties of composites. Time dependence of the physical properties is manifested by their sensitivity to the heating rate (as already mentioned above for T_g), while the mechanical properties are sensitive to the strain rate. The heating rate dependence is further discussed in Section 3.9.2 concerning the fire behaviour of composites.

The mechanical responses of polymers and composites depend not only on the strain rate, but also on the temperature to which the material is subjected, as schematically shown for the stress vs strain responses in **Fig. 3.20** (left). The material remains in the elastic state at a higher strain rate (since the reaction time to deform is too short) or at a lower temperature (as the molecular motion is restricted). The secondary bonds remain intact in both cases. When the strain rate is decreased, or the temperature is increased, the viscous component becomes more significant. The reaction time is sufficient to adapt the network to the stress direction due to the low strain rate, or the motion of the molecular network increases with increasing temperature. Most of the secondary bonds are lost in both cases; more energy is

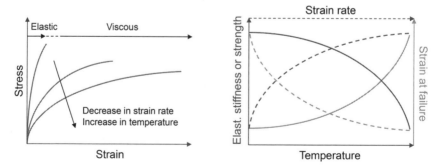

Fig. 3.20 Strain rate and temperature effects on polymer/composite stress vs strain response **(left)**; elastic stiffness, strength, and strain at failure **(right)**

3. Composites – Overcoming Limitations

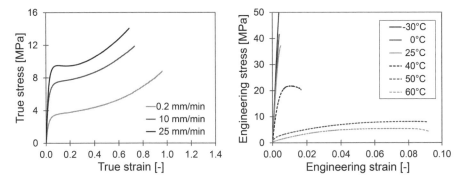

Fig. 3.21 True tensile stress vs strain curves of acrylic (elastomer) adhesive at different displacement (strain) rates, based on data in [44] **(left)**; engineering tensile stress vs strain curves of epoxy (thermoset) adhesive at different temperatures, based on data in [39] **(right)**

thus dissipated, and deformations may increase significantly. Decreasing the strain rate and increasing the temperature normally decrease elastic stiffness and strength, while the strain at failure is increased, as schematically shown in Fig. 3.20 (right) [32]. Further time-dependent phenomena are creep, recovery, and relaxation, as discussed in the next Section 3.4.5.

The rate and temperature dependencies are demonstrated in two examples shown in **Fig. 3.21**, based on the two above-mentioned adhesives. The tensile stress vs strain responses of the acrylic elastomer adhesive at different displacement (i.e., strain) rates and constant laboratory temperature, and of the epoxy thermoset adhesive at different temperatures and constant strain rate are depicted. Due to the much larger strains of the acrylic than of the epoxy adhesive, true stress vs strain values are selected in the acrylic case, which consider the large specimen deformations. Engineering strains, which refer to the initial geometry, are applicable in the epoxy case, however. Since strains are comparably small, engineering and true values do not differ notably. The responses of the acrylic adhesive depend significantly on the strain rate, as described above. The adhesive is in the rubbery state and significant strain hardening occurs in all cases due to the alignment of the molecules to the applied load direction [44]. The epoxy adhesive exhibits an elastic response clearly below T_g (which is 46°C, obtained from DSC in this case), while the viscous component increases when the temperature approaches T_g and above this value [39].

A further example, shown in **Fig. 3.22** (left), consists of a novel concept for angled joints of composite frame structures, which are semi-rigid and hybrid, and where the applied load induces shear and axial forces and a bending moment in the joint [45]. The experimental angled double-lap joint is composed of (i) pultruded laminates (of 100×10 mm cross section), (ii) 100×100 mm adhesive connections of 2.5 mm or 5 mm thickness, consisting of the acrylic elastomer adhesive, and (iii) a steel bolt of 14 mm diameter in the centre of the joint. Relative rotations can occur around the bolt, which prevents relative displacements and thus transmits shear and axial forces and assures the basic joint integrity. Consequently, the external bending moment is transmitted by pure in-plane torsion in the adhesive layers. This torsion

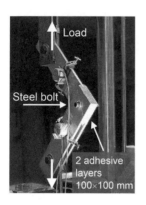 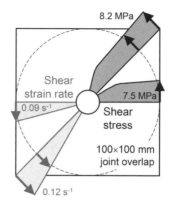

Fig. 3.22 Angled double-lap joint consisting of pultruded laminates, acrylic elastomer adhesive layers, and a central bolt, subjected to torsion [13] **(left)**; internal radial shear strain rate and shear stress distributions at peak load under a certain external displacement rate, based on data in [45] **(right)**

moment causes linear radial shear strain and strain rate distributions, which produce tangential (rate-dependent) shear stresses, as shown in Fig. 3.22 (right). The shear stresses in the centre of the joint remain in the elastic range while softening occurs towards the joint edges through the loss of the secondary bonds. The composition of these elastic and viscous components of the viscoelastic stress response depends on the strain rate, i.e., higher rates increase the elastic component and decrease the viscous component and vice versa. The integral of the product of the shear stresses (over the areas where they act) and lever arms gives the torsional moment that can be transmitted by the joint. The transmittable torsional moment thus also depends on the strain rate, as depicted in **Fig. 3.23** (left), and increases nonlinearly with an increasing rate for the same reason as mentioned above.

The joint can provide pseudo-ductile torsion vs relative rotation responses and dissipate inelastic energy under static and cyclic loading, as is the case for the linear double-lap joints shown in Figs. 3.16 and 3.17, which exhibit pseudo-ductile axial load vs displacement responses. Both joint types have the same 100×100 mm overlap geometry and acrylic adhesive material. Their pseudo-ductility can thus be compared using the ductility index, as defined in Section 3.6.2 (Eq. 3.4). Accordingly, Fig. 3.23 (right) depicts the ductility index vs strain rate relationships of both joint types. With increasing strain rate, the pseudo-ductility of both joint types decreases because the elastic component of the viscoelastic response increasingly overrides the viscous component. The pseudo-ductility of the linear joints is significantly higher than that of the angled joints, however, because the strain rates are constant in the axial direction in the linear joints, while the radial strain rate distributions are linear in the angled joints and thus limit the development of a comparable pseudo-ductility [45].

The effects of temperature on the mechanical properties of composite materials, adhesives, and sandwich core materials can be emulated by using a conversion factor in the structural design, if the temperature remains in the range of service temperatures, where the effects are reversible and do not cause permanent damage, see details in [1, 40].

3. Composites – Overcoming Limitations 125

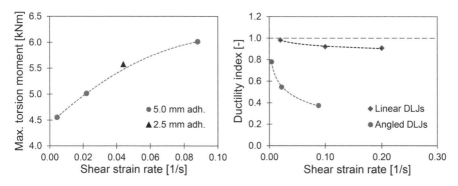

Fig. 3.23 Maximum torsion moment vs internal strain rate of composite angled joint **(left)**; comparison of ductility index vs internal strain rate of angled and linear double-lap (DLJ) joints (index of 1.0 indicates full pseudo-ductility) **(right)**, based on data in [45]

3.4.5 Creep, recovery, and relaxation

3.4.5.1 Overview

Creep, recovery, and relaxation are time-dependent phenomena related to the viscoelastic nature of the polymer network; they are defined as shown in **Fig. 3.24** (left). Creep develops in a structural member under permanent/sustained stress, i.e., the deformation or strain increases over time under these conditions until creep rupture occurs; unloading during this sustained stress results in total or partial recovery of the deformations. Conversely, relaxation occurs under sustained deformation or strain, i.e., the stress decreases over time under this condition. Based on the engineering model shown in Fig. 3.14, the basic mechanisms of creep, recovery, and relaxation will be explained in the following, also addressing the influencing parameters relevant for the structural design. More details about modelling and design are provided in [1, 40].

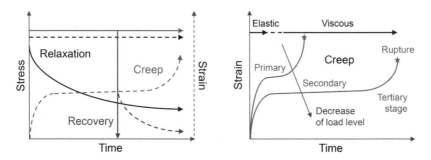

Fig. 3.24 Creep, recovery, and relaxation definitions in schematic stress and strain vs time diagram **(left)**; creep stage definitions and dependence on sustained load level **(right)**

3.4.5.2 Creep

Since composite structures are normally lightweight and the permanent/sustained stresses due to self-weight are correspondingly low, creep deformations are often not critical, particularly in the case of bridges with lightweight composite decks where, furthermore, traffic loads are considered as being short-term [1]. In buildings, however, permanent high service loads or snow loads on large composite roof overhangs may cause creep deformations, which can become relevant in the serviceability state verification. Creep deformations can also increase eccentricities in composite members subjected to axial compression and thus significantly reduce resistances against instability phenomena, such as those against buckling of members or wrinkling of sandwich panel face sheets. In contrast to reinforced concrete, where creep is relevant only under compression, significant creep may develop in composites under compression and tension (and shear, which is a combination of compression and tension).

If a sustained stress is applied to the polymer network, it will react in three stages, as shown in Fig. 3.24 (right). In the primary stage, at the beginning, the network deforms almost elastically at a high rate. With increasing time, the molecular chains can align to the stress direction in the secondary stage, as shown in Figs. 3.14 (right) and 3.15 (creep case). The secondary bonds diminish, and the deformations grow at an almost constant rate (but much slower than in the primary stage). Damage in the form of microcracks can already be initiated in the matrix in the later phase of this stage, as shown in **Fig. 3.25**, where damage in a translucent composite laminate under sustained tension is manifested by the reduction of light transmission through the specimen. In the tertiary stage, the damage level increases, i.e., matrix and fibre-matrix interface damage fully develops. Local stresses become non-uniformly distributed among the fibres, which then progressively fail until complete creep rupture occurs; the deformations rapidly increase at this stage. With the decreasing level of the sustained load, the viscous component of the response (including damage formation) develops at a lower rate over a longer time interval, until creep rupture occurs, see Fig. 3.24 (right). Furthermore, at lower load levels, the viscoelastic response remains linear, i.e., the resulting strain is proportional to the sustained stress at a given time and the linear (Boltzmann) superposition principle is applicable (as explained in Section 3.4.1).

In structural design, creep deformations can be considered by reducing the short-term elastic or shear moduli by a "creep coefficient", as described in [1]. Creep deformations in composite members mainly concern the polymer matrix or core materials of sandwich panels; glass, basalt, and carbon fibres do not exhibit notable creep deformations. In the case of composites, creep deformations thus remain small if the response is fibre-dominated (e.g., in a UD laminate under tension), while in the case of a matrix-dominated response, they may rise to significant values (e.g., in a woven $0/90°$ laminate under shear), as shown in **Fig. 3.26** (left). Polymeric foam and balsa sandwich core materials are sensitive to creep; sandwich panel creep deformations due to shear may thus be more significant than those due to bending. Furthermore, creep deformations depend on the environmental conditions and may significantly increase with increasing temperature and relative humidity. Creep deformations can be modelled based on Findley's power law, and accelerated characterisation methodologies, such as time-temperature-stress superposition principles, can be applied to extrapolate test results in time [1, 40].

3. Composites – Overcoming Limitations

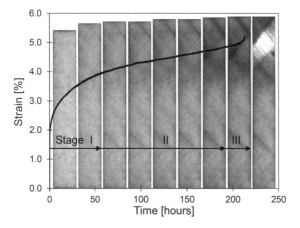

Fig. 3.25 Tension creep response of an angle-ply glass/epoxy laminate with $[\pm 45]_{2s}$ architecture and loaded in 0° direction, showing damage initiation and propagation in the secondary and tertiary creep stages, manifested by reduction of light transmittance over time through specimen under load, reconstructed from [46]

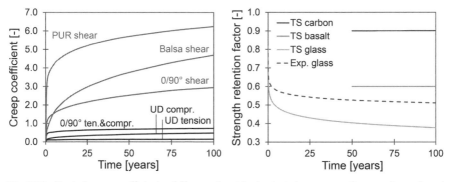

Fig. 3.26 Typical creep coefficients of fibre- and matrix-dominated responses of composites and sandwich core materials under ambient temperature and linear viscoelasticity conditions, based on data in [1] **(left)**; strength retention factors for tension creep rupture in fibre direction, experimental values for glass fibre rebars (Exp. glass) from [47] and glass, basalt, and carbon fibre values according to [1] **(right)**

An illustrative example of how creep can influence the structural responses are web-core sandwich panels where a foam core is reinforced by webs, as is the case in the Novartis Building roof, for instance (see Appendix A.4). A grid of thin composite webs of 3–6 mm thickness, spaced at 925 mm, reinforces the PUR foam core of the 70–620 mm thick sandwich panel member. In such members, the individual contributions of webs and core to the total shear stiffness and, accordingly, the distribution of the shear forces between webs and core are time-dependent, since the composite webs and foam core exhibit different creep behaviours (see Fig. 3.26, left). The core contribution to stiffness and core shear forces decreases with time, while the shear forces in the web increase, as shown in **Fig. 3.27** (left) up to a building design service life of 50 years. The shear resistances of the individual webs (due to shear wrinkling) and core components and their summation slightly decrease, as depicted in Fig. 3.27 (right).

Taking into account the continuous redistribution of shear forces between core and webs shown in Fig. 3.27 (left), effective shear resistances of the components in the hybrid core can be obtained (i.e., by dividing the individual resistances by the shear force distribution coefficients of Fig. 3.27, left). As can then be seen in Fig. 3.27 (right), the shear resistance of the hybrid web-core is governed by the effective resistance of the webs, whose curve (denominated "Hybrid web") is below (i) the curve of the effective resistance of the core ("Hybrid core") and (ii) the summation curve of the individual web and core resistances ("Indiv. web+core"). Considering the summation of the individual resistances would thus result in an unsafe design [2].

In the project of a similar composite-PUR foam web-core sandwich roof, for the CLP building (see Appendix B.3), two options were studied regarding the appearance of the roof top: (i) a white-coloured coating of the sandwich panel top face sheet and (ii) a green roof. The added weight in the latter case of 170 kg/m^2 increased the permanent loads by a factor of 3.4. The associated significant effects of creep on stiffness and resistances of webs and face sheets required a 1.8–2.8-fold increase in their thicknesses. As a result, the self-weight of the sandwich slab increased by up to 60% [48].

In contrast to the insignificant creep deformations of fibres, regardless of the fibre type (glass, basalt, or carbon), the stress limit for creep rupture depends significantly on the fibre type. Glass fibres are very sensitive to creep rupture, i.e., the strength retention factor, defined as the ratio of long- to short-term strength, is small, see Fig. 3.26 (right). Carbon fibres, however, are almost insensitive to creep rupture, while basalt fibres exhibit a sensitivity between that of glass and carbon fibres. Cables with permanent prestress, for instance, should thus be composed of carbon or basalt fibre composites, while prestressed glass fibre composites can only be economically used for temporary (short- to medium-term) purposes. The stress limits for creep rupture also depend on the loading type, being higher for fibre- than for matrix-dominated

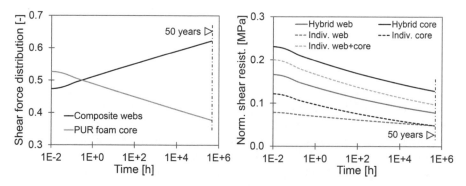

Fig. 3.27 Time-dependent distribution of shear force on individual webs and core (**left**); time-dependent normalised (by hybrid core unit area) shear resistance of hybrid web-core, delimited either by web (Hybrid web) or core (Hybrid core), and of individual web and core and their summation (Indiv. web+core) (**right**), derived from materials and geometry of composite-PUR sandwich roof of Novartis Building, based on data in [2]

responses, see [1, 40]. Creep rupture stress limit or retention vs time curves, such as shown in Fig. 3.26 (right), normally become linear in linear-log or log-log representations, which can then be used to extrapolate experimental data [47].

3.4.5.3 Recovery

If the polymer network is unloaded after having experienced sustained stress, e.g., during the secondary creep stage, the stretched molecules are released and can move back, and the secondary bonds can re-establish themselves. The deformations thus recover, either completely, if damage has not yet been initiated, or partially if damage has already occurred.

The recovery after unloading of composites can be modelled by Weibull distribution functions, as demonstrated, for instance, in [49] for composite web-flange junctions. Another example is given in [44] concerning the recovery behaviour of an acrylic adhesive after having been subjected to tension and compression. Phenomenological models can also be used, as introduced above and are applied in [50], for instance, to model the recovery behaviour during interrupted fatigue loading of (matrix-dominated) composite laminates (see Section 3.8.5). In that case, recovery after repeated fatigue loading interruptions increases the fatigue life considerably [51], which is a result that can be relevant in the case of composite bridges, for instance, where reversible creep-fatigue effects can recover during periods of reduced or absence of traffic, e.g., during the night.

3.4.5.4 Relaxation

If sustained strain or deformation is applied to the molecular network, the internal structure will re-arrange itself as time increases, i.e., the molecular chains adjust to the strain direction and thus relax. The initially arising, mostly elastic stress thus decreases with increasing relaxation time.

Stress relaxation has so far almost exclusively been studied for pretensioned carbon, basalt, or aramid cables, i.e., for highly fibre-dominated cases, and linear-log fitting relationships are used to extrapolate experimental data over time. Depending on the investigation, carbon and basalt cables exhibit only small relaxation rates of 6–10%, extrapolated to approximately 100 years, while smaller/higher values are obtained for lower/higher sustained strain, respectively, and carbon values are slightly lower than those of basalt [52]. An example is shown in **Fig. 3.28**, where post-tensioned carbon fibre cables even exhibit a slight increase in cable forces over a period of 21 years, strictly following the course of temperature (Verdasio Bridge, see Appendix A.1).
Relaxation results under imposed permanent bending deformation to form bending-active (prestressed) glass fibre composite arches are furthermore reported in Section 4.3.4.3.

3.4.6 Moisture effects

The stress vs strain responses of polymers and composites do not only depend on temperature and time, but also on their moisture content, as shown in **Fig. 3.29** for the

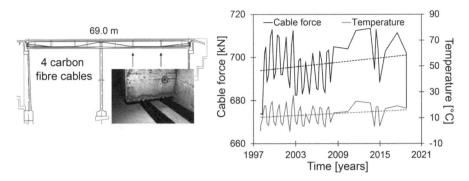

Fig. 3.28 Strengthening of Verdasio Bridge with four carbon fibre cables inside box girder [53] **(left)**; cable force and temperature vs time, over 21 years (longer measuring intervals after 2009) [54] **(right)**

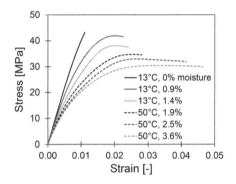

Fig. 3.29 Tensile stress vs strain responses of an epoxy adhesive at different temperatures and moisture contents (in [wt%]), according to [55]

epoxy adhesive introduced in Section 2.2.3. If water molecules penetrate the molecular network, they reduce or interrupt the secondary bonds and thus increase the molecular mobility and deformability of the molecular chains. The water molecules plastify the network, the reason for this mechanism being called plasticisation. A similar mechanism thus occurs as under increasing temperature, which has similar effects, i.e., the stress vs strain responses become nonlinear, strength and stiffness decrease, but strains at failure increase, see Fig. 3.29. In line with these responses is a decrease in glass transition temperature with increasing moisture content, as mentioned above and shown in Fig. 3.42, where the effects of moisture on the physical and mechanical properties are further discussed. Plasticisation is a physical degradation mechanism, which normally is reversible after drying. If the moisture exposure is sustained, however, chemical degradation mechanisms, such as hydrolysis, may cause irreversible damage to the chain network, see Section 3.10.2.7 [40, 55].

As is the case for temperature, the effect of moisture on the mechanical properties of composite materials and adhesives can be taken into account by including a conversion factor for moisture in the structural design for certain moisture exposure classes, see details in [1, 40]. This factor includes both physical and chemical degradation of mechanical properties.

3.4.7 Conclusions

Viscoelastic effects in composites are caused by the polymer matrix. They mainly include time (strain and heating rate), temperature, and moisture dependence of material properties, and related phenomena such as softening, creep, recovery, and relaxation. These effects depend significantly on the fibre architecture and volume fraction. If the material behaviour is clearly fibre-dominated, the effects may remain small, while in clearly matrix-dominated cases, they may become significant. Viscoelastic effects in sandwich panel core materials, such as foam or balsa, can also reach significant levels; the same applies to adhesives (for example, elastomers), particularly if they are in the rubbery state during service conditions.

Viscoelastic effects can be limited by (i) imposing a fibre-dominated response through an appropriate fibre architecture in the design (see also Section 3.2.3), (ii) limiting the levels of service temperature, permanent load, and moisture, (iii) assuring that the glassy state clearly covers the service temperature range, e.g., by extending the former to higher temperatures through fully cross-linking (i.e., fully curing or post curing the composite matrix), and (iv) using carbon or basalt fibres instead of glass fibres under high permanent loads to prevent creep rupture.

The effects of the curing degree on the mechanical behaviour and creep-fatigue interaction effects are further discussed in Sections 3.5.3 and 3.8.4, respectively.

3.5 CURING AND PHYSICAL AGEING

3.5.1 Introduction

During curing of thermoset polymers, a cross-linked molecular chain network emerges, driven by an exothermic polymerisation reaction and, consequently, the physical and mechanical material properties develop. Concurrent processes may take place during curing, such as physical ageing and plasticisation, the latter caused by moisture uptake [43, 55].

While composite components and members can be hot cured under indoor factory conditions in many cases, i.e., cured at elevated temperatures to accelerate the polymerisation reaction (e.g., in the case of pultruded profiles), applying heat is normally difficult outdoors. Examples are adhesive joints of on-site assembled structures, where hot curing is often impossible due to the frequently limited accessibility of the joints after assembly or the large scale of the members. Such joints are thus normally cold cured, i.e., cured at ambient temperature, which is made possible by specific chemical compositions of cold curing adhesives [43].

If the curing temperature is low at the early age, however, during winter for instance, the reaction progress and associated development of the mechanical properties is delayed; this should be considered in the design, e.g., for the specification of

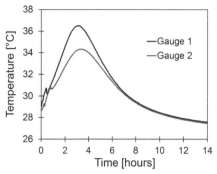

Fig. 3.30 Temperature development in 6 to 10 mm thick bridge deck-to-girder adhesive connection at two measurement points due to exothermic curing reaction [58]

construction stages or dates of inauguration. The glass transition temperature may also not yet have been fully developed, and the adhesive material may thus remain sensitive to higher ambient temperatures, during the first summer, for example. These higher temperatures, on the other hand, also accelerate the curing and increase the glass transition temperature at the same time and may thus mitigate the problem [56, 57].

Since the thermal conductivity of polymers is low, the heat generated during curing can be trapped in thick laminates or thick adhesive joints between insulating adherends, e.g., in sandwich panels with insulating foam cores. The released heat can accelerate the curing process but can also cause thermal degradation if excessive. In bridge deck-to-girder adhesive connections described in *Example-3* of Section 4.6, for instance, relatively thick bond lines of 6–10 mm have been implemented to compensate for member tolerances. The temperatures in the adhesive layer increased to 34–36°C during the first four hours of cold curing, see **Fig. 3.30**, which in this case accelerated the curing process. Excessive heating was mainly prevented by the high thermal conductivity of the bottom steel girder flange.

In the following discussion, the basic physical background of cold curing, physical ageing, and the effects of moisture uptake will be described, to make these complex processes and their interactions understandable from an engineering point of view. Since adhesive joints are particularly subjected to cold curing, the discussion will then focus on how physical and mechanical properties of cold curing adhesives develop under these concurrent processes, at the early age (i.e., during the first year), and at low temperatures in particular; both together may represent the most critical conditions for an adhesive joint design. Finally, the long-term development of the properties will be addressed, and conclusions will be drawn about related aspects that must be considered in the design of adhesive joints and composite structures. The content of this section comes mainly from two PhD theses [59, 60], completed in 2011 and 2017 at the CCLab; further and related references can be found within them.

3.5.2 Processes of curing, physical ageing, and moisture uptake

The progress of molecular cross-linking during curing of thermoset polymers can be traced by Differential Scanning Calorimetry (DSC). In the DSC method, the heat

3. Composites – Overcoming Limitations

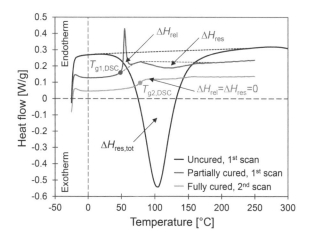

Fig. 3.31 DSC results for an uncured, partially, and fully cured cold curing epoxy adhesive, derived from [43] (ΔH_{res} = residual heat, ΔH_{rel} = relaxation enthalpy, $T_{g,DSC}$ = glass transition temperature obtained from DSC)

difference between a small material and a reference sample is measured for a simultaneous increase in temperature of both samples, see **Fig. 3.31** (for the epoxy adhesive introduced in Section 2.2.3). Three metrics can be extracted from a typical DSC curve of a partially cured material sample, i.e., residual heat (ΔH_{res}, exhibiting an exothermic valley), relaxation enthalpy (ΔH_{rel}, exhibiting an endothermic peak), and DSC glass transition temperature ($T_{g,DSC}$, at the midpoint temperature according to [61]). Residual heat is the exothermic heat released as curing is completed. Endothermic relaxation enthalpy is a metric that characterises physical ageing, as described below [43].

In a polymer, since the curing temperature (T_c) is initially higher than the glass transition temperature (T_g), the cross-linking reaction is driven by chemical kinetics and proceeds rapidly in the liquid state, and T_g increases accordingly. Once T_g reaches T_c, vitrification takes place, i.e., the material solidifies. Subsequently, when T_g exceeds T_c, the reaction rate decelerates and becomes diffusion-controlled, and the material enters the glassy state [56]. Increasing the curing degree increases the cross-link density and the specific volume, and thus decreases the mass density [43].

The curing degree (α) of a partially cured polymer can be obtained from two DSC or two Dynamic Mechanical Analysis (DMA) scans as follows [1, 43]:

DSC: $\quad \alpha = \dfrac{\Delta H_{res,tot} - \Delta H_{res}}{\Delta H_{res,tot}}$ (3.1)

DMA: $\quad \alpha = \dfrac{T_{g,onset-1}}{T_{g,onset-2}} \quad$ (for $T_{g,onset-1} > 0$) (3.2)

where ΔH_{res} is the residual heat of the partially cured polymer, $\Delta H_{res,tot}$ is the residual heat of the uncured/fresh polymer, and $T_{g,onset-1}$ and $T_{g,onset-2}$ are the onset values of the glass transition temperature in [°C], obtained from two consecutive DMA

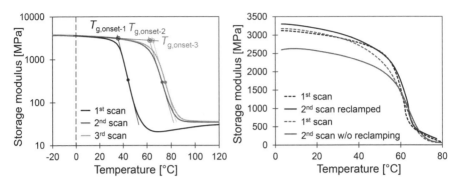

Fig. 3.32 Three consecutive DMA scans on same epoxy adhesive to define curing degree (storage modulus in log-scale according to [1]) **(left)**; effect of specimen reclamping before 2nd DMA scan on storage modulus of epoxy adhesive **(right)**

scans, 1 and 2, on the same specimen, see **Fig. 3.32** (left) and Section 3.4.3. The third DMA scan, almost overlapping the second scan, confirms that the material is already fully cured after the second scan. In this respect, it is important that the specimen is reclamped before repeating a scan. Without reclamping, the storage modulus curve may start at a much lower level, as shown in Fig. 3.32 (right). The resulting curing degree obtained from Eqs. 3.1–2 varies from 0 (uncured) to 1.0 (fully cured).

Driven by a thermodynamic disequilibrium, physical ageing occurs under an ageing (service) temperature in the glassy state. During this process, the molecular network densifies, i.e., the mass density increases (also designated as volumetric relaxation), and the molecular configurational energy decreases (enthalpy relaxation). The magnitude of these effects depends on the material type and thermal history and is higher if the ageing temperature (T_a) is low, i.e., if the difference T_g-T_a is high. The effects may be erased, i.e., the material de-aged, if T_a approaches or exceeds T_g, as shown in Fig. 3.31, where the endothermic peak of the first scan disappears in the second scan (orange curve) after heating up to 250°C during the first scan (blue curve). After cooling down below T_g, however, physical ageing begins again, and it may be active for several years. To quantify physical ageing, the relaxation enthalpy is an appropriate metric, since it considers both volume changes and molecular configurational changes [43, 62].

Possible moisture uptake causes plasticisation of the molecular network, i.e., the absorbed water breaks the intermolecular secondary bonds and thus increases the segmental mobility (see Section 3.4.6). Depending on the extent of moisture uptake, a densification of the molecular network is inhibited, i.e., physical ageing may not occur [55].

3.5.3 Early-age development of adhesive properties

3.5.3.1 Overview

The early age development of the physical properties (relaxation enthalpy, curing degree, and glass transition temperature) and mechanical properties (elastic modulus and strength) of two commercial cold curing two-component epoxy adhesives will be analysed in the following section, during curing and physical ageing. Furthermore, the effect of moisture uptake on these properties and processes will be discussed. The mechanical investigation is limited to tension properties since a complete dataset is available only for this case.

The first adhesive is filled with silica quartz particles (approx. 55% by weight, to increase the stiffness), and thus is denominated "filled" in the following. The adhesive is exposed to isothermal curing at different temperatures between 5 and 70°C during the first three to ten days. A minimum temperature of 5°C was selected based on experimental results at 0°C, which showed that curing is significantly inhibited or does not start at all. On the other hand, 70°C may be reached in adhesive joints under dark asphalt layers of bridges, for instance, and was thus selected as the maximum temperature [56, 57].

The second epoxy adhesive (introduced in Section 2.2.3) also contains some silica quartz fillers, but at a much lower content (< 20% by weight), and is thus denominated "unfilled" in the following for the ease of reading. For approximately one year, this adhesive is exposed to continuous curing at a low 13°C or moderate 21°C – or to continuous ageing at 13 or 21°C, after having been post cured for three days at 60°C after the fifth day [43]. The terms used in the following are thus "curing" and "partially cured" for the non-post cured adhesive where curing is still ongoing, and "ageing" and "fully cured" for the previously post cured adhesive, which is primarily subjected to physical ageing. Furthermore, partially and fully cured adhesive specimens are exposed to curing and ageing, respectively, under immersed conditions, i.e., in demineralised water of 13°C up to 754 days (i.e., 25 months), to study the effect of moisture uptake on curing and physical ageing [55].

The physical properties are obtained from DSC (as described above), with the exception of the glass transition temperature of the immersed materials, which is obtained from the DMA peak of the loss modulus (and represents the onset of segmental motion at the molecular level). Tension mechanical properties are obtained from standardised dog-bone-shaped specimens under laboratory conditions (i.e., at 21±3°C and 40±10% RH). Only average values of experimental results are indicated in the following; standard deviations can also be found in references [43, 55–57].

The two studies on the "filled" and "unfilled" adhesive were performed consecutively and without targeting the effect of the filler content on curing and physical ageing. The effect of the filler content can only be quantified for the glass transition temperature.

3.5.3.2   Early-age development of physical properties

The development of the relaxation enthalpy during curing of the partially cured, and ageing of the fully cured unfilled epoxy adhesive, is shown in **Fig. 3.33**; the relaxation enthalpy indicates the presence of physical ageing (see above). The relaxation enthalpy peaks shift to higher temperatures and the residual heat (exothermic valley) decreases during curing, see Fig. 3.33 (left). On the time scale, the relaxation enthalpy of the partially cured adhesive and thus the associated physical ageing exhibits a peak at approximately 20 days, while post curing after five days partially de-ages the

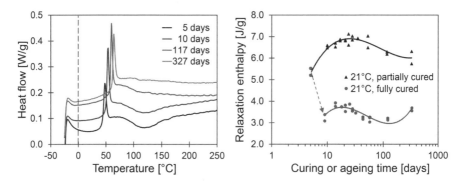

Fig. 3.33 Development of DSC relaxation enthalpy peaks **(left)**; values **(right)** of partially **(left and right)** and fully cured **(right)** unfilled epoxy adhesive during first year at 21°C, based on data in [43]

material and results in a significant drop of the values, as indicated by the arrow in Fig. 3.33 (right). In the following, physical ageing restarts, and the values of the fully cured material exhibit a similar peak as the partially cured material, but at a much lower level; after 120 days, they slightly increase again due to a new cycle of physical ageing [43]. The theory of physical ageing described above is thus confirmed by these data.

The development of the curing degree during the first three days, at different curing temperatures, is shown in **Fig. 3.34** (left), together with the vitrification curve, which separates liquid from solid material states (i.e., where $T_c = T_g$, see above). While full curing can be obtained after a few hours at a high curing temperature of 70°C, a relatively high curing degree (of approx. 0.8) is only achievable after several days at a low curing temperature of 5°C (e.g., during winter). At 25°C (e.g., during summer), such high values are already obtained after 12 hours. Vitrification furthermore initiates later at lower curing temperatures since the curing reaction is decelerated [56]. After vitrification, the curing progress slows down significantly, as also shown in Fig. 3.34 (right), which is the reason a logarithmic timescale is selected (where the progress curve becomes linear in this time interval). During curing at 21°C, a high curing degree of 0.94 results after only five days, while full cure ($\alpha = 1.0$) is not yet achieved after one year. Also shown is a specimen which reaches and maintains full curing after post curing on the fifth day [43].

The DSC-based glass transition temperature (T_g) vs curing time relationship during the first three days, at different curing temperatures, and the corresponding vitrification curve, are shown in **Fig. 3.35** (left). At 70°C curing temperature, maximum T_g-values > 50°C are reached after a few hours, while only approximately 20°C is obtained after three days at 5°C. It can be noted, however, that although the 20°C is low, it may not be critical, since this scenario takes place during winter, where the service temperatures are normally lower. Later, during spring, the values can rapidly increase, as the 25°C curve demonstrates [56]. As with the curing progress, the glass transition temperature development of the partially cured material slows down after vitrification and, at 21°C, values larger than 60°C are obtained only after one year of exposure, see Fig. 3.35 (right); however, higher fully cured values are not yet reached. Post curing increases the T_g-values

3. Composites – Overcoming Limitations   137

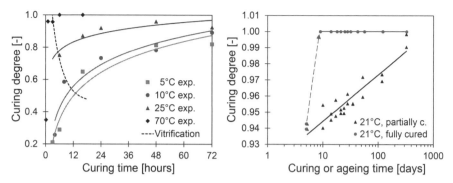

Fig. 3.34 DSC-derived curing degree vs curing time of filled epoxy adhesive during first three days and dependence on curing temperature, log-fit of experimental data in [56] (dots are average values from three samples) **(left)**, and of partially and fully cured unfilled epoxy adhesive during first year at 21°C, based on data in [43] **(right)**

to > 70°C, but they subsequently decrease slightly due to an inhomogeneous thermoset morphology caused by cooling after post curing [43].

The increase in glass transition temperature until approximately 50 days can be attributed to the combined effects of curing and physical ageing, the latter process also occurring during this period, see Fig. 3.33 (right). The subsequent increase is then mainly caused by the completion of curing, i.e., cross-linking [43].

The development of the DSC glass transition temperature vs curing degree is shown in **Fig. 3.36**, for the filled and unfilled (left) and the partially and fully cured unfilled (right) epoxy adhesive. The curves demonstrate that the glass transition temperature almost exponentially increases with increasing curing degree [63]. The reasons for the slight decrease in glass transition temperature of the fully cured material (Fig. 3.36 right) is discussed in [43].

The exponential increase in the glass transition temperature with increasing curing degree may be explained by the fact that the molecular cross-link density is not

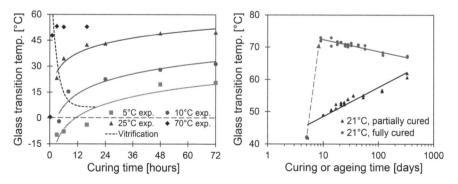

Fig. 3.35 DSC glass transition temperature vs curing time of filled epoxy adhesive during first three days and dependence on curing temperature, log-fit of experimental data in [56] (dots are average values from three samples) **(left)**, and of partially and fully cured unfilled epoxy adhesive during first year at 21°C, according to [43] **(right)**

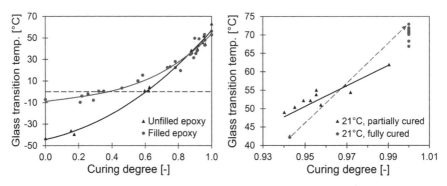

Fig. 3.36 DSC glass transition temperature vs curing degree of filled and unfilled epoxy adhesive, polynomial fit of experimental data in [63] **(left)** and of partially and fully cured unfilled epoxy adhesive at 21°C, based on data in [43] **(right)**

homogenously distributed in the material, with "clusters" of high density being surrounded by lower density zones, and the former increase while the latter decrease during curing. The glass transition temperature is governed, however, by the lower density zones, which control the overall molecular mobility. Even in the case of very high curing degrees of almost 1.0, few and small zones of lower cross-link density may still exist between large clusters of high density, and relative motions of the latter are thus still possible. Towards the end, only a limited number of additional cross-links are necessary, i.e., a minimal increase of the curing degree, to definitively "fix" all the large high-density clusters among each other, which results in a still significant and thus overall exponential increase of the glass transition temperature [13].

Also shown in Fig. 3.36 (left) is the effect of the silica quartz filler in the filled epoxy adhesive on the glass transition temperature. With an increasing curing degree, the effect disappears, and filled and unfilled values approach each other [63]. This result may also be explained by the "engineering model" developed in the previous paragraph. At lower curing degrees, the rigid fillers can reduce the molecular mobility and thus increase the glass transition temperature. At higher curing degrees, however, only the cross-links between the large high-density clusters, which include the embedded fillers, govern the overall molecular mobility and thus the glass transition temperature, which therefore becomes independent of the filler content [13].

3.5.3.3 Early-age development of mechanical properties

During the early days, significant development of the tension modulus occurs only after vitrification and is delayed at lower curing temperatures, as shown in **Fig. 3.37** (left). After approximately one week, however, similar maximum values are reached, and the elastic modulus becomes almost independent of the curing conditions (i.e., curing temperature, partially or fully cured), see Fig. 3.37 (right). The modulus increases up to a peak, but subsequently drops at approximately 180 days, and then slightly increases again [43, 57].

The increase of the tension modulus up to a peak can be attributed to a predominant effect of physical ageing (i.e., increase of mass density) over curing (increase of

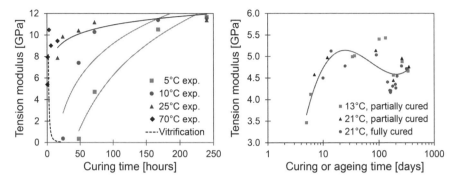

Fig. 3.37 Tension modulus vs curing time of filled epoxy adhesive during first ten days and dependence on curing temperature, log-fit of experimental data in [57] **(left)** and of partially and fully cured unfilled epoxy adhesive during first year at 13 and 21°C, with fully cured values trend line, based on data in [43] **(right)** (dots are average values from 3–4 samples)

cross-link density). This interpretation is supported by the coinciding peaks of the tension modulus and the corresponding relaxation enthalpy at approximately 20 days of curing at 21°C, see Figs. 3.33 (right) and 3.37 (right). Subsequently, curing becomes dominant since the tension modulus drops, which can be attributed to a decreasing mass density. The slight increase after 180 days may be caused by a reactivation of physical ageing, a similar increasing trend can be seen in Fig. 3.33 (right) [43].

Tensile strength develops differently during the early age if compared to the tension modulus. During the early days, the strength development also depends on the curing temperature and is delayed at lower temperatures, but in contrast to the tension modulus, the values at lower temperatures clearly remain below those at higher temperatures after ten days, see **Fig. 3.38** (left). The development of the tensile strength is thus delayed compared to that of the tension modulus [57]. After ten days, the responses of the partially and fully cured materials are and subsequently remain different, as shown in Fig. 3.38 (right), which again contrasts with the tension modulus, where they become almost independent of the curing conditions (see above) [43].

The tensile strength of the partially cured material exhibits a valley in the period where the tension modulus has a peak due to physical ageing, see above. Physical ageing, on the one hand, increases the material stiffness, but also causes embrittlement (i.e., a reduction in toughness), and tensile strength thus decreases [43, 62]. After the phase of physical ageing, the strength increases again, now driven by the predominant progress in cross-linking, and approaches the fully cured values. The latter remain at the same high level reached after post curing, until a drop occurs after 180 days, which may be attributed to a new cycle of physical ageing (see above) [43].

The early age relationships between tension modulus and glass transition temperature (T_g) are shown in **Fig. 3.39**. Initially, a significant increase in T_g occurs while the tension modulus does not yet notably develop, as shown in Fig. 3.39 (left). Subsequently, however, the tension modulus increases rapidly, while the T_g development clearly falls back. It should be noted that the time periods of these developments are

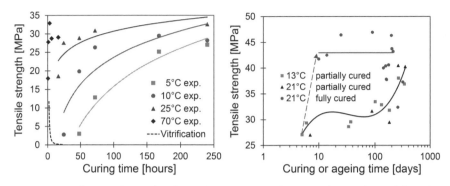

Fig. 3.38 Tensile strength vs curing time of filled epoxy adhesive during first ten days and dependence on curing temperature, log-fit of experimental data in [57] **(left)**, and of partially and fully cured unfilled epoxy adhesive during first year at 13 and 21 °C, with partially cured values trend line, based on data in [43] **(right)** (dots are average values from 3–4 samples)

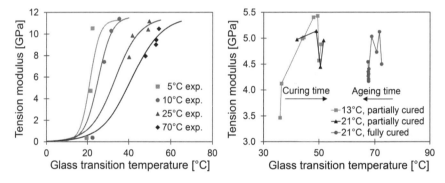

Fig. 3.39 Tension modulus vs glass transition temperature of filled epoxy adhesive during first ten days and dependence on curing temperature, derived from [57] **(left)**, and of partially and fully cured unfilled epoxy adhesive during first year at 13 and 21 °C, according to [43] (corrected) **(right)** (dots are average values from 3–4 samples)

very different, from a few hours at 70 °C to approximately ten days at 5 °C curing temperature [57].

Later, during the first year, tension modulus and T_g increase together with curing time up to a maximum during predominant physical ageing, see Fig. 3.39 (right). Subsequently, the tension modulus decreases while T_g continues to increase due to predominant cross-linking. The fully cured adhesive exhibits an inverse trend, i.e., the tension modulus exhibits a minimum at lower and a maximum at higher T_g values (note the direction of the ageing time from right to left). T_g decreases with ageing time (according to Fig. 3.35, right), but the values remain much higher than under curing conditions (see also Fig. 3.35, right) [43].

On average, the tension modulus is almost constant and independent of T_g above approximately 45 °C, i.e., at a curing degree > 0.94 (see Fig. 3.39, right); the variations

3. Composites – Overcoming Limitations   141

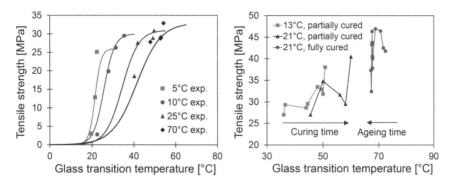

Fig. 3.40 Tensile strength vs glass transition temperature of filled epoxy adhesive during first ten days and dependence on curing temperature, derived from [57] **(left)**, and of partially and fully cured unfilled epoxy adhesive during first year at 13 and 21°C, based on data in [43] **(right)** (dots are average values from 3–4 samples)

around an average plateau level are caused by the either predominant physical ageing or curing (as discussed above) [43].

During the first ten days, the relationships between tensile strength and T_g are very similar to those of the tension modulus vs T_g, see **Fig. 3.40** (left) and compare to Fig. 3.39 (left); the main difference is that the strength development falls behind that of the elastic modulus, as already discussed above [57]. Relationships later in the first year are characterised by the individual processes, as discussed above for tensile strength and glass transition temperature. Valleys in the strength development of the partially cured material occur due to physical ageing and values subsequently increase due to curing, and approach the fully cured values, while T_g is continuously increasing. On average, an exponential increase in strength is exhibited with increasing T_g, while the variations around this trend occur due to either predominant physical ageing or curing (as discussed above) [43].

3.5.3.4 Effect of moisture on early-age development of physical and mechanical properties

The water uptake of the partially and fully cured specimens immersed at 13°C during approximately 25 months, represented by the weight increase of the specimens, is shown in **Fig. 3.41**. They can be considered fully saturated after this period according to [2] (although the curve trend is still directed upward); partially and fully cured specimens do not exhibit a notable difference. Both immersed and dry partially cured specimens continue to cure and reach almost full cure during the 25 months (i.e., α increases from 0.87 to 0.98 in both cases); the curing degree thus remains the same during the different exposures, as concluded in [55].

Depending on the extent, moisture uptake, i.e., plasticisation, may inhibit or prevent a densification of the molecular network and thus physical ageing, as explained above. This effect is confirmed in **Fig. 3.42** (left), where the initially existing relaxation enthalpy peak disappears in the immersed case after 433 days, while a smaller peak remains under dry conditions. Moisture uptake also influences the glass transition

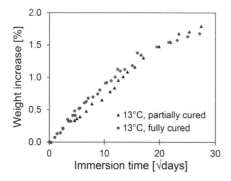

Fig. 3.41 Weight increase of immersed partially and fully cured unfilled epoxy adhesive specimens vs immersion time (specimen size 35×10×3 mm^3) based on data in [55]

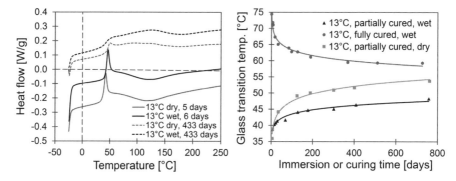

Fig. 3.42 Effect of water immersion on relaxation enthalpy, based on [55] **(left)**, and glass transition temperature, logarithmic fits of data in [55] **(right)**, partially cured **(left and right)** and fully cured **(right)** unfilled epoxy adhesive, immersed (wet) and dry conditions at 13°C

temperature, as shown in Fig. 3.42 (right). Compared to the dry condition, the T_g values of the immersed partially cured material remain approximately 5°C lower, and the values of the immersed fully cured specimens drop significantly. These lower values can be explained by plasticisation, which increases the molecular mobility (see above, [55], and Section 3.4.6).

The effect of moisture uptake on the mechanical properties and the curing and physical ageing processes is shown in **Fig. 3.43**. The peak of the tension modulus under dry conditions, caused by physical ageing, does not appear under immersed conditions, see Fig. 3.43 (left), which confirms that physical ageing is inhibited by plasticisation. The partially cured immersed values rapidly reach those of the fully cured immersed material and, subsequently, both decrease at almost the same rate. The elastic modulus of the dry condition also decreases after physical ageing (due to predominant curing, see above), but at a lower rate; the decreases in the immersed cases can thus also be attributed to plasticisation. Looking at the tensile strength, the partially cured immersed values slightly increase but remain below the dry values

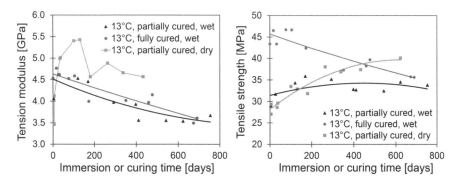

Fig. 3.43 Tension modulus **(left)** and strength **(right)** vs immersion or curing time of partially and fully cured unfilled epoxy adhesive, immersed (wet) and dry conditions at 13°C, polynomial fits of data in [55], except one curve with polygonal line (dots are average values from 4–5 samples)

during the continuation of curing, and the strength of the fully cured specimens decreases, see Fig. 3.43 (right). After 500–600 days, the partially cured immersed values approach the fully cured immersed ones since the former are now almost fully cured [55]. The immersion does thus not affect the cross-linking process.

3.5.4 Long-term development of adhesive properties

As has been shown in the previous Section 3.5.3, curing of cold curing adhesives at ambient temperature is not complete after one year and thus continues, and the glass transition temperature, accordingly, is not yet fully developed. Experimental data for the development of T_g over seven years of curing under laboratory and outdoor conditions is available in [64], for the unfilled epoxy adhesive, see **Fig. 3.44** (with linear- and log-time axis). The lab values increase by approximately 10% from the first to the seventh year. The outdoor values are a further 10% higher, which may be attributed to summer temperatures (in the material), which were higher than laboratory temperatures. As can be seen, the increasing trend follows a logarithmic law and may thus be extrapolated, as shown in Fig. 3.44 (right). High values, above 70°C, as they are obtained after post curing (see Fig. 3.36, right), may thus only be reached after 50–100 years of continuous cold curing; the advantages of post curing (if feasible) are thus evident.

Based on this development of the glass transition temperature, it can be assumed that the mechanical properties of cold curing adhesives and composites also continue to slightly develop according to a still increasing cross-link density, i.e., that tensile strength still increases over time, while the tension modulus decreases. The rates of these variations continuously decrease with time, however, and become very low. Furthermore, the variations also depend on the fibre architecture and are much higher in matrix-dominated cases (including adhesives) than in fibre-dominated cases. In the long term, such developments of physical and mechanical properties, based solely on the completion of curing, may be affected by degradation due to sustained environmental exposure, however, as discussed in Section 3.10.

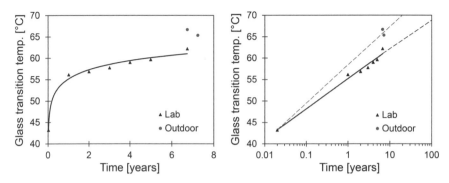

Fig. 3.44 Development of glass transition temperature of unfilled epoxy adhesive under laboratory and outdoor conditions over seven years, log-fit and extrapolation (dashed) of experimental values from [64], linear- **(left)** and log-time axis **(right)**

3.5.5 Conclusions

During the early age, the development of the physical and mechanical properties of thermoset polymers is determined by the concurrent processes of curing and physical ageing. During curing, the molecular cross-link density increases, which increases the glass transition temperature and tensile strength. The tension modulus, however, decreases, as summarised in **Fig. 3.45**. On the other hand, once the polymer has entered the glassy state, i.e., after vitrification, physical ageing takes place, which also alters the properties since the mass density increases and the material embrittles, glass transition temperature and tension modulus thus increase, while tensile strength decreases. The effects of physical ageing may be erased if the ageing (service) temperature approaches or exceeds the glass transition temperature. After cooling down, physical ageing may be re-activated, however. The properties thus develop according to the predominant processes, curing or physical ageing, which may also alternate during time. Moisture uptake increases the molecular mobility through plasticisation and thus lowers the glass transition temperature and mechanical properties and may inhibit or prevent physical ageing.

During the early days, the physical and mechanical properties depend on the curing conditions, i.e., the curing temperature, and their development is significantly delayed at low temperatures. This delayed development should be considered, particularly in the case of adhesive bonding during winter. Significant initiation and progress of curing occurs only at temperatures above 5°C.

After one year, the glass transition temperature is still not yet fully developed under cold curing conditions. Reaching the high values attainable by post curing may take decades, emphasising the advantages of post curing (when feasible). The tension moduli and tensile strengths, however, become almost independent of the curing conditions, i.e., partially of fully (post-) cured during this period.

The effects of concurrent curing and physical ageing, as described above, concern thermoset polymers used as adhesives or as matrix in composites. In the case of composites, the magnitude of the effects depends significantly on the fibre architecture and is greater in matrix-dominated than in fibre-dominated cases.

3. Composites – Overcoming Limitations   145

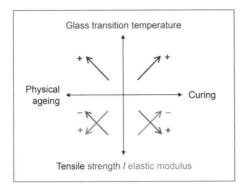

Fig. 3.45 Early age development of glass transition temperature, tensile strength, and tension elastic modulus, under increasing curing or physical ageing [13]

Similar interpretations of experimental results and conclusions as made and drawn above were obtained in other works too, as discussed in [43]. The conclusions derived in this Section 3.5 are applicable to tension mechanical properties and cannot be transferred to compression or shear properties. Epoxy polymers, for instance, may exhibit an increase in compression and shear strength due to physical ageing [62].

3.6 DUCTILITY

3.6.1 Introduction

Composite materials and structures are commonly associated with a brittle material or structural behaviour. This classification, however, applies mostly to a fibre-dominated response, as schematically shown in **Fig. 3.46**. As will be explained later, a matrix-dominated behaviour may exhibit a kind of ductility.

If composites exhibit brittle behaviour, they cannot benefit from the advantages of ductility, which are normally offered by conventional structural materials such as

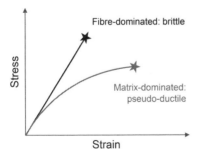

Fig. 3.46 Fibre- vs matrix-dominated stress vs strain behaviour

steel and reinforced concrete. Ductility may provide warning of an imminent failure through increasing (inelastic) deformations in overloaded sections. Furthermore, stress redistributions from overloaded to less loaded sections can occur and thus increase the structural safety. Brittle behaviour, on the other hand, is characterised by a sudden and often catastrophic failure without warning, and engineers thus aim to prevent such behaviour in structural design. For the acceptance of composites in engineering structures, it is thus crucial to provide and apply strategies for how brittle composites can offer a ductile overall behaviour.

Due to the importance of this problem, the question of how brittle composites can provide a ductile structural response has been addressed in research at the CCLab since 2000. Amongst other things, in 2004, the concept of system ductility was introduced, which demonstrates how the lack of ductility at the material level can be compensated at the structural system level. Two different sub-concepts are distinguished: (i) ductility in the proper sense, referring to the combination of ductile and brittle components, and (ii) pseudo-ductility, comprising only brittle components [65, 66].

To characterise ductility, the commonly used ductility index, which is tailored to the yielding behaviour of steel, is not applicable for composites. In the following, a more advanced definition of ductility, based on energy dissipation, is thus first introduced, and concepts are subsequently presented of how ductility can be implemented at the composite material, member, and system levels. In a second part, examples are given to discuss influencing factors on the type of failure mode, i.e., brittle or ductile, and further elucidate these concepts.

3.6.2 Energy-based definition and pseudo-ductility

Finding an appropriate definition of "ductility" in a field wider than that of steel and reinforced concrete is not straightforward. According to [67], ductility describes the ability of a structural material, member, or system to sustain inelastic deformation without significant loss of resistance prior to collapse. The corresponding ductility characterisation adopted for steel and reinforced concrete is schematically shown in **Fig. 3.47** (left), and ductility in these cases is normally expressed by the ductility index (μ) as follows:

$$\mu = \frac{\delta_u}{\delta_y} \tag{3.3}$$

where δ_u and δ_y are the ultimate and yield displacements, respectively [67].

The question regarding the ductility of a structure under static loading is often answered based on load vs displacement responses obtained from mechanical experiments. In many cases, particularly for composites, static experiments are performed under displacement control, and specimens exhibit a typical softening behaviour after the peak load before they fail, as schematically shown in Fig. 3.47 (right); a ductile behaviour is often deduced from the existence of this softening branch. If the behaviour of the actual structure is controlled by the load, however, e.g., the traffic load on bridges, the structure may, nevertheless, respond in a brittle manner, as a load-controlled test would show, see also Fig. 3.47 (right). The test control method should thus be in line with the actual structural behaviour to draw conclusions about

3. Composites – Overcoming Limitations   147

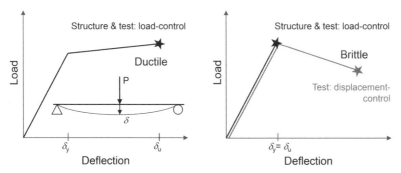

Fig. 3.47 Schematic relationships between structural behaviour and test control method: ductile behaviour (of steel) **(left)** vs brittle behaviour **(right)** (P = load, δ = deflection) [13]

the ductility of a structure, as shown in Fig. 3.47 (left). Furthermore, as demonstrated in the next paragraph, at least two load cycles should be performed in experiments, since an unloading path is required (before reloading to failure), to be able to conclusively decide on ductile behaviour.

The above ductility definition based on nonlinear responses due to yielding is inappropriate for composites, where nonlinear deformations are mainly caused by progressive damage, as will be discussed hereafter. The ductility definition used in this work is thus modified and based on energy, and the corresponding ductility index, μ, is defined as follows [49]:

$$\mu = \frac{E_{\text{inel}}}{E_{\text{inel}} + E_{\text{el}}} = \frac{E_{\text{inel}}}{E_{\text{tot}}} \qquad (3.4)$$

where E_{inel} is the inelastic or dissipated energy, E_{el} the stored elastic energy, and E_{tot} the total energy. The dissipated and elastic energies are defined by the areas under the stress vs strain curve, which must include, in addition to the loading path, an unloading path, as schematically shown in **Fig. 3.48**. If loading and unloading paths are identical, only elastic energy is stored in the system and released during unloading, and the behaviour is designated "brittle". This also applies if both overlapping paths are nonlinear, e.g., in the case of elastic buckling (top middle case in Fig. 3.48). If the unloading path is missing in such a nonlinear case, it is, strictly speaking, not possible to decide on a ductile behaviour (although this is frequently done). If the unloading path deviates from the loading path, the area between the two paths corresponds to the inelastic or dissipated energy (top right case). This behaviour is designated "ductile", and the degree of ductility depends on the ratio of the areas or energies involved, according to Eq. 3.4. The values of μ vary between 0.0 (brittle behaviour) and 1.0 (fully ductile); the latter value applies if the elastic energy is negligible.

In addition to these general cases of brittleness and ductility, two more specific cases can be defined for composites, as schematically shown in the bottom row of Fig. 3.48. To compensate for the basic material brittleness, composites can be used in redundant (statically indeterminate) configurations, at the material, member, or system level, where a certain amount of damage or local failures can occur until the configuration becomes statically determinate, see Fig. 3.48 (bottom left case, System 3).

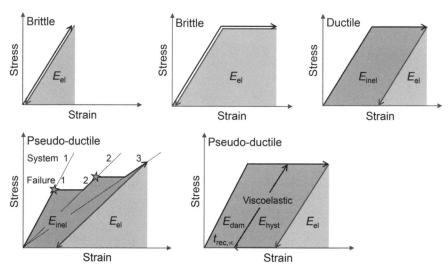

Fig. 3.48 Energy-based definition of brittle, ductile, and pseudo-ductile behaviour based on simplified stress vs strain relationships and loading and unloading paths (E_{inel} = inelastic/dissipated energy, E_{el} = elastic energy, E_{dam} = damage energy, E_{hyst} = hysteretic energy, $t_{rec,\infty}$ = recovery time after full recovery) [13]

This damage also dissipates energy, but is irreversible due to fracture, in contrast to the ductile case, where plastic deformations do not result from fracture and can be recovered by applying counteracting stress. The associated behaviour is designated "pseudo-ductile" to express this difference from the ductile case, as was already introduced in [2] to differentiate damage-caused deviations from linear responses in composites from the classical definition of ductility based on yielding. Depending on the degree of pseudo-ductility, the structural response may provide, however, the same advantages as ductility, as described above.

Further pseudo-ductility may arise from the viscoelastic response of matrix-dominated composites or adhesives, as shown in Fig. 3.48 (bottom right) and the corresponding example of tensile stresses vs strain in **Fig. 3.49**, for the acrylic elastomer adhesive introduced in Section 2.2.3. The adhesive specimens are unloaded at the same strain and then reloaded after different recovery periods (t_{rec}). After unloading, the material exhibits strain recovery, which depends on the recovery period (Fig. 3.49, left). Subsequent reloading paths approach each other above the unloading strain. The relationship between the ratio of recovered strain to strain after unloading and the recovery period is shown in Fig. 3.49 (right); the ratio levels off after a certain period. A potentially remaining permanent strain can then be attributed to damage [68].

Since the stress vs strain unloading path deviates from the loading path (as in the previous cases), energy is dissipated, which can be characterised after full recovery ($t_{rec,\infty}$). The reloading path then separates the area of dissipated energy into two sub-areas of (i) the hysteretic energy (E_{hyst}), and (ii) the damage energy (E_{dam}), as schematically shown in Fig. 3.48 (bottom right). The former is dissipated through the breaking of the secondary bonds between the molecular chains and associated friction between the chains (see Section 3.4.2), and the latter through the formation of microcracks and

3. Composites – Overcoming Limitations   149

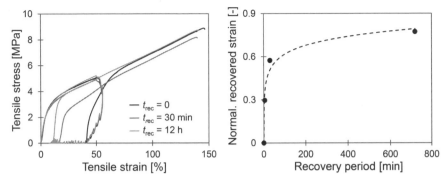

Fig. 3.49 Nominal tensile stress vs strain responses of acrylic elastomer adhesive with reloading cycles after different recovery periods (t_{rec}) **(left)**; recovered strain normalised by strain after unloading vs recovery period, with logarithmic fit **(right)** (loading and unloading strain rate 5 mm/min), based on data in [68]

associated friction in the crack flanks. If only hysteretic energy is dissipated and the strain is fully recoverable (i.e., the secondary bonds can be re-established), the behaviour can be designated as ductile. Irreversible progressive damage cannot normally be prevented at large strains, however, and the behaviour is thus pseudo-ductile in most cases, following the definition introduced above. The ductility index defined above (Eq. 3.4) can then be extended, by analogy, for viscoelastic materials as follows [49]:

$$\mu = \frac{E_{dam} + E_{hyst}}{E_{dam} + E_{hyst} + E_{el}} = \frac{E_{dam} + E_{hyst}}{E_{tot}} \qquad (3.5)$$

Depending on the strain or stress at unloading, the index varies and normally increases with increasing strain or stress at unloading [49].

Based on these definitions of ductility and pseudo-ductility, the failure modes and behaviours of structural composites can be classified at the material, member, and structural system levels, as shown in **Table 3.3**; joints between members are included at the structural system level. At the material level, the behaviour may be brittle or pseudo-ductile in the listed cases; a ductile failure mode does not exist. Whether the behaviour of adhesives is brittle or pseudo-ductile depends on the strain rate and temperature, i.e., on the dominance of either the elastic or the viscous component of the viscoelastic response, see Section 3.4.4. At high rates and low temperatures, the response is normally brittle, while energy may be dissipated over time at lower rates and elevated temperatures. At ambient temperature, thermoset adhesives (such as epoxies) normally show a rather brittle behaviour, in contrast to elastomer adhesives (such as acrylics), which are often in the rubbery state and thus exhibit a pseudo-ductile response, as shown in Fig. 3.48 (bottom right). Polymeric foams and end-grain balsa wood in cores of sandwich panels normally exhibit a pseudo-ductile behaviour, according to Fig. 3.48 (bottom left), if subjected to shear [3, 69], see also Fig. 3.3 (right) in Section 3.2.

At the structural member level, stability failures are brittle or almost brittle in most cases. Pseudo-ductile failure may occur if member components exhibit a

Table 3.3 Brittle, pseudo-ductile, and ductile failure modes of composites at material, member, and structural system levels [13]

Failure mode	Material level	Member level	Structural system level
Brittle	- Fibre-dominated failure - Fibre-matrix interface failure - Cohesive failure in adhesives (thermosets)	- Stability failure	- Brittle member or joint failure in statically determinate composite system
Pseudo-ductile	- Matrix-dominated failure - Cohesive failure in adhesives (elastomers) - Sandwich core shear failure	- Progressive failure of member component(s)	- Brittle or pseudo-ductile member or joint failure in redundant composite system - Pseudo-ductile member or joint failure in statically determinate composite system
Ductile	- N/A	- Brittle or pseudo-ductile component failure during yielding of ductile component in hybrid built-up member	- Brittle or pseudo-ductile member or joint failure during yielding of ductile member or joint in statically determinate or redundant hybrid system

pseudo-ductile response, e.g., the core of a sandwich panel. In hybrid built-up members, ductile components may provide an overall ductile response. At the structural system level, brittle failure occurs if a brittle member or joint fails in a statically determinate system. The response is pseudo-ductile if a brittle or pseudo-ductile member or joint fails in a redundant (statically indeterminate) system or if a pseudo-ductile member or joint fails in a statically determinate system. A ductile failure occurs if ductile members or joints are implemented in hybrid systems.

A condition for producing a targeted pseudo-ductile or ductile failure mode as specified in Table 3.3 is that these complex systems are also designed accordingly, i.e., that sufficient deformation capacity exists. To only combine members of certain characteristics is not sufficient; if a basically ductile component cannot enter the yielding stage at system failure, for instance, the response remains brittle.

A term also used in the context of large deformations and which often causes confusion is "flexible". Flexibility is not related to brittleness or ductility – a flexible response linked to large deformations is related to low stiffness and can be brittle, ductile, or pseudo-ductile, according to the definitions given above.

3.6.3 Examples of pseudo-ductility

Four examples of pseudo-ductile behaviours at the material, member, and structural system levels are discussed in the following. The examples demonstrate that not only material, component, or member combinations but also factors such as member geometry, fibre architecture, and material imperfections can influence the failure

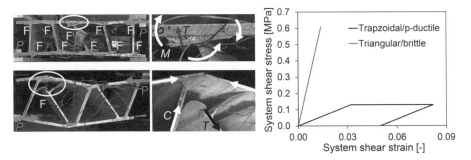

Fig. 3.50 In-plane shear failure modes (**left and middle**), and system shear stress vs strain responses (**right**), of trapezoidal/pseudo-ductile Vierendeel frame (comprising bending moments, M), and triangular/brittle truss (comprising mainly axial forces, T = tension, C = compression), derived from [71–73] (P = load, F = failure at peak load)

mode and lead to either brittle or pseudo-ductile responses. Furthermore, the different possible responses of hybrid (bonded and bolted) joints, which depend on the adhesive deformation capacity, are discussed, and it is shown how 100% joint efficiency can be obtained. The last example demonstrates how pseudo-ductile adhesive joints can lead to a moment redistribution in a redundant structural system. References are also made to pseudo-ductile angled joints in Section 3.4.4, and the hybrid construction *Examples-2* and *-3* in Section 4.6.2 of Chapter 4. In *Example-2*, inelastic energy is dissipated through progressive shear failure in a lightweight concrete (LC) sandwich core and friction in the unbonded LC-composite interface, while ductility is provided in *Example-3* by combining composite bridge decks with ductile steel main girders.

- Example-1: Effect of geometry and fibre architecture on failure mode

Different types of pultruded web-core sandwich panels were developed in the past for bridge deck applications [70] (see also Section 2.3.3). Two such panels are shown in **Fig. 3.50** (in deformed state), exhibiting differences in the shape of the cells and fibre architecture. According to their cell shapes, they are denominated "trapezoidal" and "triangular" in the following discussion. Both panel types are composed of E-glass fibres and isophthalic polyester resin. The fibre architecture of the trapezoidal panel is mainly composed of alternating layers of rovings (R) in the pultrusion direction and multidirectional (MD) fabrics, while the rovings are concentrated in the core and the fabrics are arranged on the outer sides in the triangular architecture, see **Fig. 3.51** [71].

The panels are normally installed with the pultrusion (or webs) direction transverse to the bridge main girders, see *Example-3* in Section 4.6.2. If the panels are rigidly connected to the girders, they exhibit composite action and are thus subjected to in-plane shear loading, as indicated in Fig. 3.50 by the opposed forces P in the flanges. The resulting shear stress vs strain responses of the two panels at the system level are schematically shown in Fig. 3.50 (right) [71, 72]. Triangular panels exhibit high shear stiffness and strength, but the response is brittle. On the other hand, trapezoidal panels have much lower stiffness and strength, but dissipate a large amount of inelastic energy and thus exhibit a pronounced pseudo-ductile behaviour.

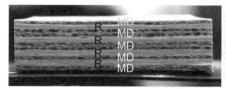 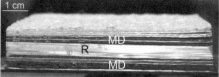

Fig. 3.51 Fibre architectures of trapezoidal **(left)**, and triangular **(right)** panel flanges (R = roving, MD = multidirectional fabric), [71]

This significant difference in the responses is mainly caused by the different cell shapes. The trapezoidal configuration acts as a highly redundant Vierendeel frame, exhibiting significant bending moments (M) in all web-face panel junctions. These moments cause tensile stresses (T) transverse to the fabrics and thus progressive local failures at all these locations (F, in Fig. 3.50), which therefore produce the pseudo-ductile behaviour. In addition, as can also be seen in Fig. 3.50 (top middle), pronounced fibre bridging occurs in the failure plane, which can be associated with the layered fibre architecture; the dissipated fracture energy is significantly increased by this failure mode (see Section 3.7.2).

The triangular configuration, in contrast, acts like a statically determinate truss (if the small eccentricities in the junctions are neglected), and diagonals and face panels are mainly loaded in axial compression (C) or tension (T). At the peak load, one of the tension diagonals is torn away from the junction, causing a brittle failure. The other local failures shown in Fig. 3.50 occur after the peak load. The tearing away of the diagonal is facilitated by the fibre architecture; the failure plane in the curvature of the junction (where transverse stresses are highest) is in the interface between the roving core and the outer fabrics and does not exhibit any significant fibre bridging, as can also be seen in Fig. 3.50 (bottom middle). The fracture energy of such interfaces is normally low, see Section 3.7.2.

The example is continued as *Example-3* in Section 4.6.2, where the two panels are used in hybrid combination and full composite action with steel girders.

- Example-2: Effect of imperfections on failure mode

The energy dissipation in the web-face panel junctions (WPJs) of the trapezoidal and triangular panels of *Example-1* is further investigated. The WPJs are subjected to bending and shear in the web-cantilever set-up shown in Fig. 3.52 (left); the inclined face panels are fully fixed by plates and screws. The failure modes and load vs deflection responses of the trapezoidal WPJs depend significantly on the imperfections in the WPJ area, as can be seen in **Figs. 3.52–54**. WPJs with wrinkles in the multidirectional fabrics (W in Fig. 3.52, middle) exhibit multiple cracks and large deformations at the peak load and an associated pronounced pseudo-ductile behaviour (Figs. 3.53–54, left). On the other hand, WPJs with a resin pocket (RP) and fabric concentrations (C in Fig. 3.52, right) reach the peak load at the first crack and thus small deformations; the behaviour is brittle accordingly (Figs. 3.53, middle, and 3.54, left). Similarly, triangular WPJs fail after the first (principal) crack in a brittle manner (Figs. 3.53, right, and 3.54, left). The response is matrix-dominated in all cases.

3. Composites – Overcoming Limitations 153

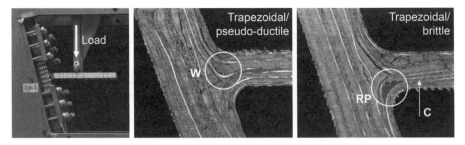

Fig. 3.52 Experimental set-up of web-face panel junction (WPJ) cantilever experiments **(left)**; main imperfections of trapezoidal/pseudo-ductile WPJ **(middle)**; trapezoidal/brittle WPJ **(right)**, photos after experiments, according to [49] (W = wrinkle, RP = resin pocket, C = fabric concentration)

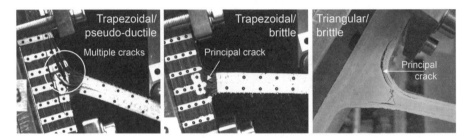

Fig. 3.53 Failure modes of WPJs at peak load from left to right: trapezoidal/pseudo-ductile WPJ (multiple cracks), according to [49]; trapezoidal/brittle WPJ (one principal crack) [49]; triangular/brittle WPJ (one principal crack)

Several load cycles are performed in the pseudo-ductile WPJ experiments, as can be seen in Fig. 3.54. The damage caused in each cycle reduces the stiffness of the next cycle, as is also schematically shown in Fig. 3.48 (bottom left). The cycles allow the hysteretic to total energy ratio (E_{hyst}/E_{tot}) to be calculated, as a function of the deflection at unloading, see Fig. 3.54 (right). The reversible hysteretic energy remains small during the cycles, and, since the irreversible damage and thus total energy increase significantly with increasing cycle number, the ratio decreases. On the other hand, the ductility index increases with increasing deflection, as also shown in Fig. 3.54 (right).

- Example-3: Hybrid (bonded and bolted) joints

The usefulness of hybrid joints, i.e., joints with both adhesive and bolted connections, is controversially discussed in the literature, mainly regarding the question as to whether the resistances of the bonded and bolted connections in the joint can be summed [74].

To clarify this question, axial tension experiments are discussed in the following, performed on hybrid double-lap joints, composed of basalt fibre-epoxy adherends and two types of adhesives, i.e., the epoxy thermoset and acrylic elastomer-thermoset adhesives presented in Section 2.2.3. The joint overlap lengths vary between 64–136 mm, the adhesive layer thickness is 2 mm, and one steel bolt of 8 mm diameter is used (in most cases). To compare, bonded-only and bolted-only joints of the same

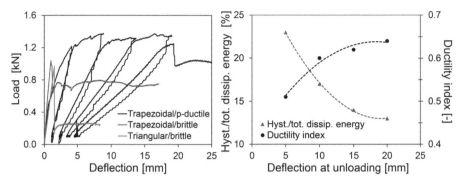

Fig. 3.54 Load vs deflection responses of trapezoidal pseudo-ductile WPJ (60 min recovery time after each unloading), trapezoidal/brittle and triangular/brittle WPJs **(left)**; hysteretic/total dissipated energy (E_{hyst}/E_{tot}) ratio and ductility index (μ) vs deflection at unloading of trapezoidal/pseudo-ductile WPJ **(right)**, based on data in [49]

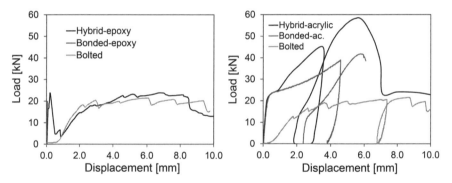

Fig. 3.55 Load vs joint displacement responses of hybrid-epoxy joints **(left)** and hybrid-acrylic joints **(right)**, and comparisons to corresponding bonded-only and bolted-only joints, based on data in [74] (double-lap joints with 64 mm overlap length, one bolt ø 8 mm, 60 min recovery after unloading)

dimensions and materials are also investigated. More details about all the parameters involved are provided in [74].

The responses of hybrid, bonded-only, and bolted-only joints with the much stiffer epoxy adhesive are shown in **Fig. 3.55** (left). The stiff adhesive connection in the hybrid joint fails in a brittle manner at small deformations, in the range of the bolt hole clearance, thus before the bolted connection can share the load. Subsequently, the bolt bears the entire load and exhibits a pseudo-ductile behaviour due to a typical pin-bearing failure (i.e., progressive crushing of the composite adherend ahead of the bolt). In the case of a comparably stiff adhesive, the resistances of bonded and bolted connections can thus not be summed. The bolt, however, can act as a back-up system to prevent joint collapse in the case of adhesive bond failure (if designed accordingly, see Section 5.3.3) [74].

Using the more flexible acrylic adhesive changes the behaviour of the hybrid joint significantly. The much larger adhesive deformations of the bonded connection

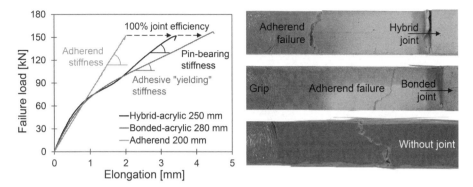

Fig. 3.56 Joint efficiency of 100% for bonded and hybrid double-lap joints, composed of glass fibre composite pultruded adherends of 100×10 mm cross section, 280 and 250 mm overlap length, and 100 mm width, with flexible acrylic adhesive, comparison to adherend response **(left)**; tension failure in adherend outside of hybrid and bonded joints and in adherend itself **(right)** [13, 58]

now overlap those of the bolted connection, and the resistance of the hybrid joint increases accordingly, as demonstrated in Fig. 3.55 (right). The individual connection resistances can thus be basically summed if the adhesive is sufficiently flexible. The bolt furthermore limits potential creep deformations of the adhesive connection and thus of the hybrid joint [74]. Also shown are individual loading cycles in each case, confirming a pseudo-ductile behaviour, which includes both cases shown in Fig. 3.48, i.e., case (bottom left) for the bolted and (bottom right) for the bonded connections. The corresponding ductility indices are 0.94, 0.94, and 0.97 for the hybrid, bonded, and bolted joints, respectively.

In cases of a stiff adhesive, as defined in Section 2.4.3, the resistance of joints is normally significantly lower than that of the composite members in between, i.e., the joint efficiency, defined as the joint to member resistance, is much lower than 100%, see Section 2.4.3, Fig. 2.16 (left), and the structural response is brittle. Similarly low is the efficiency of bolted joints [74]. Hybrid and bonded-only joints with flexible adhesives can exhibit, however, much higher efficiencies and pronounced pseudo-ductility. The efficiency can be almost linearly increased by increasing the overlap length (since the shear stress distribution is uniform, see Section 2.4.3, Fig. 2.15, left), and 100% efficiency can be reached. An example is shown in **Fig. 3.56**, where both joint types reach the failure load of the adherend and exhibit adherend failure outside of the joint and not joint failure. The hybrid joint can, furthermore, be conceived as more compact, i.e., with shorter overlap length, thanks to the contribution of the bolt. The responses also show the increased pin-bearing stiffness of the hybrid joint, which becomes effective once the bolt hole clearance is exceeded.

- *Example-4: Moment redistribution in composite two-span beams with pseudo-ductile joints*

The same pseudo-ductile acrylic adhesive as used in the previous example (and shown in Fig. 3.49) is selected to connect two pultruded glass fibre composite hollow

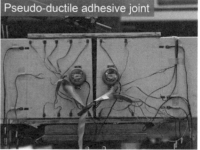

Fig. 3.57 Five-point bending experiment on (deformed) continuous two-span pseudo-ductile beam (spans 2×3.60 m) **(left)**; pseudo-ductile adhesive joint exhibiting differential rotation at mid-support **(right)**, according to [66]

beams of 3.60 m length and 240×240×12 mm cross section, with lap shear joints over the mid-support in a five-point bending set-up (three supports and two loads), see **Fig. 3.57**. The bending behaviour of these two-span beams with pseudo-ductile joint is compared with that of continuous two-span (brittle) beams (without joint) and two single-span (brittle) beams (not connected) in **Fig. 3.58** (left) [66].

The beams with pseudo-ductile joints experience a significant redistribution of the bending moments from the mid-support (M^-) to the span (M^+) thanks to the pseudo-ductile deformations in the joints. The redistribution capacity increases with decreasing overlap length of the lap shear joints (from 300 to 100 mm), as shown in Fig. 3.58 (right). This redistribution increases the ultimate failure loads by up to 15% and 22% compared to the single-span and continuous brittle beams, respectively, since the location of (web-crippling) failure can be moved from the mid-support in case of the continuous brittle beams, to the span, where failure also occurs in the single-span beams, as shown in **Fig. 3.59** [66].

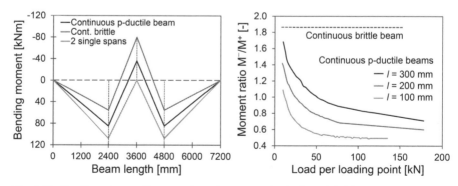

Fig. 3.58 Bending moment diagrams of continuous pseudo-ductile and brittle and two single-span beams at 135 kN per loading point **(left)**; moment redistribution in continuous pseudo-ductile beams from mid-support (M^-) to span (M^+) as a function of joint overlap length (l) **(right)**, based on data in [66]

Fig. 3.59 Web crippling failure above mid-support (continuous beam) **(left)**; in span below load (single-span beam) **(middle)**; in span below load (pseudo-ductile beam) **(right)**, based on [66, 75]

3.6.4 Conclusions

The failure behaviour of composites is complex and can vary from brittle to pseudo-ductile at the material, member, or structural system level. Small variations in geometry, fibre architecture, or imperfections of members can change the behaviour fundamentally from brittle to pseudo-ductile or vice versa. Reduction of through-thickness tensile stresses (through appropriate design), activation of fibre bridging (through appropriate fibre architecture), and implementation of redundancy (to produce progressive failure), at all levels (material-member-system), can foster pseudo-ductility and thus offer the same advantages as ductile materials. Adhesively bonded or hybrid joints can reach 100% joint efficiency if pseudo-ductile adhesives are used. Full ductility can be provided in combinations of composites with ductile components or joints in hybrid built-up members or hybrid systems.

3.7 FRACTURE

3.7.1 Introduction

Most of the design methods for engineering structures are based on stress analysis and material strength. Stress-based methods also represent the core of most structural design codes, e.g., [1]. However, these methods reach their limits when local stress concentrations are involved due to, for instance, geometric discontinuities (e.g., at sharp corners or holes), or at existing defects or cracks, where the stresses theoretically become infinite. Such stress concentrations may locally exceed strength, and the subsequent initiation of cracks and their rapid propagation cannot always be excluded; they can even lead to structural collapse, as several examples have demonstrated [76].

To base the design of such cases on energy, instead of stress and strength, is an alternative. In an energy-based approach, the energy required to propagate a crack is compared with the strain energy, which is stored in a member under load and released around the crack if such a crack initiates and propagates. If the released strain energy is smaller than the energy required to propagate the crack, the latter will not propagate, and vice versa. This approach also allows for a damage-tolerant design to be implemented, i.e., a certain crack length may be tolerated if the crack remains stable [1, 76].

Table 3.4 Approximate values of fracture energy and nominal tensile strength of common solids, according to [76]

Material	Fracture energy [J/m^2]	Tensile strength [MPa]
Glass, pottery	1–10	35–175
Cement, brick, stone	3–40	4
Polyester and epoxy resins	100	50
Bone	1 000	200
Timber (along grain)	10 000	100
High strength steel	10 000	1 000
Mild steel	1E5–1E6	400

The different nature of fracture energy-based and stress-based design approaches also becomes obvious if comparing required fracture energies to propagate a crack and tensile strengths of common solids, as done in **Table 3.4**. No clear relationship exists between these two metrics, and just basing the design on stress and strength may thus cause non-conservative solutions in cases of comparably low fracture energy. Furthermore, it should be noted (as is also shown later) that crack initiation and onset of propagation often occur long before the peak of (engineering) stress is reached in the bulk material. Depending on the problem, a fracture energy-based approach for design may thus be more appropriate than a stress-based one.

Different concepts and metrics can be found in the literature to describe the fracture behaviour of composites. Basically, they define material resistances to crack propagation, such as (i) "fracture toughness", defined as the critical value of the stress intensity factor (K_c), and (ii) "critical strain energy release rate" (G_c). Under Mode-I fracture, for instance (as defined below), stress intensity factor (K_I) and strain energy release rate (G_I) are related by $G_I = K_I^2/E$, under plain stress conditions (e.g., in thin plates), where E is the elastic modulus. In the following, the concept of "strain energy release rate" (G), henceforth denominated "SERR", with its critical value (G_c) representing the fracture resistance, will be used since it allows the effects of fibre bridging to be comprehensively described (see below). The term "rate" does not refer to a derivative with respect to time, but to the rate of change in potential energy with increasing crack area [77].

Depending on the orientation of a crack in a composite laminate with regard to the fibre architecture, three different types of crack or fracture planes can be differentiated, i.e., interlaminar, intralaminar, and translaminar crack planes, see **Fig. 3.60**. The crack plane of interlaminar cracks coincides with the planes between the different plies of the laminate, i.e., the crack is located "in-plane". The planes of intralaminar and translaminar cracks are oriented transversely to the ply planes, i.e., "out-of-plane", while the former cracks propagate parallel to the fibres and the latter transverse to the fibres.

3. Composites – Overcoming Limitations

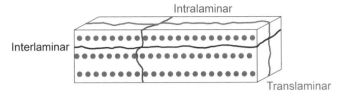

Fig. 3.60 Interlaminar, intralaminar, and translaminar crack planes in a three-ply laminate

Interlaminar fracture, also known as "delamination", is directly related to the layered material structure of composites, i.e., to the "weaker" resin-rich region at the ply interfaces, where crack propagation requires the lowest amount of energy. Intralaminar and translaminar fractures require much more energy, see **Table 3.5**, while fracture through the fibres of the translaminar orientation normally needs the largest amount. Interlaminar fracture occurs under out-of-plane tension or shear stresses, e.g., at locations (i) of significant stiffness variations through the laminate thickness, (ii) in laminate curvatures, (iii) in adhesive joints (known as "fibre-tear failure"), or (iv) in sandwich panels (e.g., debonding of face sheets from the core) [1]. Furthermore, in members manufactured by 3D-printing (with current technology), the interlaminar shear strength between the printed layers, and associated interlaminar fracture resistance, are much lower than in members manufactured with conventional processes [82]. Since interlaminar fracture is the most frequent and critical failure mode of composites, the following discussion focuses on interlaminar fracture.

Most of the work on the fracture of composites focuses on one-dimensional (1D) crack propagation, and fracture properties are based on standardised beam-like specimens, which are used to produce specific modes of fracture, as will be discussed later. In real structures, however, defects and cracks often propagate in multiple directions and with changing contours; real crack propagation is thus two-dimensional (2D). 2D crack propagation in composite laminates and sandwich panels was recently investigated at the CCLab, and results demonstrate that several significant 2D effects exist that are not considered in 1D analyses. The following discussion thus also focuses on 2D crack propagation, after having summarised the current knowledge about 1D propagation, as far as it is relevant for the structural design of composite structures.

Table 3.5 Experimental Mode-I critical strain energy release rates (G_{Ic}) [J/m²] of glass fibre composite pultruded profiles, depending on crack plane orientation

Reference	G_{Ic} [J/m²]		
	Interlaminar	Intralaminar	Translaminar
Shahverdi et al. [78]	400–2 200	-	-
El-Hajjar et al. [79]	-	8 900	23 700
Liu et al. [81]	-	7 700	-
Almeida-Fernandes et al. [80]	-	9 000–27 000	-

The work on fracture mechanics of composites at the CCLab started in 2004 and four consecutive PhD theses have been completed since then, the last two of which focus on 2D crack propagation [83–86]. The following content is mainly based on these works, and as they summarise the state-of-the art knowledge, further information and references can be found within them.

3.7.2 One-dimensional crack propagation

3.7.2.1 Experimental insights

Fracture mechanics basically considers three fracture modes, see **Fig. 3.61**: (i) Mode-I, opening mode, caused by tensile stresses normal to the crack plane, (ii) Mode-II, sliding (or in-plane shear) mode, caused by shear stresses parallel to the crack plane in the longitudinal direction, and (iii) Mode-III, tearing (or out-of-plane shear) mode, also caused by shear stresses parallel to the crack plane but in the transverse direction.

Mode-I is normally the most critical failure mode and requires the lowest amount of energy to propagate a crack. According to these failure modes, standardised 1D experimental procedures were developed to derive corresponding values of critical SERRs (G_c), e.g., double cantilever beam (DCB) for Mode-I (G_{Ic}), end-loaded-split (ELS) for Mode-II (G_{IIc}), or mixed-Mode I/II bending (MMB), as shown in **Fig. 3.62** (left). Based on load vs displacement and crack length measurements, G-values can be obtained from different data reduction methods, e.g., the Experimental Compliance Method, see [87]. Based on such experimental data, mixed-mode failure criteria can be derived, as shown in Fig. 3.62 (right), which in this case considers fibre-tear failure in adhesively bonded glass fibre composite joints [78]. A similar failure criterion is implemented in the European Technical Specification [1] for the design of adhesively bonded composite joints, see Section 5.3.3.

As can be seen in Fig. 3.62 (right), the SERR of composites (G) is normally composed of two components: (i) G_{ini} (G-initiation), which is the SERR required to initiate a crack at the crack tip (sometimes also denominated G_{tip}), and (ii) G_{br} (G-bridging), which is the SERR during crack propagation needed to break or pull-out fibres (or even rovings) that cross the crack. The sum of these two energies is the total SERR (G_{tot}) (sometimes also denominated G_{prop} for propagation).

Fibres that cross a crack and are still active or already broken are shown in **Fig. 3.63**; the active fibres are distributed along the "fibre bridging length". The initiation value (G_{ini}) is normally considered as a material property since it is mainly dependent on the matrix strength, while the fibre bridging contribution (G_{br}) depends on the curvature and thus stiffness of the two separated laminate parts (denominated

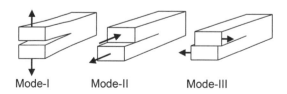

Fig. 3.61 Fracture modes of one-dimensional specimens

3. Composites – Overcoming Limitations   161

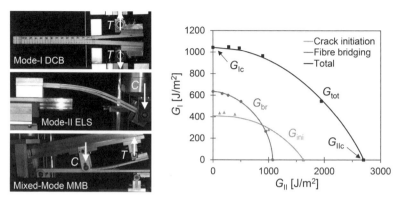

Fig. 3.62 Standardised 1D fracture experiments on adhesively bonded glass fibre composite joints: Mode-I, opening, double cantilever beam (DCB); Mode-II, sliding, end-loaded-split (ELS); mixed-Mode I/II bending (MMB) **(left from top to bottom)** (T = tension, C = compression); composition of mixed-Mode I/II failure criterion **(right)**, for G_{tot} (total SERR from Mode-I & II), based on G_{ini} (initiation component) and G_{br} (fibre bridging component), with critical values for each mode (G_{Ic} and G_{IIc}), polynomial fitting of experimental data, according to [78]

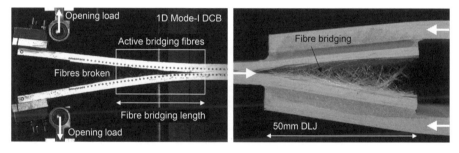

Fig. 3.63 Fibre bridging in laminate of 1D Mode-I DCB experiment, according to [90] **(left)** and in adhesively bonded double-lap joint (DLJ) under compression, [91] **(right)**

"arms"), and therefore on the specimen, member, or joint geometry [88]. The stiffer the arms, the longer the fibre bridging length, and, normally, the higher is G_{br}, since the angles of the crossing fibres, with respect to the crack plane, are smaller and their resistance against bending and decohesion is thus higher [89].

The zone where these crack initiation and propagation mechanisms take place is called the "fracture process zone" (FPZ), as shown in **Fig. 3.64** (for Mode-I), which is composed of an FPZ of initiation (FPZ$_{ini}$) and an FPZ of fibre bridging (FPZ$_{br}$), the latter being identical to the fibre bridging length. In the FPZ$_{ini}$, the traction stresses normal to the crack plane exhibit a peak at the damage initiation point (σ_c in Fig. 3.64). Matrix plasticisation and micro-cracking then occur under G_{ini}, and the traction stresses decrease to σ_{max}, which is the stress at the crack tip and also the maximum bridging traction in the bridging zone (FPZ$_{br}$). The traction stresses in the bridging zone (σ_{br}) further decrease until reaching zero at the end of the fibre bridging length, where the maximum crack opening displacement (δ_{max}) is

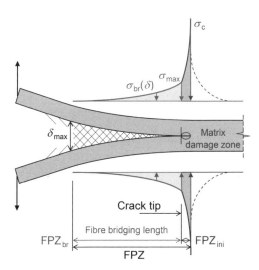

Fig. 3.64 Mode-I fracture process zone (FPZ) with schematic fibre bridging and traction stress (σ) vs crack opening displacement (δ), derived from [88] and complemented (subscripts "ini" = initiation, "br" = bridging)

obtained. The crack tip is thus located between FPZ$_{ini}$ and FPZ$_{br}$ (derived from [88] and complemented).

The resistance of composites against crack initiation and propagation is normally expressed by crack growth resistance curves, or R-curves, which show the SERR vs crack length relationship, see **Fig. 3.65**. The basic crack growth resistance is provided by the matrix where a crack initiates through matrix damage in the FPZ$_{ini}$ at G_{ini}, which remains constant with increasing crack length (and thus can be regarded as a material property, see above). Fibre bridging, which can also be manifested as roving (or fibre bundle) bridging, can increase the resistance significantly, by G_{br}, and a steady state of stable propagation can be reached under G_c. The crack length increase, during which G_{br} increases up to a steady state, corresponds to the fibre bridging length (or length of the FPZ$_{br}$). As mentioned above, the G_{br} contribution is dependent on the arm stiffness and thus geometry and cannot be considered a material property. Subsequently, the resistance may decrease, and the crack thus becomes unstable, which may lead to rapid failure (derived from [92, 93]).

The fibre bridging contribution to the critical SERR not only depends on geometry, but also on the fibre architecture. An example is shown in **Figs. 3.66–67**, where fibre-tear failure occurs at different laminate depths along different crack paths in adhesively bonded glass fibre composite DCB joints, as are shown in Fig. 3.62 (top left) [84, 94]. The joints are composed of two pultruded laminates of 6 mm thickness each, bonded together with the epoxy adhesive introduced in Section 2.2.3. The fibre architecture of the laminates consists of two combined outer mat layers (towards the adhesive layer), followed by rovings, which form the core of the laminates, see Fig. 3.66 (left). The combined mats consist of woven 0°/90° fabrics, sandwiched by two chopped strand mats (CSM); all three layers are stitched together. Fibre-tear failure occurs either within

3. Composites – Overcoming Limitations   163

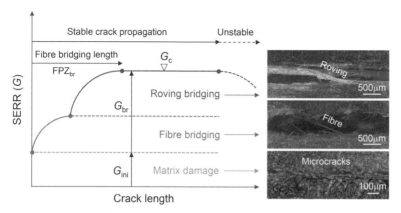

Fig. 3.65 *R*-curve (schematic): SERR (*G*) vs crack length relationship of composite laminates exhibiting matrix damage, and fibre and roving bridging (assuming roving bridging initiates after fibre bridging) [13]

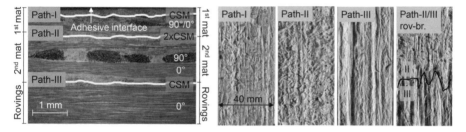

Fig. 3.66 Different crack paths producing fibre-tear failure in adherends of adhesively bonded glass fibre composite DCB joints **(left)**, view on 40 mm wide laminate crack surfaces of Paths-I–III and Path-II/III with roving bridging (rov-br.) **(right)**, based on [84, 94]

the first mat layer (Path-I), or between the first (thinner) and second (thicker) mat layers, i.e., within the two adjacent CSMs (Path-II), or between the second mat layer and the roving core, i.e., mainly in the CSM of the second mat layer towards the roving core (Path-III). In some cases, a crack of Path-II stops, and a new crack initiates in Path-III, just below, and rovings from the second mat bridge the two crack tips, i.e., providing "roving bridging"; this case is denominated "Path-II/III rov-br." The respective fracture surfaces of the three paths are shown in Fig. 3.66 (right). Path-I exhibits the lowest level of fibre bridging, while the level is highest in Path-II, and an intermediate value is exhibited in Path-III. The level of fibre bridging is directly related to the thickness (or weight) of the CSM, whose fibres penetrate the crack plane. The fracture surface of Path-II/III with roving bridging shows the failed roving bridges at the marked intersection between the two paths ([94], complemented).

Load and total crack length vs opening displacement relationships (obtained under displacement-control) are shown in **Fig. 3.67** (left) for Path-II and Path-II/III with additional roving bridging. It should be noted that crack initiation and propagation start from an implemented pre-crack in the specimens. As mentioned above,

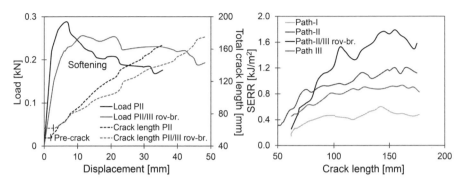

Fig. 3.67 Effect of fibre architecture on 1D Mode-I load and total crack length vs opening displacement **(left)** and SERR vs total crack length **(right)**, based on [84, 94]

cracks initiate before the load peak is reached and, subsequently, a softening behaviour is exhibited with increasing crack propagation, in both cases. Roving bridging, however, increases the fracture resistance and thus significantly decreases the softening and leads to an almost pseudo-ductile response (see Section 3.6.2). A similar case is shown in *Example-1* of Section 3.6.3, where significant fibre bridging contributes to a pseudo-ductile response, while the absence of fibre bridging causes a brittle behaviour. The *R*-curves of the four cases involved are shown in Fig. 3.67 (right). The higher the level of fibre bridging, the higher the fracture resistance; the latter reaches a maximum in the case with additional roving bridging, as is also schematically shown in Fig. 3.65. Stable crack propagation is obtained in all cases, the critical SERR (i.e., the peak of the curves) varies from 600 to 1790 J/m^2, thus by a factor of almost 3.0. An appropriate fibre architecture, which fosters fibre or even roving bridging, can thus significantly increase the fracture resistance ([94], complemented).

Further experimental fracture results are discussed in Section 3.8.8. They concern the debonding of glass fibre composite face sheets from end-grain balsa cores in sandwich beams, in an MMB set-up. Since fatigue loading was also applied, the static and fatigue results were not split and kept in Section 3.8.8. Among other things, a static fracture failure criterion is derived, similar to that shown in Fig. 3.62 (right).

3.7.2.2 Modelling

In the previous section, it has been shown how the 1D fracture resistance (or critical SERR, G_c) is defined and can be obtained experimentally. In a structural design based on fracture mechanics, the SERR induced by the external actions (G) must also be known, however, in order to apply design criteria as shown in Fig. 3.62 (right), or in [1] for adhesive joints.

Such SERR induced by external actions can be derived from linear elastic or nonlinear fracture mechanics modelling (LEFM or NLFM modelling, respectively). LEFM is applicable if the FPZ is concentrated at the crack tip, and all mechanical relationships remain linear. If significant fibre bridging is involved, however – i.e., in cases of large-scale bridging in particular, where the bridging length (and FPZ$_{br}$) is

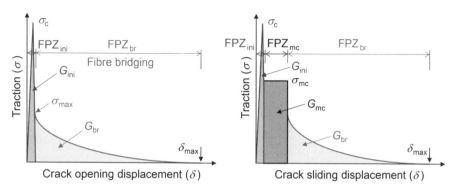

Fig. 3.68 Mode-I traction vs opening displacement relationship considering fibre bridging, based on [88] **(left)**; Mode-II traction vs sliding displacement relationship with additional micro-cracking (mc), based on [95] **(right)**.

comparable to one of the specimen dimensions – the responses become nonlinear and NLFM must be applied.

SERR values induced by external actions are normally obtained from finite element modelling; depending on whether LEFM or NLFM applies, different methods are available. In the LEFM case, the virtual crack closure technique (VCCT) can be used to model crack initiation; the method also allows to determine the mode-mixity at the crack tip [1, 78].

If NLFM applies, cohesive zone models (CZM) can be used to model crack initiation and propagation [1, 88]. Cohesive elements are introduced along the crack path, which follow cohesive "laws", such as the one shown in **Fig. 3.68** (left) for Mode-I opening. The traction stress vs opening displacement relationship corresponds to that shown in Fig. 3.64 and considers crack initiation and propagation with (large-scale) fibre bridging. The fracture energies required for initiation (G_{ini}) and propagation (G_{br}) are represented by the areas below the curve. A more complex CZM is used for Mode-II crack propagation in [95], where the FPZ of propagation is subdivided into an FPZ for pronounced micro-cracking (FPZ$_{mc}$) and an FPZ for fibre bridging (FPZ$_{br}$), with the associated fracture energies G_{mc} and G_{br}, respectively, as shown in Fig. 3.68 (right), see also Section 3.7.3.4. Such CZM depend on the specimen geometry and can thus not be assumed to be material properties; they need to be calibrated experimentally [88]. The degree of complexity of such models can be increased to be a better fit to experimental results. However, a mechanical background and justification of the introduced parameters should exist.

3.7.3 Two-dimensional crack propagation

3.7.3.1 Experimental design

As introduced above, crack propagation normally occurs in a crack plane in two-dimensions (2D) and thus significantly deviates from the current one-dimensional (1D) approach. To derive and quantify the differences in 1D vs 2D crack propagation

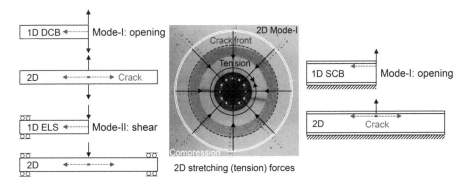

Fig. 3.69 One-dimensional (1D) standardised fracture set-ups (Mode-I, DCB and SCB, and Mode-II, ELS) and derived 2D set-ups for laminates **(left)** and sandwich panels **(right)**; with radial and circumferential stretching/tension stresses and compression ring caused by Mode-I opening, load application in black centre **(middle)**, all [13]

and fracture resistance, the standardised 1D experimental set-ups for delamination in laminates and face sheet debonding in sandwich panels can be transformed into 2D set-ups, as shown in **Fig. 3.69**.

In the case of Mode-I opening in laminates, the 1D DCB set-up (Fig. 3.62, top left) can be mirrored and extended in the second dimension. The crack thus initiates in the centre of a plate and then propagates radially in all directions. The boundary conditions at the load application point are different, however. While the DCB arms can rotate, as shown in Fig. 3.63 (left), a rotation of the plate parts above and below the crack is not possible due to the laminate continuity, as is visible in Fig. 3.69. In contrast to the 1D case, this different 2D boundary condition is consistent, however, with real cases of 2D crack opening.

The 2D Mode-II set-up for laminates can be similarly derived as the Mode-I set-up, i.e., the 1D ELS set-up (Fig. 3.62, left middle) can be mirrored and extended in the second dimension. The support boundary conditions are kept identical, while those at the load application point again differ. The 1D arm rotations and differential sliding displacements of the two arms are hindered by the 2D plate continuity. Again, the new 2D set-up better represents real Mode-II boundary conditions than the 1D set-up.

In sandwich panels, due to the elastic property mismatch at the face sheet/core interface, debonding of face sheets from the core always occurs under a certain mode-mixity. In this respect, the 1D sandwich single cantilever beam (SCB) set-up, which provides a Mode-I dominant loading [88], can also be mirrored and extended in two dimensions (Fig. 3.69, right). The boundary conditions at the bottom are identical, i.e., fully fixed, while those at the load application point differ in the same way as they do under Mode-I of laminates.

The 2D plate opening under Mode-I and 2D plate deflection under Mode-II activate additional in-plane stresses in the plates, which do not exist under 1D conditions. In both cases, radial stretching (tension) stresses are activated as soon as the plate deforms out-of-plane, and a compression ring thus forms ahead of the crack front,

3. Composites – Overcoming Limitations   167

in which the stretching stresses are anchored, as schematically shown in Fig. 3.69 (middle) for Mode-I. Furthermore, in addition to the radial stresses, circumferential tension stresses are also activated. These radial and circumferential tension stresses develop in the mid-plane of each of the two separated plate parts under Mode-I, while they are induced in the mid-plane of the whole plate under Mode-II. The compression rings only form because of the limited plate specimen dimensions. In larger plates, the radial stresses simply level off ahead of the crack front.

3.7.3.2 Mode-I 2D delamination in laminates

An opened 2D laminate under the above introduced Mode-I set-up and displacement control is shown in **Fig. 3.70** (left). The translucent laminate allows the crack front to be observed and measured since the delaminated plate parts turn opaque. The measured load and total radial crack length (including a pre-crack) vs the opening displacement responses of an orthotropic laminate are shown in **Fig. 3.71** (left). In contrast to the 1D case, i.e., Fig. 3.67 (left), the load continues increasing during the crack propagation process until a maximum is reached. The corresponding measured compliance vs total crack area response is depicted in Fig. 3.71 (right). The compliance, which is the opening displacement divided by the load (or the reciprocal of

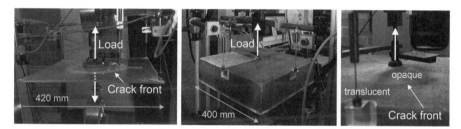

Fig. 3.70 Mode-I 2D opening set-ups for laminates [88] **(left)**; sandwich panels **(middle)**; face sheet debonding of sandwich panels [97] **(right)**

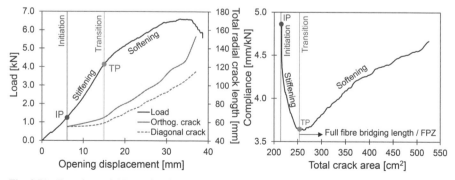

Fig. 3.71 Experimental 2D Mode-I load and total radial crack length vs opening displacement **(left)**; compliance vs total crack area **(right)**, orthotropic laminate with 50/50% glass fibre woven architecture (IP = crack initiation point, TP = transition point, separating stiffening and softening branches), derived from [96] and complemented

stiffness), first decreases, and subsequently increases with increasing crack area. The lowest point marks the transition point (TP) which separates the stiffening from the softening branch and further indicates that the full fibre bridging length is reached. The two branches can also be recognised on the load vs opening displacement curve by a marked change, i.e., decrease, of the slope at the TP. In contrast to the 1D case, where the load decreases during softening, the load still increases in the 2D case during softening [96].

These different 1D vs 2D responses can be explained by the crack front length, which increases in the 2D case (in contrast to the 1D case where it remains constant), and the different mechanisms at work, i.e., softening and stiffening. While the crack area increases proportionally with the crack length in the 1D case and the load decreases (Fig. 3.67, left), a disproportionate growth of the crack area occurs in the 2D case, which always forces the load to increase (Fig. 3.71, left) to overcome the fracture resistance and further propagate the crack [96].

Crack propagation normally softens the response; the softening is counteracted, however, by fibre bridging in the FPZ_{br}, and micro-cracking and plasticisation in the FPZ_{ini}. The higher the resistance due to fibre bridging, the more the softening is hindered and the higher the fracture resistance. As discussed above, the resistance provided by fibre bridging can be increased by reducing the curvature in the FPZ_{br}, thus by an increase in stiffness. In contrast to the 1D case, such stiffening mechanisms are activated through the 2D behaviour, i.e., through (i) the higher plate stiffness (compared to the 1D beam stiffness), (ii) the persistence of the continuity at the loading point, and (iii) the in-plane stretching (as described above). Since the resistance obtained from fibre bridging is thus increased, the 2D fracture resistance is higher than the 1D resistance, which contributes to an increase in the load. Up to the TP, stiffening mechanisms prevail over softening mechanisms, while the softening mechanisms dominate the stiffening mechanisms beyond the TP.

The increased 2D fracture resistance is confirmed by numerical Mode-I analyses performed in [88, 98], where the fracture behaviour of an isotropic composite plate is modelled using CZM and a cohesive law such as that shown in Fig. 3.68 (left). The 2D critical SERR (G_c) is up to 50% higher than in the 1D case, see **Fig. 3.72** (left). An assumption made in the modelling is that the cohesive law remains constant during propagation and the 2D R-curve also exhibits a plateau, as in the 1D case. This is confirmed by analysing the curvature in the 2D FPZ during a propagating crack, see Fig. 3.72 (right); the isotropic simplified square model in [98] is applied, using a constant law. The results reveal that the curvature in the FPZ, and thus bridging area, does not notably change during propagation, thus indicating that the cohesive law and 2D fracture resistance may remain constant; this conclusion still needs further confirmation, however.

CZM modelling in case of anisotropic plates is more complicated since the stiffness in the radial direction is not uniform and bending/opening also occurs in the circumferential direction [99], as is the case in the example shown in Fig. 3.71 (left), where the stiffness in the orthogonal directions is significantly higher than in the diagonal directions. Accordingly, the crack propagates faster in the orthogonal direction. Fibre bridging thus is influenced by varying stiffnesses in radial and circumferential directions, and the cohesive law can thus not be assumed to be constant.

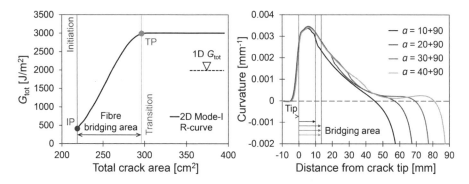

Fig. 3.72 Numerical 2D Mode-I R-curve, i.e., total SERR vs total crack area, isotropic laminate with long continuous glass filament mat architecture (IP = crack initiation point, TP = transition point), derived from [88] **(left)**; plate curvature in bridging area at four propagation states (total crack lengths, a, including 90 mm pre-crack) [13] **(right)**.

Real cracks do not propagate in one, but in two dimensions, as discussed above, but also, they are not normally perfectly circular, as assumed so far. The effect of the shape of an initial defect from which a crack develops, on the fracture behaviour, is therefore further investigated in [98]. Numerical simulations reveal that, independent of the initial shape of the defect, the crack always develops towards a circular shape under a circular loading block shape, as shown in **Fig. 3.73**. The crack initiates and develops first in the direction where the stiffness is the highest, i.e., where the defect length is the smallest (in 90° direction in Fig. 3.73, left), until full fibre bridging is reached in this direction. In the meantime, the crack initiates at the longest length of the defect (in 0° and 180° directions), where stiffness is smallest, and also develops until full fibre bridging is reached. The circular shape results when full fibre bridging occurs in all directions. The G-values develop accordingly, as shown in Fig. 3.73 (right).

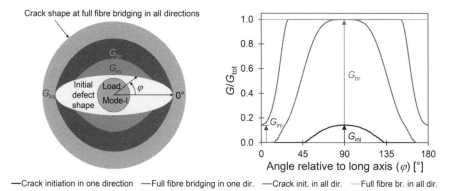

Fig. 3.73 Crack shape development from elliptical (initial defect shape) to circular at four different fracture states **(left)**; numerical 2D G-distribution along crack perimeter **(right)**, derived from [98]

3.7.3.3 Mode-I 2D debonding in sandwich panels

The experimental 2D sandwich panel face sheet debonding set-up, as described above, and a debonding face sheet from a balsa core under displacement control, are shown in Figs. 3.70 (middle and right); the circular crack front can be clearly seen. The corresponding load and total radial crack length vs opening displacement responses of two selected sandwich panels are depicted in **Fig. 3.74** (left). The difference in the panels is that one panel was manufactured with a continuous filament mat (CFM) at the face sheet/core interface to increase fibre bridging in the debonding/crack plane, while this mat layer is missing in the reference panel. The positive effect of the CFM layer can be clearly seen, the load significantly increases compared to the reference panel without (w/o) CFM layer and a crack arrest finally occurs [97].

The compliance vs total crack area relationship, shown in Fig. 3.74 (right) is similar to the laminate case (Fig. 3.71, right), exhibiting branches with prevailing stiffening and softening mechanisms, separated by the transition point (TP) at the lowest compliance level. The same softening and stiffening mechanisms are active as in the laminate case (Fig. 3.71) but are complemented by the fact that the stretching in the debonding face sheet also increases the mode-mixity, i.e., the contribution of shear fracture modes, which further increase the fracture resistance. Due to these differences, the TP does not mark the slope change in the load vs opening displacement curve and the achievement of full fibre bridging, as done in the laminate case (Fig. 3.71) [97].

Since sandwich panels composed of composite face sheets and balsa core are also used as bridge deck slabs of vehicular bridges (see Appendix A.5), the 2D debonding behaviour of the top face sheet under fatigue loading is also investigated and compared to the 1D case. The same 2D set-up is used as in the quasi-static investigations (see above), and the experiments are performed under load control and at different R-ratios, i.e., minimum-to-maximum fatigue load ratios, with values of 0.1, 0.3, and 0.5 selected. Furthermore, as in the quasi-static case, the effect of a CFM layer in the face sheet/core interface is analysed [100].

A first difference to 1D Mode-I fatigue loading already resides in the fact that load control can be applied under 2D Mode-I fatigue loading, as further demonstrated below. Load control is also the way that fatigue loads are implemented in bridges. 1D experiments under load control are difficult to perform, however, since the crack often becomes unstable with increasing propagation, which is the reason why experiments are normally performed under displacement control, where the load decreases with increasing propagation [100]; the significance of such results for real cases may thus be questioned. Only 1D results under displacement control are thus available to be compared to 2D results under load control in the following.

The 2D fatigue responses strongly depend on the R-ratio and the existence of the CFM layer. Surprisingly, and in contrast to the quasi-static experiments, the sandwich panel with CFM exhibits the lowest fatigue resistance, at $R = 0.1$. The crack propagates rapidly, without exhibiting a significant reduction in the crack growth rate, as shown in **Fig. 3.75** (left), and in contrast to the panel without (w/o) CFM, where the crack growth rate (dA/dN) decreases towards the end. Due to the high fatigue amplitude at $R = 0.1$, the cyclic crack closure crushes most of the CFM bridging fibres and thus inhibits a positive effect of fibre bridging. At the higher R-ratios with lower amplitude,

3. Composites – Overcoming Limitations   171

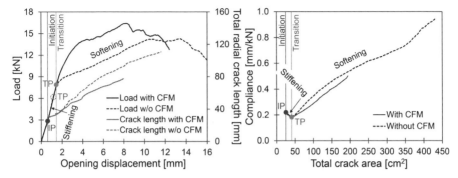

Fig. 3.74 Effect of continuous filament mat (CFM) between sandwich panel face sheet and core on 2D Mode-I load and total radial crack length vs opening displacement **(left)**; compliance vs total crack area **(right)**, experimental data derived from [97] and complemented

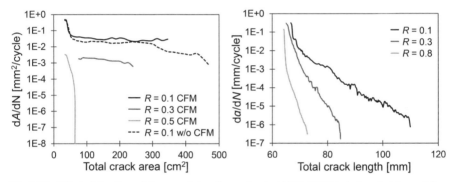

Fig. 3.75 Effect of R-ratio on fatigue crack (area) growth rate dA/dN of 2D Mode-I sandwich panel face sheet debonding with and without (w/o) continuous filament mat layer (CFM) (under load control), derived from [100] **(left)**; crack (length) growth rate da/dN of 1D Mode-I adhesively bonded glass fibre composite joints (under displacement control), according to [101] **(right)**, both experimental data

however, the crack growth rate significantly decreases, even up to an arrest ($R = 0.5$), or remains stable ($R = 0.3$), until the end of the experiments at 12.5 million cycles. In general, and in contrast to 1D experiments, the crack growth rate (da/dN) decreases under load control since an increasing load is required to propagate the 2D crack (as explained above for the quasi-static case). A stable propagation and final arrest of a 2D crack can thus normally be achieved under load control conditions [100].

1D Mode-I crack growth rates at different R-ratios, obtained under displacement control, however, are shown in Fig. 3.75 (right). The rates also continuously decrease with increasing crack length since the load decreases, as mentioned above. Steeper curves are also obtained with increasing R-ratio [101].

3.7.3.4 Mode-II 2D delamination in laminates

The 2D Mode-II set-up, as derived in Fig. 3.69 (left bottom), and a plate specimen with visible crack front, are shown in **Fig. 3.76**. Furthermore, the measured 2D Mode-II

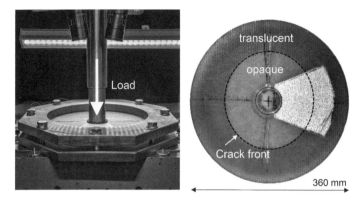

Fig. 3.76 Mode-II 2D set-up for laminates with 40 mm radius pre-crack (**left**); crack front top view (**right**), derived from [102]

load and total radial crack length vs opening displacement, and compliance vs total crack area, are depicted in **Fig. 3.77**, for two different pre-crack sizes of 40- and 80-mm radius. As can be seen, the 2D Mode-II laminate responses are similar to the 2D Mode-I laminate results, exhibiting initial stiffening and subsequent softening branches, separated by the TP. The stiffening and softening mechanisms and their relative predominance can thus be assumed to be similar under both fracture modes. However, much more plastic matrix deformation and micro-cracking occur under Mode-II, which is the reason why this zone is attributed to the FPZ for propagation in [102], and the cohesive law is modified accordingly, as shown in Fig. 3.68 (right). The pre-crack size has a notable effect on the results, i.e., the branches exhibiting stiffening are more pronounced in the larger pre-crack case since the deflection and thus stretching effects are larger at the same load.

The location of the crack front (or tip) in the above described 2D Mode-I and Mode-II experiments can be identified due to the translucency of the undamaged laminates and face sheets, and their subsequent whitening in damaged and delaminated zones (see Section 4.4.2). The crack front is thus at the separation between the translucent and the opaque laminate zones. Laminates are not always translucent, however, e.g., if fibres other than glass fibres are used, and the crack front is difficult to determine, since in 2D specimens – unlike in 1D specimens – no lateral edges exist where crack propagation can be monitored; normally only measurements on the top or bottom surfaces are possible.

Such measurements, i.e., of the radial strains on the top surface, are shown in **Fig. 3.78** (left) for different crack length (a), obtained from the Mode-II 2D experiments shown in Fig. 3.76 (with 40 mm pre-crack radius). Most of the plate is subject to negative bending moments throughout the loading process, and tensile strains thus occur. The curves basically exhibit three segments: (i) an inner delaminated segment with increasing strain and higher slope, (ii) an outer intact segment with strain increases and lower slope, and (iii) an intermediate segment where the strains drop, and which contains the crack front. The derivatives of the intermediate strain segment

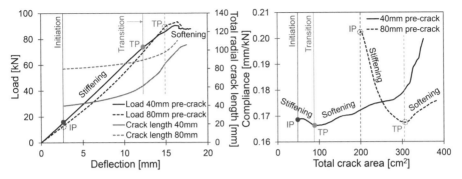

Fig. 3.77 Experimental 2D Mode-II load and total radial crack length vs deflection **(left)**; compliance vs total crack area **(right)**, isotropic laminate with CFM in centre and multidirectional fabrics on outer sides, and two different pre-crack sizes (IP = crack initiation point, TP = transition point, separating branches of stiffening and softening), derived from [102] and complemented

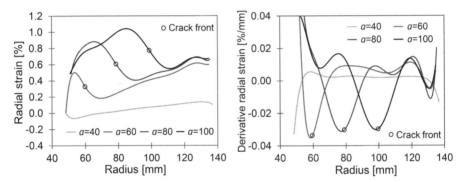

Fig. 3.78 Localisation of 2D Mode-II crack front based on radial strain measurements on top surface (fitted with ninth-order polynomial curves) **(left)**; minima of derivatives of radial strains vs crack front location, based on data in [95]

curves, shown in Fig. 3.78 (right), exhibit minima, which closely match the crack front location; differences of only about 2 mm are recognisable. This method thus provides promising results to identify the crack front but needs further verification in other cases.

3.7.4 Multiple cracks

In the previous Section 3.6.3, *Example-2* shows that almost parallel, multiple cracks can appear in curved web-face panel junctions (WPJs) of pultruded cellular profiles, see Fig. 3.53, and a magnified case in **Fig. 3.79** (left), where three main cracks develop on the side exposed to curved tension (T). The cracks consecutively initiate at the location of the highest curvature, where the radial deviation tensile stresses (t) are greatest, and then extend to both sides towards the straight arms. Depending on

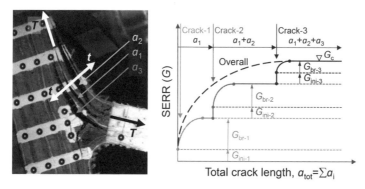

Fig. 3.79 Multiple crack development under curved tension (*T*) and corresponding radial tension (*t*), according to [103] **(left)**; composition of overall (schematic) *R*-curve with critical SERR (G_c), consisting of SERR for initiation (G_{ini}) and fibre bridging (G_{br}) of three sequentially initiating parallel cracks [13] **(right)**

the sequence and growth rate of the individual cracks, an overall *R*-curve can be constructed, which is composed of the *R*-curves of the individual cracks, as schematically shown in Fig. 3.79 (right) and derived in the following. Initiation of the individual cracks may overlap, however, and not be as sequential, as schematically shown in the diagram.

The SERR (*G*) and the belonging compliance (*C*) can be derived from linear elastic fracture mechanics, for a certain crack length (*a*), as follows [103, 104]:

$$G = \frac{P^2}{2 \cdot b} \cdot \frac{dC}{da} \quad \text{and} \quad C = (\alpha \cdot a + \beta)^\chi \tag{3.6}$$

where *P* is the load, *b* is the specimen width, and α, β, and χ are constants fitted to the experimental compliance curve.

Assuming that the total crack length (a_{tot}) is composed of the individual crack lengths (a_i) as follows:

$$a_{tot} = \sum_{i=1}^{N} a_i = a_{tot} \cdot \sum_{i=1}^{N} k_i \quad \text{with} \quad k_i = a_i / a_{tot} \tag{3.7}$$

where k_i is a constant, *N* is the number of cracks, and furthermore assuming that α, β, and χ are identical for all individual cracks i, the total SEER ($G(a_{tot})$) is then [103]:

$$G(a_{tot}) = \frac{P^2}{2 \cdot b} \cdot \sum_{i=1}^{N} \frac{dC}{da_i} = \frac{P^2}{2 \cdot b} \cdot \chi \cdot k_i \cdot a_{tot} \cdot \sum_{i=1}^{N} (\alpha \cdot k_i \cdot a_{tot} + \beta)^{\chi-1} \tag{3.8}$$

The above procedure is applied for a WPJ, similar to that in Fig. 3.79 (left), under the bending set-up shown in Fig. 3.52. The load vs deflection response, depicted in **Fig. 3.80** (left), shows a softening after the peak load, which is mainly caused by three cracks; they represent approximately 95% of the total crack length. Crack-1 and -2 appear at the same time, while Crack-3 initiates later; the total length of these three cracks is also shown. The compliance vs total crack length relationship results from fitting Eq. 3.6 to the experimental results, i.e., the load vs deflection relationship,

3. Composites – Overcoming Limitations

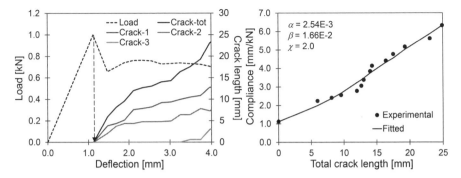

Fig. 3.80 Load and individual and total crack length vs deflection **(left)**; compliance (deflection/load) vs total crack length **(right)**, according to [103]

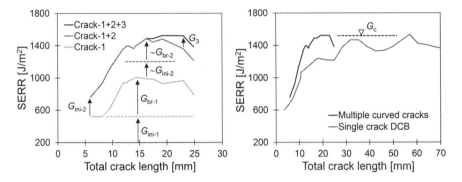

Fig. 3.81 Multiple crack R-curve: superposition of SERR vs total crack length relationships of three parallel cracks **(left)**; comparison to DCB single crack R-curve **(right)**, based on data in [90, 103]

see Fig. 3.80 (right), and the constants of this equation are obtained. The total SERR is finally derived according to Eq. 3.8 and depicted in **Fig. 3.81** (left). The results demonstrate that the contributions of new cracks to the total SERR diminish. These decreasing contributions of new cracks may be explained by the fact that their initiations overlap with previous cracks, and the local stiffness decreases with additional cracks, which reduces the fibre bridging contribution, see Section 3.7.2.1.

Although the different cracks always occur between the same materials, i.e., in the interfaces between rovings and triaxial fabrics, the hypothesis that the compliance constants are identical for all cracks still needs further validation. Furthermore, the effects of potential interactions between cracks, e.g., that one crack closes because of a newly opening adjacent crack, and the associated effects on fibre bridging, need further investigations.

In a case similar to the one discussed above, it has been shown that the SERR in a curved crack mainly originates from Mode-I fracture [105]. The derived multiple crack R-curve can thus be compared with a single crack R-curve obtained from Mode-I DCB experiments on the same laminates, where failure occurs between the

176 Composites in Structural Engineering and Architecture

same materials [103]. The two curves compare quite well, i.e., the critical SERRs (G_c) are similar, as demonstrated in Fig. 3.81 (right), and thus support this approach.

3.7.5 Conclusions

The fracture behaviour of composites has a significant effect on the load-bearing response of composite structures. If the fracture process zone (FPZ) is small and only small amounts of strain energy can thus be dissipated, rapid crack propagation may occur and lead to sudden and brittle failure. Larger FPZs can be achieved by activating fibre bridging, which increases the dissipated strain energy and thus fracture resistance and may even provide an advantageous pseudo-ductile response behaviour.

The fibre architecture plays an important role in this respect. Arranging unidirectional fibres/rovings or nonwoven fabrics only in the principal stress directions may be very efficient in terms of laminate strength. However, the interlaminar fracture resistance of such architectures may be low in comparison since fibre bridging may only be sparsely activated. Interspersing "non-structural" layers, such as chopped strand mats (CSM) or continuous filament mats (CFM), into such architectures, may increase fibre bridging significantly and thus interlaminar fracture resistance. Such additional mat layers are particularly efficient at locations which are sensitive to interlaminar fracture, i.e., where laminate out-of-plane tensile stresses occur, as is the case in laminate curvatures, adhesive joints, or sandwich panel face sheet/core interfaces.

In contrast to the current one-dimensional (1D) approach to characterise fracture resistances, real two-dimensional (2D) crack propagation exhibits significant differences, as summarised in **Table 3.6**. The increasing crack front and thus disproportionally increasing 2D crack area require a progressively increasing load to propagate a crack, in contrast to 1D propagation, where the load drops shortly after crack

Table 3.6 Summary of similarities and differences in 1D vs 2D crack propagation [13]

Effects	1D Mode-I & II	2D Mode-I & II
Load during crack propagation	Decrease	Increase (due to increasing crack front)
Softening	Crack propagation	Crack propagation
Stiffening (increases fibre bridging contribution)	None	Plate bending Plate continuity Stretching
Fracture resistance	Benchmark	Increased (if fibre bridging occurs)
Fatigue crack growth rate under load control (only Mode-I)	Unstable	Crack arrest (due to increasing crack front length)

initiation. Significant stiffening mechanisms are activated under 2D propagation, which decrease the curvature in the FPZ and thus may significantly increase the level of fibre bridging and fracture resistance compared to a 1D propagation. Without the occurrence of fibre bridging, however, no differences between 1D and 2D propagation exist, since the fracture resistance is mainly provided by the matrix, particularly under Mode-I fracture, while micro-cracking under Mode-II may still make some differences.

3.8 FATIGUE

3.8.1 Introduction

Fatigue failure in composite structures may represent a relevant ultimate limit state because, in many cases, the self-weight is low compared to the variable actions, which is particularly the case in lightweight composite bridges subjected to heavy traffic loads. Low permanent/variable action ratios may have negative consequences, such as high fatigue stress amplitudes and associated low mean stress levels, both of which may shorten the fatigue life. While the fatigue resistance of steel, for instance, only depends on the stress amplitude, that of composites also depends on the mean stress level, as is the case for viscoelastic materials in general, where higher mean stress levels may induce creep and thus creep-fatigue interaction. In contrast to steel, furthermore, the fatigue resistance of composites also depends on the loading type, i.e., compression or tension.

The basic triggers of fatigue failure in composites are, however, the same as in other materials, i.e., stress concentrations, which may occur at defects and geometrical or material discontinuities, as is the case, for instance, in connections or sandwich panels. Such stress concentrations can be prevented by appropriate geometrical detailing and material selection. The fatigue resistance furthermore depends on temperature and moisture conditions.

Fatigue design verification, together with fatigue resistance, requires relevant and appropriate fatigue load models, which cause damage similar to that caused by real fatigue actions, and from which the action effects can be derived (e.g., stress amplitudes). In this respect, a difficulty exists in composite bridge design since appropriate fatigue load models do not yet exist. The current fatigue load models (e.g., in Eurocode EN 1991-2 [106]) were calibrated for the damage rate of steel in steel and reinforced concrete bridges, and are partially based on variable amplitude loading. In the latter case, the application of the Palmgren-Miner linear damage accumulation assumption is involved, which is not reliably applicable to composites, however [40].

In the following, after summarising the basic representation of fatigue resistance in stress vs number of cycle relationships (*S-N* curves), mean stress dependence and creep-fatigue interaction will be first discussed, before addressing variable amplitude loading. Effects of temperature and humidity on the fatigue resistance are subsequently analysed, fatigue modelling will be discussed, and fatigue-tailored detailing addressed. Finally, a first approach to establish a fatigue load model for composite bridges is presented. Methods to determine stresses due to fatigue actions (e.g., by finite element analysis) are normally independent of the structural material and are thus not further addressed. The following content is mainly based on the fatigue research on composites accomplished at the CCLab since 2002, including four PhD theses [83, 84, 107, 108].

3.8.2 Basic representation of fatigue resistance

Experimental fatigue data are normally represented in stress vs number of cycles to failure (S-N) curves, which generally follow a power law; stress range, stress amplitude, or maximum stress can be indicated on the S-axis. The N-axis is generally in log-scale, while the S-axis can be linear or in log-scale too. In double logarithmic scale, the fatigue response is normally linear, i.e., it exhibits a constant slope. These relationships can be represented as follows:

$$\sigma = \sigma_0 \cdot N_f^{-1/m} \quad \text{or} \quad \log \sigma = \log \sigma_0 - m^{-1} \cdot N_f \tag{3.9}$$

where σ is the stress (range or amplitude or maximum value), $\log \sigma_0$ is the ordinate value, m is the slope coefficient, defined as the reciprocal of the slope of the straight line in log-log-scale (i.e. the damage rate), and N_f is the number of cycles to failure. The m-values are independent of the selected stress metric, i.e., stress amplitude or maximum stress. Furthermore, experimental fatigue data normally refers to a specific fatigue stress ratio (R) and corresponding mean stress (σ_m), which depend on the maximum stress (σ_{max}) and minimum stress (σ_{min}) as follows:

$$R = \sigma_{min}/\sigma_{max} \quad \text{and} \quad \sigma_m = (\sigma_{max} + \sigma_{min})/2 \tag{3.10}$$

where tensile and compressive stresses have positive and negative signs, respectively. Typical S-N curves of composites profiles and joints are shown and compared to those of structural steel in **Fig. 3.82**. The former are normally flatter than the latter, i.e., they have higher m-values, and, most importantly, do not exhibit fatigue limits, as steel does (horizontal lines beyond 5E+6 cycles for constant amplitude, CA). In steel, the m-value is constant, and only the fatigue limit depends on the type of structural component or connection and decreases with their sensitivity to stress concentrations. In composites, however, the slope is not constant and depends on several factors: (i) the R-ratio – where higher R-ratios, i.e., smaller amplitudes, reduce the damage rate and thus increase m, (ii) the loading type, i.e., compression or tension – where tension fatigue normally exhibits a higher damage rate, and (iii) the viscous response of the polymer – where dominating viscous responses, e.g., in elastomeric polymers or at elevated temperatures, increase the creep-fatigue interaction (see below) and thus increase the damage rate. As can be seen throughout the following examples, at $R = 0.1$ for instance, the lowest m-values were observed for acrylic adhesives (elastomers) and epoxy adhesives at around T_g (i.e., both in rubbery state, $m \approx 7$–8), while epoxy adhesives in the glassy state (≈ 10–12), glass fibre-polymer laminates and epoxy-adhesive composite joints (≈ 10–15) exhibit intermediate values. Carbon fibre-polymer laminate values (≈ 17–20 [109]) are normally the highest, mainly due to a high carbon fibre-matrix interface strength.

In the two examples shown in Fig. 3.82, the tension-tension fatigue loading of the pultruded glass fibre composite profiles continuously damaged the matrix and

3. Composites – Overcoming Limitations   179

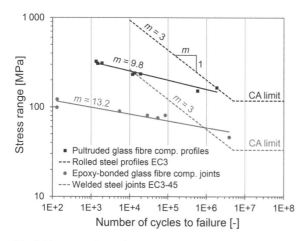

Fig. 3.82 Comparison of *S-N* curves of glass fibre composite vs steel profiles and joints (*m* = slope coefficient, i.e., reciprocal of slope of power fit), based on data in [110–112]

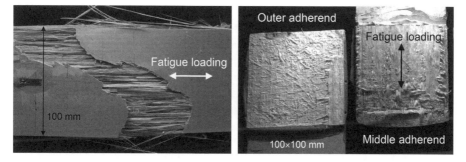

Fig. 3.83 Failure modes under tension-tension fatigue at $R = 0.1$, of pultruded glass fibre composite profile (cross section of 100×10mm²) **(left)**; epoxy-bonded composite double-lap joint composed of same profiles **(right)**, based on [111, 112]

fibre-matrix interfaces, which led to the large opening of the diagonal crack shown in **Fig. 3.83** (left) [111, 112]. In the epoxy-bonded joints composed of the same profiles, however, no effects of damage were observed until brittle fibre-tear failure occurred in the profiles, depicted in Fig. 3.83 (right) [112].

Occasionally, the quasi-static resistances are included in fatigue resistance curves; they can be interpreted as resistances at a half cycle, i.e., $N_f = 0.5$. This assumption is, strictly speaking, only valid at $R = 0.0$, but may be used as an approximation at $R = 0.1$. These quasi-static resistances must be obtained, however, at the same (normally high) loading rate as applied in the fatigue experiments. If this rule is not respected, the values may be too low, and their inclusion may lead to an incorrect (nonlinear) fitting of the experimental data at low-cycle fatigue. Applying a constant loading rate (r) in fatigue experiments also implies that the frequency (f) needs to be adapted to the stress amplitude (σ_a) and thus should increase with decreasing amplitude, according to the relationship $r = 4 \cdot \sigma_a f$.

3.8.3 Mean stress dependence

As composites are viscoelastic materials, they exhibit creep deformations when under constant stress and may fail through creep rupture, as discussed in Section 3.4.5. A constant stress can also be understood as a constant amplitude fatigue loading at a stress ratio which approaches $R = 1.0$ (where $\sigma_{max} = \sigma_{min} = \sigma_m$). Fatigue loading may thus also induce creep deformations and associated damage, and both of which may be approached by the corresponding values under constant static stress at the fatigue mean stress level (σ_m) (under constant amplitude). While fatigue loading at R-values approaching 1.0 may involve significant creep effects, fully reversed loading at $R = -1$, i.e., $\sigma_m = 0$, is normally not sensitive to creep. The extent of creep deformations and the sensitivity to creep-induced rupture also depend on whether the fatigue response is fibre- or matrix-dominated. In the former case, creep effects may be small, while in the latter case they may become significant, see Section 3.4.5. Creep effects furthermore increase with increasing temperature and humidity.

The effects of the R-ratio and mean stress on the fatigue life can be represented in constant life diagrams (CLDs), as shown in **Fig. 3.84** (left) for epoxy-bonded composite double-lap joints. A CLD represents the stress (or load) amplitude vs the mean stress (or mean load) relationship at selected fatigue lives (i.e., numbers of cycles to failure) and is built up of straight lines, which represent the S-N curves of the relevant R-ratios (Fig. 3.84, right). On the abscissa, the constant-life curves are normally directed to the quasi-static resistances in tension and compression (i.e., $R = 1.0$), which does not, however, consider potential creep-fatigue interaction, as will be discussed later. The CLD shown in Fig. 3.84 (left) is left-inclined and thus reveals that the joints are more resistant to fatigue under compression than under tension [113]. As introduced above and shown in **Fig. 3.85**, the m-values depend on the R-ratio and, in this case, are almost constant at low ratios, but significantly increase towards higher ratios and are higher in compression ($R = 1.1$) than in tension ($R = 0.9$), as discussed above.

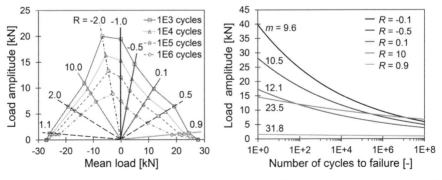

Fig. 3.84 Constant life diagram (load amplitude vs mean load at different R-ratios and fatigue lives) **(left)**; load amplitude vs number of cycles to failure relationships at different R-ratios **(right)**, both for epoxy-bonded composite double-lap joints, based on data in [113]

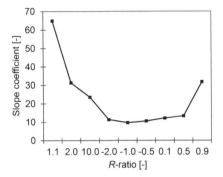

Fig. 3.85 Relationship slope coefficient (m-value) vs R-ratio, for epoxy-bonded composite double-lap joints, based on data in [113] (order of R-values on abscissa as in CLD)

3.8.4 Creep-fatigue interaction

When fatigue loading is applied to viscoelastic composites, hysteresis loops appear in the cyclic stress vs strain responses, as shown in **Fig. 3.86**. Three metrics can be derived from the progression of these loops, i.e., cyclic stiffness (slope of loop, E_i), mean strain (shift of loop, $\varepsilon_{m,i}$), and cyclic dissipated energy (hysteresis area, w_i). Cyclic stiffness is a short-term stiffness and does not include any creep effects. Mean strain is caused by creep deformations and damage under the mean stress and/or progressive damage accumulated due to the cyclic loading. At R-values approaching 1.0, creep deformations and damage prevail, while the mean strain mainly originates from cyclic damage at values around -1.0. Cyclic dissipated energy includes energy required for (i) damage formation and propagation (e.g., formation of microcracks, initiated from creep and/or cyclic loading), (ii) viscoelastic deformation (breakage of secondary molecular bonds, see Section 3.4.2), and (iii) friction among sliding

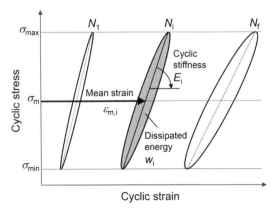

Fig. 3.86 Definitions of cyclic stiffness (E_i), mean strain ($\varepsilon_{m,i}$), and cyclic dissipated energy (w_i, hysteresis area) (N_1, N_i, N_f = 1st, ith, failure cycle), based on [109, 114]

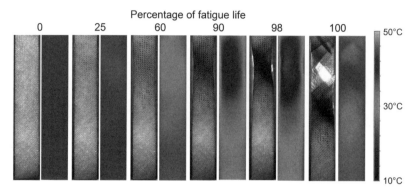

Fig. 3.87 Damage formation and propagation in translucent angle-ply composite laminate under $R = 0.1$ tension-tension fatigue, damage progression visualised by pairwise light transmittance and self-generated temperature, at different percentages of fatigue life, based on [109]

molecules and between crack flanks, which may result in self-generated heat, i.e., increases in temperature [109, 114].

An example is shown in **Fig. 3.87**, where damage formation and propagation can be observed by the increasing opacity and temperature of a translucent angle-ply laminate under tension (see also Section 4.4.2). Towards the end of the fatigue life, damage concentrates, and the temperature reaches a hot spot at the location where failure subsequently occurs (but clearly remains below $T_{g,onset}$, i.e., 78°C). It can also be seen that final damage mainly occurs diagonally, i.e., along the fibres in the fibre-matrix interfaces. Cyclic stiffness significantly decreases, and mean strain and cyclic dissipated energy significantly increase in the first approximately 20% of life, as shown in **Fig. 3.88**; subsequently, the rates of all three metrics (i.e., slopes of the curves) decrease. These changing responses can be explained by the fact that matrix micro-cracking initially develops in the whole specimen volume but does not yet notably change its translucency and temperature. In the later phase, micro-cracking in the matrix becomes saturated, and damage continues to nucleate and propagate in the diagonal fibre-matrix interfaces on a smaller volume, which decreases the rates but locally increases opacity and self-generated temperature; the latter is caused by internal friction between crack fronts and secondary bonds. Failure finally occurs through fibre breakage or fibre pull-out [109].

As has been shown above, fatigue damage normally originates from both creep and cyclic loading, thus from creep-fatigue interaction. To quantify this fatigue interaction, i.e., to establish a partition of the fatigue damage into creep- and cyclic-dominated parts, a model based on the total dissipated energy (W_{tot}), accumulated during the fatigue life, is presented in [114], where it is assumed that:

$$W_{tot} = W_{cyclic} + W_{creep} \qquad (3.11)$$

where W_{cyclic} and W_{creep} are the total dissipated energies attributed to cyclic and creep loading, and are basically defined as follows:

$$W_{cyclic} = \int_{i=1}^{N_f} w_i \quad \text{and} \quad W_{creep} = \sigma_m \cdot \varepsilon_{m,f} \qquad (3.12)$$

3. Composites – Overcoming Limitations   183

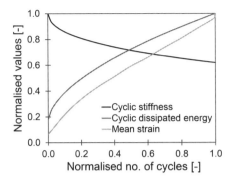

Fig. 3.88 Normalised cyclic stiffness (E/E_f), mean strain ($\varepsilon_m/\varepsilon_{m,f}$), and cyclic dissipated energy (w/w_f) vs normalised number of cycles N/N_f, for angle-ply composite laminate under $R = 0.1$ tension-tension fatigue (index-f = failure), based on data in [109]

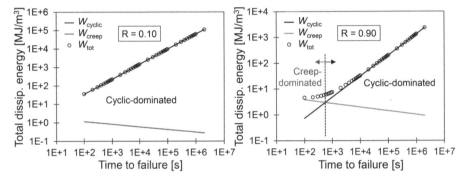

Fig. 3.89 Dissipated energy attributed to creep and cyclic loading during fatigue life (time to failure) of angle-ply composite laminate at two different R-ratios, based on data in [114]

An example of a W-partition is shown in **Fig. 3.89** for the angle-ply composite laminate already introduced above (Figs. 3.87–88). At $R = 0.1$, the creep part is almost negligible, while at $R = 0.9$, the fatigue damage is creep-dominated in the first part of the life and then changes to cyclic-dominated. Based on a linear interpolation, stress vs time (S-t) curves at selected R-ratios can be obtained or CLDs can be constructed with a more realistic representation at R-values close to 1.0, where significant creep-fatigue interaction normally occurs, as shown in **Fig. 3.90**. In contrast to the representation in Fig. 3.84 (left), cyclic- and creep-dominated regions can be distinguished based on this model [114].

3.8.5 Variable amplitude loading

Real loading conditions are normally of variable amplitude (VA), as demonstrated in the weight-in-motion (WIM) traffic data, measured on a typical highway in Switzerland, see **Fig. 3.91**. The 40-tonne weight limit of lorries in Switzerland can be clearly

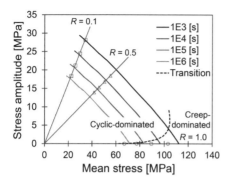

Fig. 3.90 Constant life diagram for angle-ply composite laminate showing cyclic- and creep-dominated regions, based on data in [114] (symbols are experimental data)

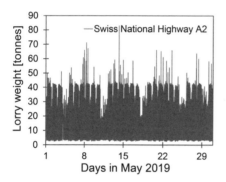

Fig. 3.91 Lorry weight distribution over one month (weigh-in-motion data from Swiss National Highway A2 across Gotthard Mountain, 01–31 May 2019, minimum weight considered is 3.5 tonnes), based on [115].

identified. However, the weights significantly decrease during nights and weekends, and high overloads occur several times during working days (up to double the allowed load in this case).

In contrast to constant amplitude spectra, VA spectra thus comprise various variations such as different load-sequences (high-low vs low-high), overloads (as mentioned above), and load interruptions. Such variations may significantly affect the fatigue life – negatively or positively [51, 116, 117]. They are also responsible for the fact that the Palmgren-Miner linear damage accumulation assumption (e.g., used in steel) is not reliably applicable to composites [40], which significantly complicates the derivation of fatigue design resistances based on experimental VA data. Since such effects on the fatigue life may result from the viscoelasticity of composite materials (e.g., in [51]), the application of the Palmgren-Miner damage assumption may be more appropriate in fibre-dominated than in matrix-dominated cases; this hypothesis needs experimental verification, however.

3. Composites – Overcoming Limitations

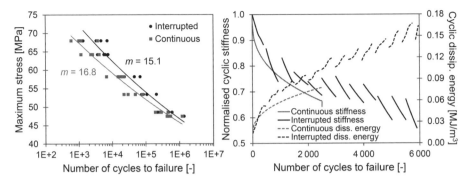

Fig. 3.92 Effect of fatigue load interruptions and associated recovery on maximum stress vs number of cycles to failure **(left)**; normalised cyclic stiffness and cyclic dissipated energy vs number of cycles to failure, at 64 MPa maximum stress **(right)**, angle-ply composite laminate at $R = 0.1$, based on data in [51]

Experimental results obtained for epoxy-bonded composite double-lap joints (as introduced above, Fig. 3.83 right) reveal significant sequence effects for tension ($R = 0.1$) and compression ($R = 10$) fatigue loading, when applying two-block sequences [116]. The transition from low-to-high load cycles is more damaging under tension loading than the high-to-low sequence, since the crack propagation rate is accelerated in the former and decelerated in the latter case. The inverse behaviour is observed for compression loading. The sequence effects diminish, however, with an increasing number of transitions. The same type of composite joints is subsequently subjected to VA loading. A sudden acceleration of the crack propagation rate is observed due to short-term overloads. However, the rate decreases to the preceding rate after a short period [117]. The damage quantification in both works [116, 117] is based on the application of the Palmgren-Miner linear damage accumulation assumption, due to the lack of alternatives. However, further references cited in [116, 117] confirm these results.

Fatigue loading may be interrupted, e.g., during nights or weekends on highways, as has been shown above (Fig. 3.91). Such interruptions allow viscoelastic materials to recover and thus extend their fatigue life, as demonstrated in **Fig. 3.92** (left) for the angle-ply composite laminate already mentioned above, where the fatigue life is now periodically interrupted for two hours, between constant fatigue loading blocks. After each unloading, the cyclic stiffness partially recovers, i.e., increases, and the damage rate and the associated cyclic dissipated energy decrease, as shown in Fig. 3.92 (right). Within the subsequent fatigue loading blocks, however, the cyclic stiffness decreases at an elevated rate (compared to the average rate), which is always similar to that in the first loading block, indicating that further damage formation and propagation occurs at a similar elevated rate too. A second mechanism intervenes during unloading, i.e., crack tip blunting, which reduces the local stress intensity at the crack tip and thus delays the crack growth and enables the initiation and growth of new cracks. This mechanism allows the new damage to be better distributed over the whole specimen volume. On average, cyclic stiffness, therefore, decreases at a lower rate and cyclic dissipated energy increases at a higher rate than in the continuous fatigue case.

Both mechanisms together, recovery and crack tip blunting (i.e., a better distribution of damage), allow the fatigue life to be extended. The life extension is significantly higher at high stress level where recovery is more influential and diminishes with decreasing level ([51], revised data interpretation).

3.8.6 Temperature and moisture dependence

Temperature and moisture may exhibit a significant effect on the fatigue life of composites. Increasing temperature and increasing moisture normally move the S-N curve downwards, in line with the quasi-static resistance. In the first case, the matrix softens, and the viscous part of the response may become dominant, while plasticisation of the matrix occurs in the latter case (see Section 3.10.2), which has a similar effect. In both cases, creep-fatigue interaction effects are increased, and the slopes of the curves, i.e., the damage rates, are also increased, thus decreasing fatigue life.

Two examples are shown in **Figs. 3.93–94**. The first case concerns epoxy-bonded composite double-lap joints (similar to those shown in Fig. 3.83, right), where the downwards shift of the curves with increasing temperature and moisture is clearly visible; the resulting slope behaviour, however, is not totally consistent since the scatter of the data is quite significant. More consistent results are shown in the second case for the epoxy adhesive introduced in Section 2.2.3. The curves shift downwards, and the slopes increase with increasing temperature and decreasing R-ratio. The $T_{\mathrm{g,onset}}$ value is 51°C, the material thus enters the rubbery state at 55°C.

3.8.7 Fatigue modelling vs fatigue design

Due to the complexity of the fatigue phenomena in composites, most of the models developed to describe fatigue resistances are phenomenological models, which are based on fitting of experimental data obtained from specific damage metrics. A hybrid exponential-power law formulation for S-N curves is derived, for instance, to better simulate the low-cycle fatigue region [119], or it is demonstrated how a polynomial constant life diagram formulation outperforms other CLD models [113]. Furthermore, fatigue failure criteria for constant crack growth rates are developed, which can be used to establish progressive damage models under variable mode-mixity fatigue loading [120].

As can be seen from these examples, quite large experimental databases are normally required to calibrate fatigue resistance models. Furthermore, the validity of such models is generally limited to the specific datasets, i.e., the materials and experimental set-ups, to which the models were fitted. New experimental datasets and models should thus be established for each new design problem. Due to cost and time constraints, however, experimental work in structural engineering design is limited in most cases, and experiments are thus specifically shaped to obtain, for a specific problem, the required fatigue design resistances directly, without any further modelling. Accordingly, design codes (e.g., [1]) instead focus on specifying the required experimental work to directly obtain specific fatigue resistances rather than referring to modelling.

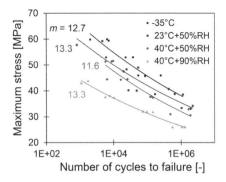

Fig. 3.93 S-N curves for epoxy-bonded pultruded composite double-lap joints under different temperature and relative humidity (RH) conditions, at $R=0.1$, power fits of exp. data, based on data in [118]

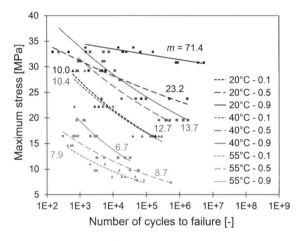

Fig. 3.94 S-N curves for epoxy adhesive under different temperatures (20, 40, 55°C) and R-ratios (0.1, 0.5, 0.9) [58]

3.8.8 Fatigue-tailored detailing

Avoiding or minimising stress concentrations through appropriate detailing of geometrical changes in sections or changes in materials can improve the fatigue life [1, 40]. Stress concentrations can be reduced by extending geometrical or material changes along the member axis, e.g., by tapering thickness changes or implementing scarfed transitions in laminate or core joints. An example is shown in **Fig. 3.95**, where different types of joints between a higher density Douglas fir and lower density balsa wood sandwich core are compared, i.e., butt, step-lap, angled-lap, and scarf joints. The stress concentrations significantly decrease in this sequence, i.e., by extending the joint along the axis. In the case of a scarf joint, the values reach almost those of a continuous balsa core without joint [121].

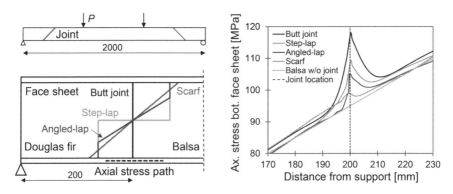

Fig. 3.95 Sandwich panel core joints between higher density Douglas fir and lower density balsa wood (**left**); numerical stress concentrations according to joint type in outer layer of bottom face sheet, based on data in [121]

Bridge decks of vehicular bridges are composite members which are subjected to significant fatigue loading by the vehicle tyre loads. Two types of commercial sandwich panel products basically exist for this application, i.e., panels with web-core or with homogeneous core, see Section 2.3.3. Vulnerable components of these decks are, in the web-core case, the upper face panels in the transverse (to the web) direction and their junctions to the webs, and in the homogeneous core case, the face sheets and their connection to the core.

A significant difference in the support conditions of the face panel/sheet exists between the two deck types, as shown in **Fig. 3.96** (left). In the web-core case, the face panel acts as a continuous beam supported by the webs at the web-face panel junctions (WPJs); the spans correspond to the web spacing (b). A typical tyre contact length is in the range of 350 mm, which is thus larger than typical web spacings, and the tyre contact pressure distribution is slightly trapezoidal [122, 123], but can be assumed to be evenly distributed. Under these conditions, the WPJ locations are subjected to combined bending and shear. While the bending moment (M) increases quadratically with the web spacing, the shear force (V) increases only linearly, see Fig. 3.96 (right). In the homogeneous core case, if assuming a sufficiently stiff core (e.g., consisting of high-density end-grain balsa), the upper face sheet can be assumed to be continuously supported and local bending and shear thus do not arise, as also shown in Fig. 3.96 (left).

Stress concentrations thus occur above the WPJs of web-core panels, which not only depend on the web spacing, however, but also on the face panel thickness, i.e., on the web spacing to face panel thickness ratio. As is shown in Section 2.3.3, this ratio can be manufactured much smaller in vacuum-infused than in pultruded web-core panels, and stress concentrations are thus smaller in the former case.

Concerning the WPJs themselves, there are differences in the transverse fibre architecture between pultruded and infused web-core panels, as discussed in Section 2.3.3. While the architecture can be better controlled and optimised regarding the risk of delamination in infused panels, significant imperfections normally exist in

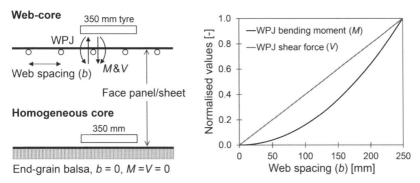

Fig. 3.96 Discontinuous and continuous support of upper face panel/sheet of composite sandwich bridge decks **(left)**; normalised bending moment (M) and shear force (V) in web-panel junctions (WPJ) vs web spacing (b) in web-core sandwich bridge decks **(right)** [13]

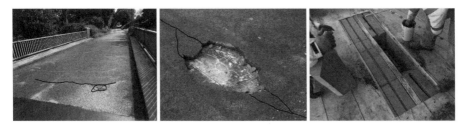

Fig. 3.97 West Mill Bridge, UK, after 13 years of use: surfacing with transverse cracks and local disintegration **(left)**; abrasion damage and cracks in upper face panel of web-core deck at location of surfacing disintegration **(middle)**; rehabilitation by removing damaged face panel section and filling cavity with foam blocks, before bonding filling plate and double-layer bypass plates (not shown) **(right)**

pultruded panels. They may affect the static and fatigue junction resistance and fracture mode, since matrix-dominated responses and stress concentrations are induced, see Sections 3.3.2 (Fig. 3.11) and 3.6.3 (*Example-2*).

A different fatigue failure mode, which may also be related to the discontinuous support conditions of the upper face panel, however, occurred in the pultruded web-core deck of the West Mill Bridge (UK, built in 2002), after 13 years of use. Multiple cracks formed and local disintegration occurred in the thin polymer concrete surfacing, see **Fig. 3.97** (left). Both may be traced back to the varying deflections of the upper face panel between the webs under tyre loading, and associated loss of interface bond between deck and surfacing. Continuous panel abrasion subsequently occurred through surfacing fragments and associated local point loads under tyre load (Fig. 3.97, middle), which finally led to a fatigue delamination failure in the panel. The rehabilitation of the deck occurred by (i) local removal of the damaged upper face panel section, (ii) filling of the cells with high-density foam (Fig. 3.97, right), (iii) closing of the cavity by an adhesively bonded plate, and (iv) adhesive bonding of double-layer bypass plates transverse to the pultrusion direction, over the entire deck [124].

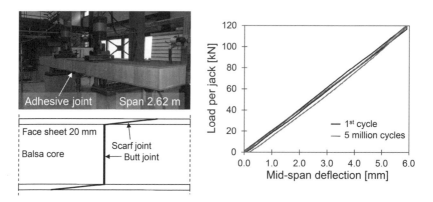

Fig. 3.98 Full-scale composite-balsa sandwich beam with adhesive joint at mid-span under four-point bending fatigue **(left)**; fatigue hysteresis loops at first and after 5 million cycles **(right)**, based on data in [5]

Concerning sandwich decks with homogeneous core, during the design of the Avançon Bridge (see Appendix A.5), two full-scale and two series of small-scale static and fatigue experiments were performed on composite-balsa sandwich beams, produced by vacuum infusion. In one full-scale beam, an adhesive joint was implemented at mid-span with butt and scarf connections in the core and face sheets, respectively, as shown in **Fig. 3.98** (left). The beam was subjected to fatigue loading in a four-point bending set-up for up to five million cycles [5]. No visible signs of damage were observed, i.e., the hysteresis loops of the first and last cycles almost overlapped, and no cyclic stiffness degradation (slope change of loop), cyclic creep (shift of loop), or damage (significant increase of loop area) appeared, see Fig. 3.98 (right).

In the first series of small-scale experiments, four sandwich beams, of 845 mm span and 37 mm thickness, were subjected to ten million fatigue cycles in a four-point bending set-up. Similarly, no signs of damage were observed [5]. The second series of small-scale experiments comprised eleven static and three fatigue experiments on beams of 600 mm length, in a mixed-mode bending (MMB) set-up (see Section 3.7.2.1), as shown in **Fig. 3.99** (left). The face sheet and core thicknesses were 5 mm and 20 mm, respectively. A pre-crack of 68 mm length was inserted between the top face sheet and balsa core at the right beam end on the photo using a thin Teflon sheet. Mixed-mode ratios G_I/G_{II} of 0.13, 3.92, and 6.47 were applied by varying the MMB set-up configuration; the R-ratio was 0.1 and the frequency 5 Hz.

Fibre bridging occurred during the debonding of the face sheet from the balsa core, as depicted in Fig. 3.99 (right), which increased the fracture resistance, see Section 3.7.2.1 (Fig. 3.65). A mixed-mode static failure criterion was derived, furthermore, as shown in **Fig. 3.100** (left). The shape of the criterion is convex, in contrast to the concave fibre-tear failure criterion for adhesively bonded glass fibre composite joints, shown in Section 3.7.2.1 (Fig. 3.62, right).

The fatigue responses during face sheet-core debonding were compared to those during fibre-tear failure of the same adhesively bonded glass fibre composite joints, as mentioned above. The crack growth curves, shown in Fig. 3.100 (right), exhibited

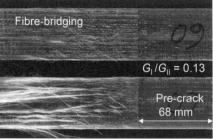

Fig. 3.99 Debonding of glass fibre composite face sheet from balsa core under mixed-mode bending fatigue, beam length 600 mm **(left)**; fibre bridging in two opposite faces of face sheet-balsa core debonding under static loading **(right)** [58]

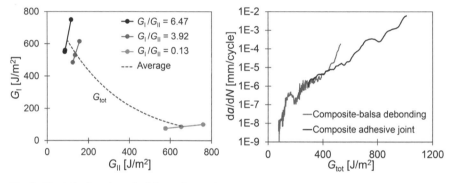

Fig. 3.100 Mixed-mode static failure criterion for debonding of glass fibre composite face sheet from balsa core **(left)**; crack growth rate (da/dN) vs G_{tot} comparison between glass fibre composite face sheet debonding from balsa core (at R-ratio R=0.1, G_I/G_{II}=3.92) and fibre-tear failure in glass fibre composite adhesive connection (at R=0.5, G_I/G_{II}=3.70) **(right)**, based on [58, 120]

a similar rate (slope) during stable crack propagation in both cases, although G_{tot} at failure was significantly smaller in the case of the sandwich beam. It should be noted, however, that the R- and G_I/G_{II}-ratios did not fully match, see caption in Fig. 3.100 (right).

Based on these comparisons and results, which admittedly are not numerous, it may nevertheless be concluded that sandwich panel bridge decks with homogeneous core, and thus continuous and uniform support of the upper face sheet, may resist traffic fatigue loading better than web-core decks with discontinuous support of the upper face panel. A precondition is adequate bond resistance between face sheets and core, however. In this respect, a good bond normally exists in composite-balsa sandwich panels produced by vacuum infusion, since the resin slightly penetrates the end-grain balsa, and fibre bridging occurs during debonding, which increases the fracture resistance. Within web-core panels, vacuum-infused panels with (i) optimised fibre architecture in the web-face panel junctions and (ii) low ratio of web spacing to face panel thickness may be less sensitive to fatigue damage than pultruded panels due to an improved resistance against delamination and reduced stress concentrations.

3.8.9 Fatigue load model for composite bridges

Current European fatigue load models (FLMs) were calibrated for structural steel; FLM3, for instance, is based on a damage rate of $m = 5$ [106, 110]. Since the damage rate of composites is significantly smaller, as discussed above (Section 3.8.2), and to safely design composite bridges against fatigue damage, the development of FLMs calibrated for composites is therefore necessary.

A first step towards a FLM for composite bridges is made in [125]. Considering that composite bridges are still designed with smaller spans, a simple two-axle fatigue load model is assumed, as is shown in **Fig. 3.101** (left). The two axle loads (Q) are calibrated using (i) weigh-in-motion (WIM) databases (i.e., static axle loads and spacings) from most loaded motorways in three European countries, recorded for periods of up to 22 years, and (ii) the two-lane Avançon Bridge, as described in the previous Section 3.8.8 and in detail in Annex A.5. The plan view of the bridge, as also shown in Fig. 3.101 (left), is simplified for this purpose, i.e., the skew angles are neglected, since they do not notably affect the calibration. As can be seen, the bridge is composed of three sandwich panels, which are adhesively bonded together, as also depicted in Fig. 3.98 (left). The bending moment transverse to one adhesive joint at the most stressed joint location, i.e., at mid-span between the steel girders (location A) in Fig. 3.101 (left), is considered the critical internal force related to the fatigue resistance. Accordingly, the influence surface for this moment is derived, as shown in Fig. 3.101 (right), taking the deck orthotropy into account.

Since the damage rate of the adhesive joint is unknown, three rates ($1/m$) are assumed, with $m = 10, 15, 20$, thus covering the range of typical m-values for composites, see Section 3.8.2. Based on the static full-scale beam experiment shown in Fig. 3.98 (left), the design moment resistance vs number of cycles to failure curves of the joint can be derived for the three damage rates, as shown in **Fig. 3.102** (left), assuming that the static resistance represents a fatigue experiment with $N_f = 0.5$. This assumption can be justified by the fact that the R-ratio in the case of this lightweight bridge deck approaches $R = 0.0$, see Section 3.8.2. The location of the experimental beam in the deck, transverse to the joint, and its dimensions and four-point bending set-up, are also shown in Fig. 3.101 (left, top and side views).

Due to the small-scale and multi-component configuration of composite bridge decks, particularly of web-core decks, the relevant number of fatigue cycles should be selected as the number of individual axle loads, and not as the number of tandem- or tridem-axle groups, as is the case in concrete bridges, for instance. The corresponding cumulative frequency of axle loads of the WIM data of the slow lanes of motorways in France (A6 near Auxerre), the Netherlands (A16 near Moerdijk), and Switzerland (A1 near Zurich and A2 Gotthard), used for the calibration, are shown in Fig. 3.102 (right). The Auxerre motorway (which was used for the calibration of the current Eurocode FMLs) exhibits the highest frequency of axle loads around the 120 kN European limit, while the Moerdijk and Switzerland motorways have similar highest axle loads.

3. Composites – Overcoming Limitations   193

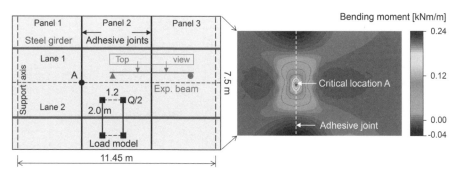

Fig. 3.101 Two-axle composite fatigue load model on two-lane calibration bridge, composed of three composite sandwich panels, and set-up of experimental beam across adhesive joint **(left)**; influence surface for bending moment transverse to adhesive joint at critical location A **(right)**, based on [125]

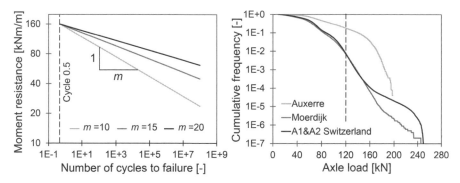

Fig. 3.102 Joint design moment resistance vs number of cycles to failure for three different damage rates ($1/m$) (static value at 0.5 cycle, double-logarithmic scale) **(left)**; cumulative frequency vs axle load on slow lanes of WIM databases used for model calibration, with European 120 kN limit indicated **(right)**, based on data in [125]

These axle loads together with their spacings are subsequently run across the influence surface and the moment range spectrum per year is derived by performing cycle count, using the Rainflow method. The cumulative damage per year can then be calculated for the three damage rates by applying the Palmgren-Miner damage assumption (due to missing alternatives for this step); the results of this calculation are shown in **Fig. 3.103** for the three motorways. The traffic from the Auxerre motorway causes the largest damage per year due to the high frequency of high axle loads, although the highest axle loads are lower than for the other motorways. The effect of the damage rate is clearly visible, i.e., the highest rate ($m = 10$) causes the highest damage.

The equivalent moment range ($\Delta M_{eq,WIM}$) for a specified number of cycles (n_{sp}), generated by the traffic records, can then be calculated based on the rule of equivalent damage, according to Eq. 3.13:

$$\Delta M_{eq,\,WIM} = \left(\frac{\sum n_i \cdot (\Delta M_i)^m}{n_{sp}}\right)^{1/m} \qquad (3.13)$$

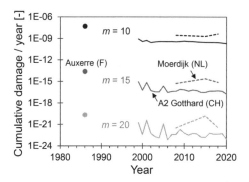

Fig. 3.103 Cumulative damage per year caused by WIM traffic records at critical location A of calibration bridge, for three different damage rates ($1/m$), with $m = 10, 15, 20$, based on data in [125] (dots, dashed and continuous lines are from Auxerre, Moerdijk, and A2 Gotthard, respectively)

where n_i is the cycle count of moment range ΔM_i. The moment range obtained from the fatigue load model (ΔM_{FLM}) is similarly calculated, as a function of the axle load (Q), and the axle load is subsequently obtained from the condition $\Delta M_{E,WIM} = \Delta M_{FLM}$.

Applying this procedure for two million axle loads and the total number of axle loads obtained from the traffic records, extrapolated to a 100-year design service life, results in the relationships of axle loads (Q_k) vs slope coefficient (m) shown in **Fig. 3.104** (left). The axle loads shown are characteristic values, i.e., they are derived as 95% fractile values from the traffic data. As can be seen, the axle loads almost linearly increase with the slope coefficient, while this increase is less pronounced for lower axle numbers. Lower axle numbers and higher slope coefficients lead to higher axle loads since the same damage as caused by actual traffic must always be produced. The corresponding dependence of the axle loads from the number of cycles (axles) is shown in Fig. 3.104 (right) and is linear in a double-logarithmic scale. These curves can be applied for designing fatigue test verifications, i.e., the pairs of axles loads to be applied on the bridge and number of cycles used for testing can be selected, as a function of the slope coefficient (m). The experimental fatigue loads can then be derived according to [1], see Section 5.3.2.

The obtained FML needs, however, further validation with other composite bridges. In contrast to the European FMLs for concrete and steel bridges, the proposed model for composite bridges does not yet contain a dynamic amplification factor, since it is based on the static axle loads obtained from the WIM databases. Dynamic amplification and damage equivalent factors may then also be derived in later stages, as available for concrete and steel bridges [106, 110]. On the other hand, the procedure described above can be individually applied for a specific bridge, as done above, and bridge specific FMLs can thus be derived.

As is the case for FMLs for composite bridges, contact surface dimensions for wheel loads are not yet specified in standards, since they depend on the type of composite bridge deck and surfacing [40]. However, information for the determination of contact surfaces is provided in [1, 40].

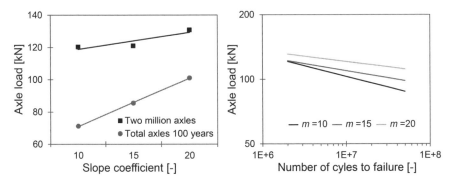

Fig. 3.104 Characteristic values of axle load of calibrated fatigue load model vs slope coefficient for two million and total measured number of axles extrapolated to 100 years (with linear fits) **(left)**; characteristic values of axle loads vs number of cycles for different slope coefficients (double-logarithmic scale) **(right)**, based on data in [125]

3.8.10 Conclusions

Fatigue resistance depends essentially on two metrics, the quasi-static resistance and the slope of the resistance curve, i.e., the damage rate. The higher the quasi-static resistance, the more the fatigue resistance curve is normally upwards shifted, i.e., the higher the fatigue resistance, at least at low-cycle fatigue. The fatigue resistance can be kept high (at high-cycle fatigue) if the slope of the curve remains flat, i.e., at a low damage rate. The damage rate can be reduced by (i) reducing stress concentrations through appropriate structural detailing, (ii) reducing the fatigue amplitude, (iii) subjecting a critical structural component to compression (instead of tension), and (iv) decreasing the viscous response to prevent detrimental creep-fatigue interaction effects (e.g., by keeping an adhesive or the matrix in the glassy state and assuring high fibre-matrix interface strength). Furthermore, the positive effects of load interruptions may be considered if fatigue resistances are conservative, i.e., obtained from constant amplitude loading (without interruptions).

Concerning appropriate structural detailing, it is particularly important that the upper face sheets or face panels of sandwich bridge decks are continuously supported by the core as much as possible, in order to reduce stress concentrations. Homogenous cores, such as end-grain balsa cores, seem to fulfil such conditions particularly well, which does not mean to say, however, that web-core panels with appropriate web-face panel geometry and fibre architecture cannot resist fatigue loading.

Fatigue design basically requires knowledge about the fatigue internal forces or stresses (induced by the fatigue loads) and the corresponding fatigue resistances. There is a complication in the determination of fatigue internal forces or stresses in composite bridges since fatigue traffic load models, calibrated for composite bridges, do not yet exist, and their assumption must thus be based on engineering judgment. A first step towards a simple two-axle fatigue load model shows that the axle loads depend on the damage rate, which is variable in composites (in contrast to steel).

Further cases where fatigue is relevant are discussed in other sections, such as (i) adhesively bonded bridge deck-to-girder connections in Section 4.6.2 (*Example-3*), where it is demonstrated that such adhesive connections can survive ten million fatigue cycles without any signs of damage, and (ii) Mode-I 2D debonding of face sheets in sandwich panels in Section 3.7.3.3, where it is shown that cracks originating from defects may arrest under certain conditions.

3.9 FIRE

3.9.1 Introduction

Fire exposure causes thermal actions to which engineering structures must resist [126]. Composites, in this respect, exhibit advantages and disadvantages. Advantages are the low thermal conductivity (except in the case of carbon fibres), which delays the temperature increase in structural members, and the significant high-temperature resistance of glass, basalt, and carbon fibres. In fibre-dominated response cases, i.e., under tension in the fibre direction (see Section 3.2.3), composites may thus exhibit a significant fire resistance. On the other hand, the combustibility of the polymer matrix, above about 300°C, represents a significant disadvantage. Even more critical, however, is the glass transition of thermoset polymers, where a significant softening of the material already occurs at temperatures around 100°C. In matrix-dominated response cases, e.g., under compression or shear, unprotected composite members may already fail in this, only elevated, temperature range, and thus exhibit a low fire resistance. In this respect, a distinction is made between elevated and high temperatures. The former range includes temperatures above T_g-20°C up to 200°C (where T_g is the glass transition temperature), and the latter represents temperatures above 200°C [1, 127]. The 200°C value is considered as a threshold beyond which material changes become irreversible [128].

The fire resistance of composite engineering structures also depends on the intensity of a fire and the type of surrounding environment. The same fire in a closed room of a building develops very different thermal actions than one on top of a bridge deck. In the first case, the released heat is trapped and rapidly affects the surrounding structure, while in the second case, most of the heat can escape and thus harms the underlying structure much less. In design, such differences may be taken into account by different nominal temperature-time fire curves, as shown in **Fig. 3.105**. The two extreme cases are the hydrocarbon and external fire curves, while the ISO 834 standard fire curve lies in between. In all curves, the temperatures rise, in only a few minutes, to high values of at least 680°C in the case of the external fire curve. The ISO 834 curve is normally applied for building interior fires, while the external curve is used for bridges, in the latter case considering the possible spread of the heat by limiting the maximum temperature. The hydrocarbon curve is applicable in the case of petroleum fires.

In fire design, fire reaction behaviour and fire resistance behaviour must be differentiated. Fire reaction behaviour is mainly relevant for buildings and depends on properties which are either related to the growth of a fire (e.g., ignitability, heat

3. Composites – Overcoming Limitations   197

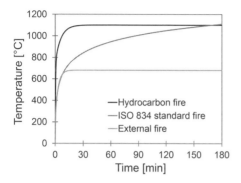

Fig. 3.105 ISO 834 standard, external, and hydrocarbon nominal temperature-time fire curves, based on [126]

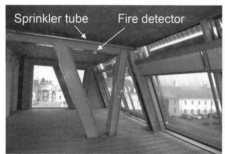

Fig. 3.106 Visible composite building structures: Eyecatcher Building interior on third floor, with fire detector and water sprinkler tube **(left)**; Novartis Building, interior view on roof with structural redundancy **(right)**

release, flame spread), or to lethality or human survivability (e.g., gas toxicity, smoke density). These properties mainly depend on the selected resin system and concern the early stages of a fire, i.e., from ignition to flashover. Fire resistance behaviour, on the other hand, concerns the structural integrity and performance during the whole fire exposure, and mainly depends on the thermophysical and thermomechanical material properties. After a fire incident, the post-fire performance is also of interest and serves as the basis for a potential rehabilitation of a fire-damaged structure [129].

Fire resistance can be improved by the installation of passive or active fire protection measures. In the first case, protective layers are applied to the members, while in the second case water sprinkler or water-cooling systems can be installed. Active fire protection measures allow the structure and composite material to be kept visual and thus be an element of an architectural concept. An example is the five-story Eyecatcher Building (see Appendix A.3), where beams and columns of the frames can be left unprotected because a water sprinkler system is installed, see **Fig. 3.106** (left). Another approach to keep the structure visible is to incorporate redundancy into the structural system, i.e., to allow the failure of individual members without causing the collapse of the whole structure. An example is the 400 m^2 composite-PUR foam web-core sandwich roof of the one-story Novartis Building (see Appendix A.4, and Fig. 3.106, right), where the design considered a potential burn-through on a roof surface of 2×2 m^2, at any location, without causing the collapse of the unprotected roof. The fire resistance cannot be improved by selecting specific resins such as phenolics or adding fire retardants to the resin. Such measures only improve the fire reaction behaviour but do not increase the critical glass transition temperature.

The fire resistance behaviour of composite structures has been investigated at the CCLab since 2001 in primarily four consecutive PhD theses [130–133] where, inter alia, full-scale combined fire-mechanical loading experiments on composite web-core slabs and walls were performed. The following discussion is mainly based on this research outcome. Thermophysical and thermomechanical material properties are addressed first, and the fire resistance behaviour of pultruded members and sandwich panels, as used in bridges and buildings, are subsequently discussed, including the post-fire performance. Unprotected and actively protected members are considered, taking water-cooling into account in the latter case. Furthermore, results of recent three-dimensional dynamic simulations of a composite building fire are presented, which consider the effects of the floor plan layout, the location of the fire, and the fresh air supply on the thermal responses. Not, or only marginally addressed, are the fire reaction behaviour and passive fire protection measures; information about these topics and further references can be found in [129].

3.9.2 Thermophysical and thermomechanical properties

3.9.2.1 Introduction

The structural responses of composite members under fire primarily depend on mechanical properties, such as stiffness (e.g., elastic modulus), strength, and coefficient of thermal expansion. Since the polymer component of a composite material undergoes different material states and transitions if subjected to elevated and high temperatures, such as glass transition (see Section 3.4.3) and decomposition (see next Section 3.9.2.2), the mechanical properties are dependent on temperature. To determine these thermomechanical properties, the temperature distributions through the thickness of a composite member, during the fire exposure, must be known.

The required through-thickness temperature distributions are derived from another set of properties, thermophysical properties, i.e., density, thermal conductivity, and specific heat capacity, which are also dependent on temperature for the same reasons as thermomechanical properties. Furthermore, it has proved to be efficient to use "effective" thermophysical properties, which implicitly include, in addition to the "true" temperature-dependent material properties, effects of physical and chemical processes on the temperature responses, such as polymer dehydration, endothermic decomposition, and delamination (radiative shielding) [134].

This already complex system of interdependent thermomechanical and thermophysical properties is further complicated by the fact that these properties are also time-dependent, i.e., they depend on the heating rate [136], see also Section 3.4.4. Depending on the type of fire, e.g., standard or external fire (see Fig. 3.105), the properties may thus be different (as will be shown hereafter).

Different methods exist for obtaining effective temperature- and time-dependent material properties. Effective thermophysical and thermomechanical properties of composite laminates were derived from a decomposition model based on chemical kinetics in [41, 127, 135–137]. Kinetic parameters were determined using thermogravimetric results at different heating rates. Effective thermophysical properties of balsa wood were derived in [138–140] using an inverse analysis based on (i) experimental data obtained from the standard and external fire curves, (ii) a one-dimensional transient heat transfer finite element model, and (iii) assuming parametrised temperature- and density-dependent piecewise linear functions across the decomposition stages (see hereafter).

In the following discussion, the effective thermophysical and thermomechanical properties of pultruded glass fibre-polymer composites and balsa wood will be compared, the latter used as core material of sandwich panels. This comparison of a fibre-polymer composite and a polymer material allows the ongoing physical and chemical processes during fire exposure, and their effects on the properties, to be clearly understood. Balsa wood was selected since it can retain relevant mechanical properties up to approximately 200–250°C, while other polymeric materials, such as adhesives or foams, normally lose stiffness and strength at much lower temperatures, i.e., already below 100°C in most cases [1, 69] (see also Fig. 2.5). Balsa wood, furthermore, in contrast to polymeric foams, can develop significant char layers, whose thermal insulation capacity can provide positive effects on the fire performance of composite-balsa sandwich panels [138–140].

Balsa wood panels, as used in sandwich panel cores, are normally composed of smaller end-grain blocks of varying densities, which are bonded together with a polyurethane adhesive, as described in Section 2.2.2.1. Regarding fire performance, it may become relevant that the bond lines are continuous along the panel in one direction, while they are staggered in the other direction, see Fig. 2.4. The shear resistance of the panels, after glass transition of the adhesive, may thus be lower transverse to the continuous bond lines. The air gap imperfections mentioned in Section 2.2.2.1 are always located transverse to the continuous bond lines, see Fig. 2.4. Subjected to fire, they may represent paths of rapid heat ingress into the core, as will be discussed hereafter (Section 3.9.3.3). Due to the complexity of this assembly, effective thermophysical and thermomechanical properties are normally determined as average properties across the panel plane, including the effects of varying density and adhesive bond lines, and the material is thus considered as isotropic in this plane [140].

3.9.2.2 Decomposition and density

Both composite materials and balsa wood normally decompose in three stages during fire exposure, which can be recognised in a thermogravimetric analysis (TGA), see **Fig. 3.107**, where remaining mass fraction and mass loss rate vs temperature relationships are measured. During the first stage, moisture evaporates from both polymers (composite matrix and balsa wood), followed by the decomposition of the polymers (pyrolysis) in the second stage, where the chemical bonds are progressively broken and decomposition products are formed during an endothermic reaction, such as residual char, various liquids, smoke, and incombustible and combustible gases.

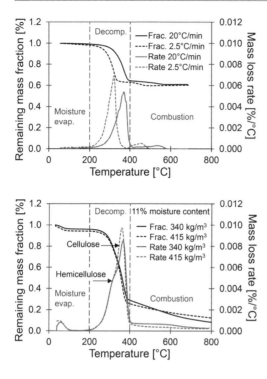

Fig. 3.107 Remaining mass fraction and mass loss rate vs temperature relationships of pultruded glass fibre composites under two different heating rates (**top**), and balsa wood at two different densities (**bottom**), obtained from TGA experiments under nitrogen atmosphere, based on data in [41, 133, 138]

Depending on the concentration of the combustible gases, temperature and availability of oxygen, ignition and flaming or smouldering combustion occur in the third stage during an exothermic reaction [138, 141]. The released heat can further accelerate the decomposition of the remaining material.

The fact that structural composites contain a significant number of fibres – which normally neither contain moisture nor decompose (except polymer-based fibres, e.g., aramid) – has two main effects on the decomposition behaviour. First, mass loss due to moisture evaporation is almost insignificant in composites since the moisture content is low (approx. 1–4 wt% at full saturation [142]), while the evaporation mass loss of balsa wood can be clearly recognised, since the moisture content is higher (approx. 9–12 wt%), and the density of balsa is lower compared to that of composites, see Fig. 3.107. Second, the mass loss rate of composites is significantly lower during decomposition and, after decomposition, the remaining mass and thus char fraction is much higher than in the case of balsa wood, because the char of composites mainly consists of fibres. The temperature ranges of decomposition and the decomposition temperatures (i.e., peaks of the mass loss rate curves) are, however, similar for both materials, as also shown in Fig. 3.107.

The time dependence of the decomposition process is demonstrated in Fig. 3.107 (top), where decomposition of the composite material starts later, and mass loss

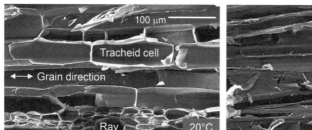

Fig. 3.108 Balsa wood structure at 20°C **(left)** and after cooling down from 250°C **(right)**, longitudinal sections along tracheid cells, based on [144]

occurs slower at a higher heating rate [41]. In the case of balsa wood, hemicellulose decomposes first (shoulder in the curves), while the highest mass loss rate is exhibited by the decomposition of cellulose, see Fig. 3.107 (bottom) [138, 143]. The whole decomposition process is, however, almost independent of the balsa wood density. The scanning electron microscope photos in **Fig. 3.108** demonstrate how the wall structure of the balsa tracheid cells is heavily damaged during decomposition [144].

The temperature-dependent density of the materials, which is required for thermal response modelling, can be derived from such TGA mass loss curves.

3.9.2.3 Effective thermal conductivity

Effective thermal conductivities of glass fibre composite laminates and balsa wood were derived in different ways in the CCLab's work on fire since 2006, as shown in **Fig. 3.109**. The first model derived for pultruded composite laminates was developed in 2006 from different references in the literature and own experimental observations [134]. The piecewise-linear curve takes delamination in the laminate during decomposition into account, which produces a radiative shielding effect and thus significantly decreases the thermal conductivity above approximately 300°C to the level of char material (Fig. 3.109, left). The loss of fibre layers after relatively long exposure to fire is considered by a steep increase in thermal conductivity above 700°C. Models for the same material exhibiting similar trends, i.e., considering the same physical phenomena but with continuous curves, were subsequently derived from chemical kinetics and a series model in [41, 127] (2007 and 2008); they also include the properties of char material. Finally, a piecewise-linear model was obtained from an inverse analysis (in 2023) for laminates produced by vacuum infusion. The model also shows the shielding effect and thus validates the previous models, although it does not exhibit the steep increase at very high temperatures [139].

The effective thermal conductivities of balsa wood in grain and transverse-to-grain directions, and for two different fire curves, were derived from an inverse analysis in [138–140] and are shown in Fig. 3.109 (right). In contrast to the composite laminate models, the balsa wood models exhibit significant peaks during moisture evaporation due to the higher moisture content (see above). The values are significantly higher in the grain than in the transverse-to-grain direction because vapor

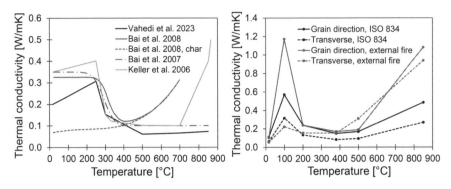

Fig. 3.109 Effective thermal conductivity vs temperature relationships of pultruded (2006/7/8) and vacuum-infused (2023) glass fibre composites and char material according to different models, for ISO 834 fire curve (left) and of balsa wood in different directions, for ISO 834 and external fire curves (average values from densities of 200–525 kg/m^3) (right), based on data in [41, 127, 134, 140]

transmission mainly occurs along the tracheid and vessel cells. The peaks are followed by similar valleys during decomposition (pyrolysis), as observed in composite laminates. This decrease in thermal conductivity is rather attributed to the insulation capacity of the forming char layer, however, than to a shielding effect due to delamination. Subsequently, the values for balsa wood increase again, which can be traced back to the cracking of the char layer (and not to a loss of fibre layers), which makes the heat ingress into the material easier. Differences in heating rates exhibit an effect during moisture evaporation and char cracking, where the values obtained from the ISO 834 fire curve are significantly lower than those from the external fire curve. In the grain direction particularly, the effective thermal conductivity of balsa wood is significantly higher than that of glass fibre composite laminates (note the different y-axis scales).

As can be concluded from the above, the derivation of such models for polymer-based materials is much more challenging than, for instance, in the cases of concrete or steel. The complexity of the processes involved and the changing material compositions during the fire exposure introduce uncertainties into the models which should be considered in the thermomechanical analysis.

3.9.2.4 Effective specific heat capacity

Effective specific heat capacities of glass fibre composite laminates and balsa wood were derived in a similar way as the effective thermal conductivities. Models for glass fibre composites based on literature (2006), derived from chemical kinetics (2007) and inverse analysis (2023), are shown in **Fig. 3.110** (left). The characteristics of the curves are similar and thus consider the same chemical and physical processes. In contrast to the effective thermal conductivity, the required heat to evaporate moisture is visible in overlapping smaller peaks (except in the continuous curve where moisture evaporation is not taken into account in the chemical kinetics approach). The specific heat capacity subsequently reaches highest values during the endothermic decomposition of the polymer matrix and then decreases to the level of the char material.

3. Composites – Overcoming Limitations   203

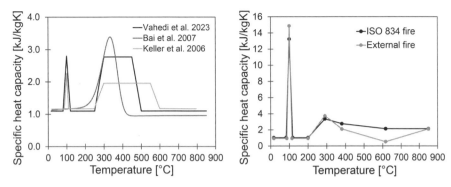

Fig. 3.110 Effective specific heat capacity vs temperature relationships of pultruded (2006/7) and vacuum-infused (2023) glass fibre composites according to different models, for ISO 834 fire curve **(left)** and balsa wood, for ISO 834 and external fire curves (average values from densities of 200–525 kg/m^3) **(right)**, based on data in [41, 127, 134, 140]

The trends of the models for balsa wood, obtained from inverse analysis, are similar to those of composite laminates, see Fig. 3.110 (right). The narrow peaks during moisture evaporation are much more pronounced, however, while the peaks during decomposition are only slightly higher (note the different y-axis scales). The values are independent of the grain direction and almost independent of the heating rate, i.e., the type of fire curve [138–140].

3.9.2.5 Elastic modulus

The temperature sensitivity of the elastic modulus of composites mainly depends on the type of the basic structural response, i.e., matrix- or fibre-dominated, since softening of the polymeric matrix already occurs at elevated temperatures, while fibres soften only at high temperatures. In the case of glass fibres, for instance, significant softening is observed above approximately 500°C, as shown in **Fig. 3.111** (left) [145], while even higher softening temperatures of 830°C are reported in [134]. If the composite response is clearly fibre-dominated, the softening may be limited to 20–30% up to a temperature in the range of 500–600°C [146]. If the response is, however, matrix-dominated, the softening behaviour approaches the DMA storage modulus curve of the pure polymer, thus exhibiting a steep drop in the elastic modulus during glass transition and a further drop during decomposition, as shown in Fig. 3.111 (left) [136]. The drop in the elastic modulus during glass transition, caused by the loss of the secondary intermolecular bonds, can recover after cooling down, however (i.e., the secondary bonds can re-establish themselves), while the drop after decomposition is irreversible, see Section 3.4.2. To compare, the elastic modulus decrease in carbon steel is also shown. The decrease is more pronounced than in fibre-dominated cases, but much lower than in matrix-dominated ones [147].

The softening behaviour of balsa wood, subjected to bending along the grain direction, at different moisture levels, was studied in [144] using DMA, as shown in **Fig. 3.112**. Two phase changes can be recognised within the measured temperature range up to 260°C (before pyrolysis initiates), which can be attributed to the glass

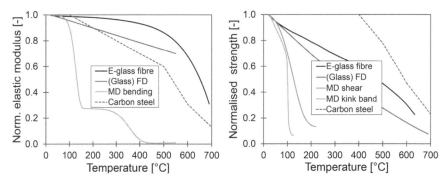

Fig. 3.111 Normalised elastic modulus (**left**) and strength (**right**) vs temperature of (i) "E-glass fibre" modulus [145] and strength [148] under axial tension, (ii) "(Glass) FD" modulus and strength (for UD glass fibre composite rebar under axial tension [146]), (iii) "MD bending" modulus (based on kinetic modelling of glass fibre composite pultruded profile [136]), (iv) "MD shear" strength (for UD glass fibre composite laminate under 10°-off-axis tension [137]), (v) "MD kink band" failure (for UD glass fibre composite laminate under axial compression [149]), and (vi) "Carbon steel" modulus and "yield" strength [147], FD = fibre-dominated, MD = matrix-dominated

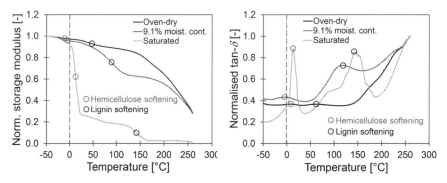

Fig. 3.112 Normalised storage modulus (**left**) and tan-δ (**right**) vs temperature relationships of balsa wood for bending along grain direction and at different moisture levels, glass transition temperatures of hemicellulose and lignin assumed at inflection points of storage modulus and maximum of tan-δ, respectively, based on data in [144] (corrected)

transition of hemicellulose and lignin. The corresponding and indicated glass transition temperatures are assumed to be the inflection points of the storage modulus (Fig. 3.112, left) and the maxima of the loss factor, tan-δ (Fig. 3.112, right). Hemicellulose softening may thus already occur at ambient temperature and the degree of softening rises strongly with increasing moisture level. Lignin, however, softens at higher temperatures, between 50–150°C. Compared to hemicellulose, the moisture level of lignin shifts the softening to higher temperatures. The increased softening due to moisture can be attributed to the plasticisation of the polymers [144].

Almost linear decreases in stiffness up to 250°C, for which less than 10% of the initial values are retained, are furthermore reported in [144] under tension, compression, and shear in all balsa wood material directions, if excluding effects of specimen dimensions on the responses.

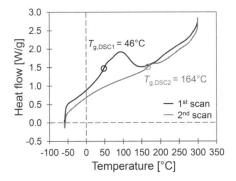

Fig. 3.113 DSC glass transition temperatures of partially cured (1st scan) and fully cured (2nd scan) polyurethane adhesive used in balsa wood panels between blocks, based on data in [144]

As described above, a polyurethane adhesive is normally used to bond the balsa blocks together, and the panels are cold cured without post curing [144]. DSC experiments performed after seven days of curing at 20°C reveal that the glass transition temperature of the adhesive is still quite low (46°C), while fully cured values increase up to 164°C, see **Fig. 3.113** (DSC first and second scan values). Accordingly, adhesive softening exhibits a clear effect on the tensile stress vs strain responses of veneered balsa wood specimens in [144], i.e., the curves become nonlinear at elevated temperatures.

3.9.2.6 Viscosity

When the temperature and load duration increase, the viscous component of the viscoelastic response of the polymer component of a composite material may become predominant over the elastic component and thus influence the mechanical responses. Such changes in the viscous component can be described by the changes in the viscosity of the material, which is a viscoelastic parameter. Accordingly, a model for temperature-dependent viscosity was developed in [136] based on DMA results and kinetic theory. As shown in **Fig. 3.114**, the viscosity exhibits a peak at the glass transition temperature and decreases to low values during decomposition. As long as the material state does not change and the material remains undamaged, the material behaviour is linearly viscoelastic, and the viscosity is independent of the stress level. Beyond these limits, the material response is characterised by nonlinear viscoelasticity, and the viscosity depends, among other things, on the stress level (see Section 3.4.1). In the case shown in Fig. 3.114, the viscoelasticity changes from linear to nonlinear in the ramp up to the glass transition temperature (T_g).

3.9.2.7 Strength

As with the elastic modulus, the strength of composites also depends on temperature and is influenced by the type of mechanical response. In fibre-dominated cases, the tensile strength reduction with increasing temperature is lowest and close to that of the fibres themselves, see Fig. 3.111 (right) for glass fibres. In matrix-dominated

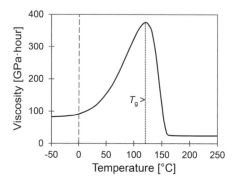

Fig. 3.114 Temperature-dependent viscosity of a pultruded glass fibre composite laminate, based on data in [150]

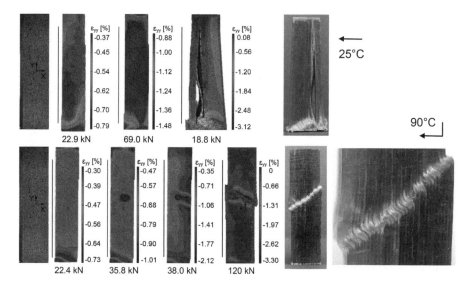

Fig. 3.115 Failure modes of compact UD glass fibre composite specimens (12.7×12.7×50 mm nominal size) under axial load at 25°C, bottom crushing and splitting **(top)**, and at 90°C onset of glass transition, kink band and fibre micro-buckling **(bottom)**; axial strains obtained from DIC measurements on specimen front side under increasing load, and experimental failure mode on specimen back side, based on [151]

cases, e.g., under 10°-off-axis tension of UD composites (to characterise the in-plane shear strength), the normalised reduction curve approaches that of the normalised matrix-dominated modulus curve, exhibiting a steep drop in strength during glass transition, compare Figs. 3.111 (right) and (left). A similar matrix-dominated response is exhibited under axial compression when kink band failure occurs at the onset of the glass transition temperature, as shown in **Fig. 3.115**, where micro-buckling of the UD fibres is clearly visible [149, 151]. Also shown is the corresponding failure mode in the glassy state, i.e., combined crushing and splitting. As a comparison, the normalised yield strength decrease in carbon steel is included in Fig. 3.111 (right).

3. Composites – Overcoming Limitations

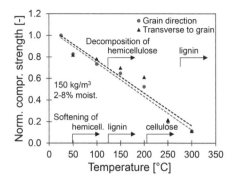

Fig. 3.116 Normalised compressive strength vs temperature relationship of balsa wood in different directions, and softening and decomposition ranges of balsa components, based on data in [152]

The decrease initiates only above 400°C but then drops rapidly to almost the level of fibre-dominated composites [147].

Models to predict strength degradation of composites under elevated and high temperatures are developed in [137] based on the mixture of materials in the glassy and leathery state. Upper and lower bounds of strength were defined based on the rule and inverse rule of mixture. The analysis revealed that shear strength degradation can be described by the rule of mixture, while the degradation of compressive strength follows the inverse rule of mixture.

The degradation of the compressive strength of balsa wood was investigated in [152]. Strength degradation occurs almost linearly with increasing temperature up to 200–300°C, for which only 10–20% of the initial strength is retained, see **Fig. 3.116**. The normalised representation shows that the rates of degradation in grain and transverse-to-grain direction are similar. The diagram also indicates the ranges of softening and decomposition of the balsa wood components, from which it is concluded that the main reason for strength degradation is the softening of these components and not their decomposition. The softening temperature of hemicellulose under compression is higher than that under bending when compared with Fig. 3.112 (left), while the softening temperature of lignin is in a similar range. A reference for the higher hemicellulose value is missing in [152], however.

Similar results are obtained in [144] under tension, compression, and shear in all material directions, i.e., almost linear decreases of strength up to 250°C, where less than 10% of the initial values are retained, if the effects of specimen dimensions are excluded. The damaged cellular structure at 250°C, shown in Fig. 3.108 (right), suggests that decomposition also contributes to strength reduction at high temperatures, in addition to the (above-mentioned) softening.

The adhesive bond lines do not normally negatively affect the mechanical properties of balsa wood at ambient temperature, since material failure always occurs in the balsa material according to [6]. Delamination and subsequent strength reduction caused by the polyurethane adhesive degradation are reported in [144], but only at the highest temperature of 250°C. At lower temperatures, failure occurs in the balsa layer with the lowest density.

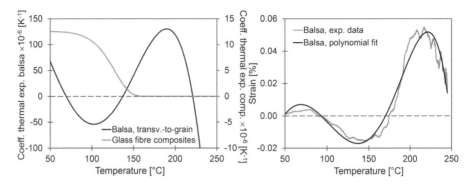

Fig. 3.117 Coefficient of thermal expansion vs temperature relationships for pultruded glass fibre composites in pultrusion direction and balsa transverse-to-grain **(left)**; typical thermal strains in balsa transverse-to-grain **(right)**, based on data in [138, 150] and complemented

3.9.2.8 Effective coefficient of thermal expansion

An effective coefficient of thermal expansion is derived in [150] for glass fibre composites based on chemical kinetics and a rule of mixture. It is considered, furthermore, that the elastic modulus decreases towards zero after glass transition and that the thermal expansion of the material at temperatures above glass transition thus no longer influences stresses, see **Fig. 3.117** (left); the effective value was thus assumed to be zero after glass transition.

For balsa wood, temperature-dependent curves of the effective coefficient of thermal expansion, transverse to the grain, are obtained as derivatives of measured strains in [138]; a typical example is shown in Fig. 3.117 (right). The transverse-to-grain direction was selected since this is the relevant direction for sandwich panel deflections with end-grain balsa core. During moisture evaporation, strains decrease, and the coefficient becomes negative (Fig. 3.117, left). Once that material is dry, both values increase until pyrolysis starts. Subsequently, the values decrease again; these last measurements may be affected by the disintegration of the material, however, as is demonstrated by the large variations.

3.9.3 Fire resistance behaviour

3.9.3.1 Introduction

The fire resistance of composite members is basically defined as the period of time during which the members can sustain a serviceability load without collapse, if subjected to a simultaneous fire. Full-scale furnace experiments are performed in most cases to confirm a required fire resistance, which, in buildings, normally depends on their category and number of stories, and varies from 30 to 120 min. Requirements regarding the fire resistance of bridges normally depend on the significance of the bridge and are defined in accordance with the bridge owner.

As will be shown hereinafter, the fire resistance depends on some basic parameters, such as (i) the structural response of members, i.e., fibre- or matrix-dominated,

(ii) the type of cross section, i.e., open or closed (cellular), (iii) the material combination, i.e., the involvement of low fire resistance materials, such as sandwich core materials or adhesives, (iv) the presence of paths of easy heat ingress, e.g., through air gaps or metallic inserts in sandwich cores, and (v) the application of fire protection systems. Examples of such parameter combinations and how these parameters can affect fire resistance will be discussed hereinafter.

The examples are complemented by modelling results. A two-step analysis is normally carried out, where temperatures are first obtained by performing a heat transfer analysis based on the thermophysical material properties, including heat conduction, convection, and radiation. Stresses and deformations then result from a subsequent structural analysis based on the thermomechanical material properties, where effects of thermal elongation and viscosity may need to be considered. The time-to-failure, i.e., the fire resistance, can finally be predicted based on temperature-dependent strength degradation, or as the time until a given maximum deflection or rate of deflection increase is reached [153].

After a fire incident, post-fire models estimating the remaining stiffness and strength are required to decide on the remaining structural safety and serviceability of composite members; these will also be discussed. Based on such evaluations, required measures to rehabilitate a fire-damaged structure can be defined and designed.

3.9.3.2 Pultruded members: furnace experiments and modelling

An overview of furnace experiments performed on pultruded glass fibre composite members is given in **Table 3.7**. To make the results comparable, only experiments performed under an ISO 834 fire curve are considered. All experiments are performed under simultaneous mechanical serviceability limit state (SLS) and fire loading. Single

Table 3.7 Beam and column members subjected to simultaneous SLS load and ISO 834 fire (CS = calcium silicate, GP = gypsum plaster, GM = glass magnesium)

Member	Span or height L [mm]	Set-up	Initial SLS deformation	Fire exposure	Fire protection	Fire resistance [min]	Structural response dominated by
Open-section beams [154]	1 500	3–point bending	L/150 (480) deflection	4–sided	None	1.5 (2.3)	Matrix
					Protected	12 (16)	Matrix
Cellular beams [155]	1 300	4–point bending	L/400 deflection	1–sided	None	36	Matrix
					CS board	83	Matrix
					Water-cooled	>120	Fibres
				3–sided	None	8	Matrix
					CS board	46	Matrix
					Water-cooled	9 [1]	Matrix
			L/250 deflection	1–sided	None	31	Matrix
					CS board	66	Matrix
Web-core panels [156]	2 750	4–point bending	L/300 deflection	1–sided	None	57	Matrix
					Water-cooled	>90, >120	Fibres

Member	Span or height L [mm]	Set-up	Initial SLS deformation	Fire exposure	Fire protection	Fire resistance [min]	Structural response dominated by
Web-core panels [157]	3 300	4–point bending	L/870	1–sided	None	54	Matrix
					GP board (2x)	83 (113)	Matrix
					GM board (2x)	103 (158)	Matrix
Polyurethane foam core sandwich beams [158]	1 400	4–point bending	L/250	1–sided	None	~25	Matrix
					CS board	~80	Matrix
					CS board + air cavity	~100	Matrix
Balsa-core sandwich beams [140]	1 100	3–point 4–point bending	L/500 deflection	1–sided	None	35 53	Matrix Matrix
Cellular columns [159]	1 500	Axial compr.	L/1 500 shortening	1–sided	None	16	Matrix
					CS board	51	Matrix
					Water-cooled	>120	Fibres
				3–sided	None	5	Matrix
					CS board	39	Matrix
					Water-cooled	20	Matrix
			L/750 shortening	1–sided	None	9	Matrix
					CS board	37	Matrix
Web-core panels [160]	2 805	Axial compr.	L/4 200 shortening	1–sided	None	49	Matrix
					Water-cooled	>60, >120	Fibres

[1] Only bottom flange was water-cooled, lateral webs were not, which explains low fire resistance.

open and closed (cellular) section profiles and multicellular web-core panels are examined, the latter assembled from single cellular profiles by adhesive bonding, as shown in **Figs. 3.118**. The members are subjected to either bending in three- or four-point set-ups or axial compression. SLS loads are implemented accordingly to produce initial SLS deformations, defined as fractions of the span or height (L). Fire exposure occurs from one, three, or all four sides. Unprotected members and members protected by various board types (passive protection) or water-cooling (active protection, see Section 4.5.3) are examined. Active protection by water sprinkler systems has not been investigated.

Looking at the fire resistances in Table 3.7 reveals that they depend significantly on the type of structural response, i.e., whether the response is fibre- or matrix-dominated. The resistances are significantly higher in fibre-dominated cases. This dependence is consistent with that of the stiffness and strength degradation shown in Fig. 3.111, where the material properties remain significantly higher in fibre-dominated cases.

Taking into account the unprotected cellular profiles and multicellular web-core panels and looking at the span to mid-span deflection ratio vs time relationship for the bending case in **Fig. 3.119** shows similar (almost overlapping), descending trends of the curves. The ratios rapidly decrease, and the members fail when the

3. Composites – Overcoming Limitations

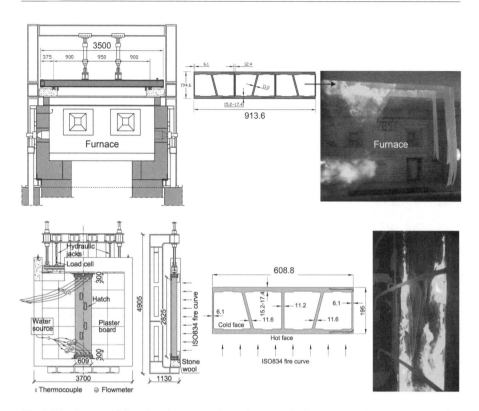

Fig. 3.118 Set-up of full-scale web-core panel experiments under four-point bending and one-sided fire, panel cross section (194.6×913.6 mm), and view through furnace on burning hot face of water-cooled panel after 45 min, showing debonded fabrics, based on [156] **(top)**; set-up under axial load and one-sided fire, cross section (194.6×608.8 mm), and view through furnace on burning hot face of non-cooled panel after 45 min, based on [160] **(bottom)** (dimensions in [mm])

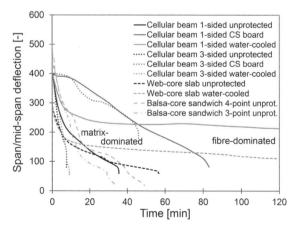

Fig. 3.119 Span/mid-span deflection ratio vs time relationships of composite cellular and balsa sandwich beams, and web-core panels, under simultaneous SLS load and one- or three-sided ISO 834 fire, with and without fire protection, based on data in [140, 155, 156]

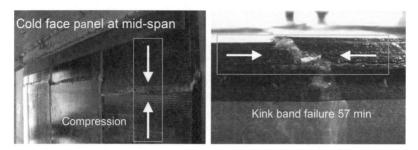

Fig. 3.120 Post-fire inspection, top view on cold face flange of unprotected web-core panel after kink band failure **(left)**; detail of kink band failure in section cut across cold face panel, based on [156]

glass transition temperature is exceeded on the cold face, where compression failure occurs in the top flanges during a matrix-dominated response, as shown in **Fig. 3.120**. In the case of the web-core panels in [156], the adhesive connections between the profiles do not negatively affect the responses, mainly because they run parallel to the member axis, and members are subjected to one-way bending. Three-sided exposure further reduces the resistances and failure already occurs in the top parts of the webs [155, 156]. Open-section profiles cannot provide any relevant fire resistance. Failure occurs as soon as the glass transition temperature is exceeded in one part of the section [154]. The resistance would therefore not be significantly higher if the fire exposure were only one-sided. Compared to the open-section profiles, the cellular section members exhibit an increased fire resistance due to the insulation capacity of the air enclosed in the cells, which protects the cold side web parts and top flanges from rapid increases of temperature. Multicellular (web-core) panels, furthermore, seem to exhibit higher fire resistances than single-cell profiles. A not fully effective lateral insulation of the webs close to the hot face may, however, have affected the single-cell results.

In the cases where water-cooling is applied, the temperatures do not exceed the glass transition temperature in webs and cold face flanges during the whole 120 min fire exposure. The temperatures on the cell-side of the hot face flanges also remain below the glass transition temperature, see **Fig. 3.121** (left), while the furnace temperature increases up to 1 050°C after 120 min. Only the matrix on the fire side of the flanges is decomposing, but the fibres remain load bearing, as shown in **Fig. 3.122**. The hot face flanges therefore remain partially functional, and the deflections thus stabilise (Fig. 3.119); structural failure is thus prevented due to this fibre-dominated response. A precondition is, however, that the fibres remain anchored in member parts which remain in the glassy state (as was the case in the experimental support regions). The required water flow velocity in the cells is low and the increase of the water temperature at the outlets small, as shown in Fig. 3.121 (right). The protection provided by the CS boards can significantly increase the fire resistances, but not prevent the glass transition temperature from finally being exceeded in the top parts of the webs; a combined compression/shear failure thus occurs during a matrix-dominated response [155, 156].

3. Composites – Overcoming Limitations   213

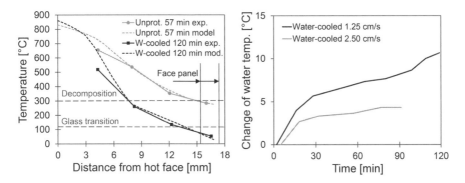

Fig. 3.121 Temperatures through hot face flange of web-core panel, unprotected at failure (57 min) and after 120 min of water-cooling, experimental and modelling results **(left)**; change of outlet water temperature as a function of water flow velocity up to 120 min **(right)**, based on data in [156, 161]

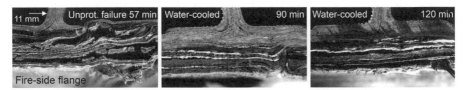

Fig. 3.122 Post-fire inspection, fire-side flange of unprotected web-core panel after failure with resin completely decomposed **(left)**; after water-cooling during 90 min **(middle)**; and 120 min **(right)** with partially functional flanges (on cell-side), based on [156]

The fire resistance behaviour under axial compression is basically comparable to that under bending, as can be derived from Table 3.7, although the temperature profiles in the compressed parts can be very different. In unprotected cases, the members collapse under local instability induced compression failure modes in flanges and webs (kink band failure, buckling after delamination) when the glass transition temperature is exceeded, as was the case under bending [159, 160]. Applying water-cooling maintains the temperatures of the webs, cold face flanges, and on the cell-side of the hot face flanges below the glass transition temperature, as in the bending case. The hot face flanges thus also remain partially functional, and the members do not fail up to a 120 min fire exposure. Although the entire cross sections are under compression in this case, the responses remain fibre-dominated, according to [1], since the compression modulus of the flanges in axial direction is based on a high fibre content, and the material remains in the glassy state. As observed under bending, three-sided fire exposure reduces and CS board protection increases the fire resistances. The fire resistances are generally slightly lower under compression than in the bending case; the values cannot, however, be directly compared since the initial stress states are different.

Modelling results of the multicellular web-core panels under four-point bending furnace experiments are shown in Fig. 3.121 (left) (hot face temperature gradients), **Fig. 3.123** (thermomechanical property distributions), and **Fig. 3.124** (mid-span deflection ratios). Temperature distributions are obtained from the thermophysical

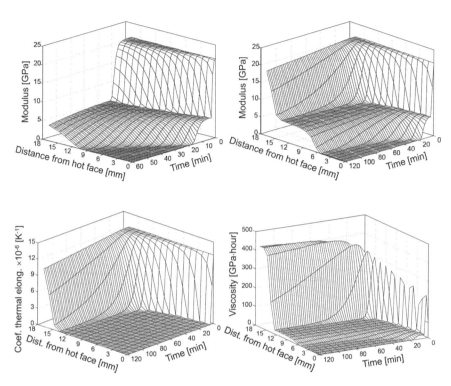

Fig. 3.123 Modelled elastic modulus distributions through hot face flange thickness of unprotected **(top left)** and water-cooled **(top right)** web-core panels; and effective coefficient of thermal elongation **(bottom left)** and viscosity **(bottom right)** distributions in water-cooled panels during fire exposure time, based on [150]

properties and one-dimensional heat transfer analysis [141]; an example is shown in Fig. 3.121 (left). The temperatures through the hot face flanges of unprotected and water-cooled web-core panels are well modelled. The thermomechanical properties are subsequently derived from these temperature distributions [136]. The temperature- and time-dependent elastic modulus degradation for unprotected and water-cooled panels, effective coefficient of thermal elongation, and viscosity for the water-cooled panel are depicted in Fig. 3.123, through the hot face flange thickness [150]. The elastic modulus degradation is based on the matrix-dominated elastic modulus curve shown in Fig. 3.111 (left). While the modulus rapidly decreases within 10 min to very low values through the whole hot flange thickness in the unprotected case, the decrease in the cell-sided part of the flange (up to 17.4 mm distance from the hot face) is slow in the water-cooled case and a certain stiffness of the flange is thus maintained up to 120 min. The response of the effective coefficient of thermal elongation is based on Fig. 3.117 (left) and, accordingly, exhibits a trend similar to that of the elastic modulus. The viscosity progression is derived from Fig. 3.114; the peak values are thus obtained in the material which is at the glass transition temperature [150].

3. Composites – Overcoming Limitations

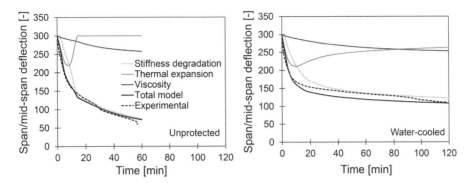

Fig. 3.124 Span/mid-span deflection ratio vs time relationships of unprotected (**left**) and water-cooled (**right**) web-core panels under bending and one-sided fire; modelling of ratio components (stiffness degradation, thermal expansion and viscosity) and total compared to experimental results, based on data in [150]

Based on the thermomechanical properties, the mid-span deflections can be calculated considering a simply supported beam in a four-point bending set-up and including shear deformations. The model can be extended to capture the effects of thermal expansion and viscosity [150]. The modelling results are shown in Fig. 3.124 for the unprotected (left) and water-cooled (right) web-core panels, i.e., the span/mid-span deflection ratio vs time relationships are depicted for stiffness degradation, thermal expansion, viscosity, and the total deflection is compared to the experimental results. In both cases, the main contribution to the deflections results from stiffness degradation. The thermal expansion contributes mainly to the first phase where the temperatures are below the glass transition temperature (see Section 3.9.2.8). The viscosity continuously and slightly increases the deflections. The total deflections obtained from modelling agree well with the experimental results [150].

As mentioned above, a matrix-dominated stiffness degradation has been assumed through the hot face flange of the water-cooled panels, although their responses are fibre-dominated. Nevertheless, the experimental deflections are modelled well. This inconsistency may be explained by the resin-rich flange-web junctions (see Section 3.3.2), which exhibit a matrix-dominated behaviour and thus may reduce the composite action between flanges and webs and thus increase the deflections.

Modelling results for the multicellular web-core panel under axial compression and one-sided fire are depicted in **Fig. 3.125** (elastic modulus) and **Fig. 3.126** (lateral mid-height deformations) for the unprotected and water-cooled cases (left and right, respectively). The temperature and time-dependent elastic modulus is obtained as indicated for the bending case (above) and shown through the whole panel depth (Fig. 3.125). It should be noted that the initial modulus (at time $t = 0$) is higher in the hot and cold face flanges than in the webs (steps in the curve). In the unprotected case, the modulus rapidly decreases in the hot face flange, webs and cell-side of the cold face flange; while in the water-cooled case the decrease in the webs is only minor and the cell-side of the hot face flange also remains partially functional [160].

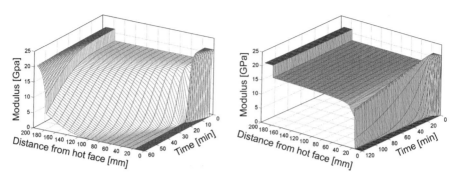

Fig. 3.125 Modelled elastic modulus distributions through whole panel depth of unprotected **(left)** and water-cooled **(right)** web-core panels, based on [160]

The lateral mid-height deformations are subsequently calculated taking second-order effects into account. The deformations are caused by two opposing mechanisms, i.e., the increasing eccentricity of the axial load due to decomposition of the hot face material and the increasing through-thickness gradient of the thermal expansions. The former mechanism produces lateral deformations away from the fire side while the latter causes deformations towards the fire side, as shown in Fig. 3.126. The total deformations are towards the fire side and in fairly good agreement with the experimental results. The discrepancies are mainly attributed to the small remaining values of each other compensating deformation components, which are thus difficult to predict accurately. The boundary conditions may also not always have been fully hinged (as assumed), which can have an influence on the deformations. The trends, i.e., deformations away from and towards the fire side, are correctly captured, however [160].

3.9.3.3 Sandwich panel members: furnace experiments and modelling

Composite-balsa core sandwich beams are investigated under three- and four-point bending and simultaneous one-sided ISO 834 fire, as also listed in Table 3.7. Reference beams with homogeneous balsa core and beams with transverse air gaps, steel plate, and steel T-shape inserts in the core are examined as potential paths of rapid heat ingress into the core, as shown in **Fig. 3.127**. Air gaps and steel plate inserts do not penetrate the hot face sheets while the webs of the T-shape inserts penetrate the fire-side face sheets and further accelerate the heat ingress due to the direct exposure of the T-shape flanges and high thermal conductivity of steel. The experimental results are complemented by numerical modelling. The two-step modelling is performed similarly as described for the pultruded members (above). The heat transfer analysis of the beams containing air gaps furthermore uses fluid-structure interaction simulation in the thermal model. In addition to radiation from the hot to the cold face inside the air gap, convection occurs, which transmits the heat to the cold face in the centre of the air gap, as depicted in **Fig. 3.128** [139, 140]. As will be shown, experimental and numerical results agree well if considering the significant possible variations of the balsa core block densities (see Section 2.2.2.1).

3. Composites – Overcoming Limitations   217

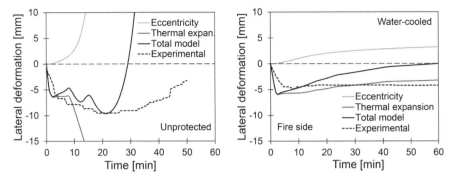

Fig. 3.126 Lateral mid-height deformation vs time relationships of unprotected **(left)** and water-cooled **(right)** web-core panels under axial compression and one-sided fire; modelling of deflection components (eccentricity and thermal expansion) and total compared to experimental results, based on data in [160]

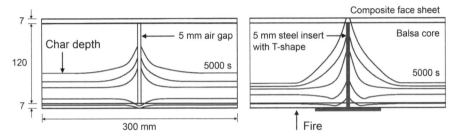

Fig. 3.127 Composite-balsa core sandwich beams with 5 mm wide air gap in core **(left)** and 5 mm steel T-shape insert penetrating hot face sheet **(right)**, and numerically obtained progress of char depth into balsa core vs time, in 1000–s steps up to 5000 s, based on [139]

The expected detrimental effects of air gaps and steel inserts are experimentally and numerically confirmed, as also shown in Fig. 3.127. The char depth around these singularities is significantly increased and the remaining undamaged section reduced accordingly. It can also be seen that air gaps and steel plate inserts have a cooling effect in the vicinity of the hot face sheet and a heating effect more inside the core, as further shown **Fig. 3.129**. The overall effects on the char depth and thus structural responses are the lowest for air gaps and the highest for steel T-shape inserts. In the latter case, the glass transition temperature in the cold face sheet is rapidly exceeded. In all cases, however, the effects remain local and are limited to the surrounding material of the singularities and are thus normally not critical for structural safety. Steel inserts should not, however, be located in the vicinity of slab supports, where they could interrupt critical load paths and thus lead to the collapse of the structure.

A much more critical failure mechanism is shown in the three-point bending experiment on the reference beam with homogeneous core, see **Fig. 3.130**. Due to the low transverse tensile strength of the end-grain balsa core and high transverse tensile stress, a transverse crack initiates at mid-span near the hot face sheet at an early stage and then rapidly opens. The crack acts as a transverse air gap, as described above, i.e.,

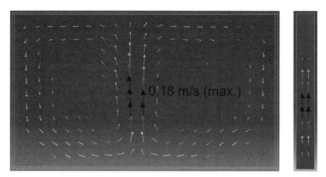

Fig. 3.128 Convection cells in 100 mm long and 5mm wide air gap, velocities 0–0.18 m/s, based on [139]

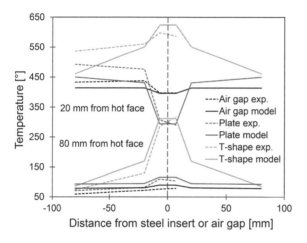

Fig. 3.129 Heating and cooling effects of 5 mm wide air gap, 5 mm wide steel plate insert, and 5 mm steel T-shape insert, at 20 mm and 80 mm depth from hot face, after 2 000 s, experimental and modelling results, based on data in [140]

as a path of accelerated heat ingress into the core. The temperature thus rapidly rises in the cold face sheet, which leads to a wrinkling failure when the glass transition temperature is reached. Since such transverse cracks may not remain local, as air gaps and inserts do, and extend over the whole slab width, this failure mechanism may lead to the collapse of the whole slab. In the four-point bending experiments on reference beams, the transverse tensile strength of the balsa core is not, however, exceeded, since the transverse tensile stresses are smaller, and the fire resistance is thus significantly higher, see **Fig. 3.131** and Table 3.7 [140]. The appropriate design of sandwich slabs in this respect is thus highly relevant. The post-fire specimen furthermore shows char peaks at the adhesive joint locations, caused by heat ingress into the core where the adhesive is decomposed – which is a similar mechanism to that observed in air gaps. The joints do not, however, open, in contrast to the mid-span crack in the three-point bending set-up.

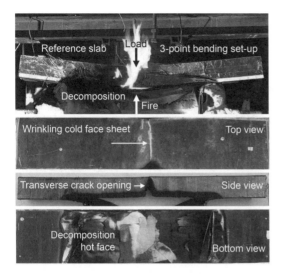

Fig. 3.130 Composite-balsa core sandwich beam under three-point bending and one-sided fire **(top)**; wrinkling of cold face sheet above mid-span crack **(middle top)**; transverse crack opening and crack-flank decomposition at mid-span (middle bottom); decomposition of hot face sheet **(bottom)**, based on [133, 140]

3.9.3.4 Post-fire assessment

The post-fire performance of composite members can be experimentally examined by comparing their structural response after fire exposure to that of "virgin" members, which were not exposed to fire. As an example, the axial compression load vs displacement responses and failure modes of post-fire and virgin web-core panels are depicted in **Fig. 3.132** (left). The post-fire panels were previously exposed to simultaneous mechanical load and one-sided fire during water-cooling, as described in Section 3.9.3.2. The responses demonstrate that the fire exposure results in significant losses of stiffness (~25%) and strength (~45%, on average). The losses during the second 60 min period of the 120 min water-cooling are, however, significantly smaller than during the first period, which coincides with the stabilisation of the deflections in the bending case, shown in Fig. 3.119. In all three panels, failure is initiated by the separation of the flanges from the webs and subsequent local buckling of webs and flanges, as shown in Fig. 3.132 (right, details) [162].

Post-fire stiffness can be estimated based on different models. A simple post-fire model is shown in **Fig. 3.133**, where the post-fire hot face flange of a previously water-cooled web-core panel is discretised into three layers, i.e., a virgin, partially degraded, and fully degraded layer [134]. The layer depths, $d(T_{g,\,onset})$ and $d(T_{d,\,onset})$, are derived visually or from the onset values of the glass transition and decomposition temperatures ($T_{g,\,onset}$ and $T_{d,\,onset}$), respectively. The virgin and fully degraded layers exhibit 100% and 0% stiffness, respectively, while a certain remaining stiffness is assumed for the partially degraded layer, e.g., 30% in Fig. 3.133.

The stiffness decrease during glass transition fully or partially recovers after cooling, depending on whether decomposition has already been initiated, as can be

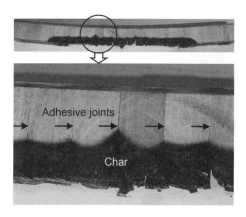

Fig. 3.131 Longitudinal mid-width cut through post-fire composite-balsa core sandwich beam under previous four-point bending and one-sided fire **(top)**; small char peaks at locations of adhesive joints between balsa blocks **(bottom)**, based on [140]

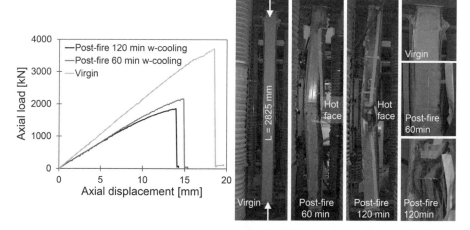

Fig. 3.132 Axial compression load vs axial displacement responses **(left)** and failure modes with details **(right)** of virgin and post-fire web-core panels after 60 and 120 min simultaneous mechanical and one-sided fire exposure with water-cooling, based on [162]

seen in Fig. 3.32 (right) and [162]. Accordingly, a continuous modulus degradation curve, similar to the one also shown in Fig. 3.133, can be derived from chemical kinetics, assuming a certain degree of recovery, as demonstrated in [163]. The post-fire elastic modulus distribution can then be derived, as depicted in **Fig. 3.134** for the case of a previously unprotected and a 120 min water-cooled web-core panel, assuming an 88% recovery of the elastic modulus after glass transition (and before decomposition), based on DMA results. A comparison with the distributions at the end of the fire experiments, shown in Fig. 3.125, demonstrates the significant modulus recovery in the hot face flanges (i.e., at 0–17.4 mm distances from the hot face).

3. Composites – Overcoming Limitations   221

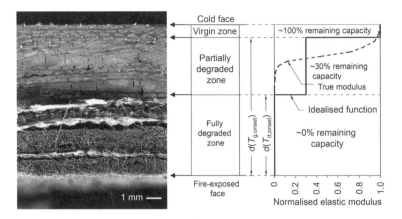

Fig. 3.133 Three-layer post-fire model of hot face flange of multicellular web-core panel, based on [134]

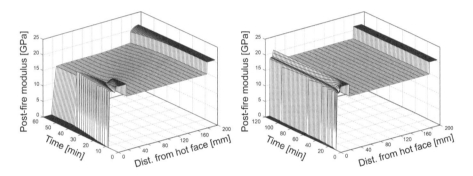

Fig. 3.134 Post-fire elastic modulus distributions through whole panel thickness of unprotected (**left**) and water-cooled (**right**) web-core panels, based on [162]

3.9.4 Fire dynamic simulation of indoor fires

Investigations concerning the fire performance of composite structures are so far limited to one-dimensional structural members in most cases, such as beams or columns, which are uniformly heated according to standard fire curves, as previously discussed. In three-dimensional buildings, however, thermal responses of composite members also depend on factors such as (i) the floor plan layout (with the arrangements of rooms, walls, doors, and windows), (ii) the location and heat release rate of the fire source (which does not necessarily follow a standard curve), (iii) the ventilation conditions (fresh air/oxygen supply), and (iv) possible fire protection measures. To obtain the thermal responses under such conditions, fire dynamic simulations based on computational fluid dynamics can be used to perform three-dimensional heat transfer analyses, as have already been successfully performed for steel and timber building structures [164].

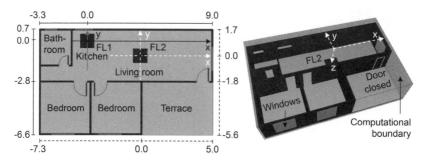

Fig. 3.135 Dimensions and configuration of floor layout of residential building with two locations of fire sources FL1 and FL2, plan view **(left)** and 3D view, with FL2 **(right)**, derived from [164]

A first example using this modelling approach for composite building structures is shown in **Fig. 3.135**. A typical single-storey residential house with walls and ceiling composed of glass fibre composite web-core sandwich panels of 116 mm thickness is investigated. The floor layout consists of a living room, two bedrooms, an open kitchen, and a bathroom, with a total area of 84 m². Doors in walls and windows are arranged as shown in Fig. 3.135. Two fire locations (FL) are considered, FL1 in the kitchen corner, or FL2 in the centre of the living room. Fresh air supply can occur through open windows or be limited by closed windows. Furthermore, fire protection of walls and ceiling with 12-mm-thick plasterboard panels is possible. The fire source is composed of wooden pallets with a volume of 1.325 m³. The heat release rate quadratically increases to a maximum value of 2400 kW within 226 s, and then remains constant if sufficient fresh air supply is provided; elsewise the rate decreases. The thermophysical properties of the composite materials up to decomposition are assumed according to Section 3.9.2. Among other things, the additional heat caused by the exothermic decomposition is considered [164].

As an output of the model, time-dependent temperature distributions on member surfaces can be obtained, as shown in **Fig. 3.136** for the temperatures on the ceiling surface above both fire locations, in x-direction (see Fig. 3.135). The temperature directly above fire location FL1 (at x=0) rises to a peak of 900°C after 600 s. Compared to the ISO 834 fire curve (which is also shown), this rise is more severe since, in the latter case, the temperature would only reach 678°C in the same interval. While the maximum temperature subsequently remains constant at 900°C up to 3600 s, the ISO-temperature slightly increases to 945°C. Above fire location FL2, the maximum temperature at the ceiling reaches only 550°C, however, which is significantly less compared to the ISO 834 curve. The higher maximum temperature above FL1 can be explained by a higher confinement of the fire in the corner of the kitchen.

Velocity fields of airflows can also be extracted from the model, as shown in **Fig. 3.137** for vertical (yz) planes through the two fire locations, delimited on the right side by the same wall. Since this wall is close to the fire location FL1, and the fire is more confined, only one airflow circulation develops, while two circulations occur in

3. Composites – Overcoming Limitations

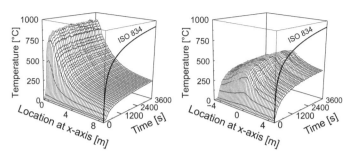

Fig. 3.136 Temperatures along x-direction on ceiling surface above fire locations FL1 **(left)** and FL2 **(right)**, with open windows, ISO 834 fire curve added, according to [164]

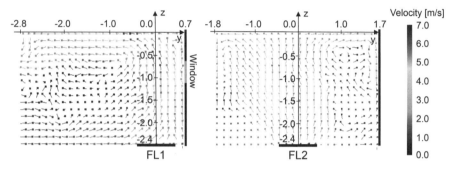

Fig. 3.137 Air velocity fields in yz-plane at 1800 s, at fire locations FL1 **(left)** and FL2 **(right)**, with open windows, according to [164]

the less confined FL2 case, where the wall is more distant. Accordingly, temperatures on the ceiling and the adjacent wall are higher in the FL1 than the FL2 case, as discussed above.

The temperature distributions on the ceiling surface, after 300 s, can be seen in **Fig. 3.138** for the two fire locations and with sufficient fresh air supply (open windows). Based on the composite glass transition and decomposition temperatures, $T_g = 117°C$ and $T_d = 300°C$, respectively, three areas can be identified: (i) the structurally intact area with temperatures $T < T_g$, most distant from the fire, (ii) the area with mostly reversible damage, $T_g < T < T_d$, and (iii) the area with irreversible damage due to resin decomposition, $T > T_d$, above the fire. These results demonstrate that temperatures and fire damage are not uniformly distributed. The ceiling locally loses resistance in both cases, but significant areas may remain undamaged, particularly in rooms separated by walls, i.e., the bathroom in this case.

For longer periods, however, the areas that undergo glass transition and decomposition significantly increase, as shown in **Fig. 3.139** (left), where 92% of the total area exceed the glass transition temperature for fire location FL1, while 50% exceed the decomposition temperature, after 3600 s. Also shown is the effect of fire protection

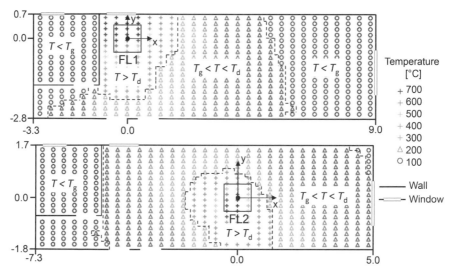

Fig. 3.138 Temperature distributions after 300 s on ceiling surface above fire locations FL1 (**top**) and FL2 (**bottom**), with open windows, areas indicated with temperatures below T_g, between T_g and T_d, and above T_d ($T_g = 117°C$, glass transition temperature, $T_d = 300°C$, decomposition temperature), derived from [164]

panels for fire location FL1, which significantly decreases these values to 62% and 8% for glass transition and decomposition, respectively.

The effect of fresh air supply, i.e., open or closed windows, is depicted in Fig. 3.139 (right) for fire location FL1. The maximum temperature over time on the ceiling surface, above the fire, is delayed and much smaller with closed compared to open windows. After 400 s, furthermore, the temperature begins to decrease due to oxygen depletion and carbon dioxide accumulation.

3.9.5 Conclusions

Depending on the composite material and structural configurations and type of fire exposure, the fire resistance of unprotected composite members may vary significantly from a few minutes up to one hour. The fire resistance can be extended mainly by (i) imposing a fibre-dominated response, (ii) selecting closed section (cellular) members, and (iii) assuring, by appropriate detailing, that only one side of the member is exposed to fire. The fire resistances can be further extended by applying fire protection systems, which can prevent or delay the temperature increase beyond the glass transition temperature in compressed member components. Selecting specific resins such as phenolics or adding fire retardants to the resin can improve the fire reaction behaviour, but not increase the fire resistance since the glass transition temperature is not notably altered.

Composite web-core and composite-balsa core sandwich panels, such as those discussed in Section 3.9.3, are frequently used for bridge decks, see Section 2.6.2. Such bridge decks are, in most cases, supported by two steel main girders, which

3. Composites – Overcoming Limitations   225

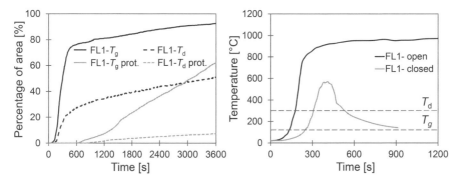

Fig. 3.139 Area percentages reaching T_g and T_d, over time on ceiling surface for fire location FL1 with open windows and without or with fire protection panels (**left**); comparison of maximum temperatures over time on ceiling surface above fire location FL1 with closed and open windows (**right**) ($T_g = 117°C$, glass transition temperature, $T_d = 300°C$, decomposition temperature), based on data in [164]

are not fire-protected. In the case of a fire below the bridge (e.g., on a crossed road), and considering that (i) the temperature exceeds 600°C already after 6 min in the case of an external fire curve (Fig. 3.105), (ii) the high thermal conductivity of steel (Table 4.6), (iii) the significant modulus and yield strength degradation of steel at 600°C (Fig. 3.111), and (iv) a fire resistance of unprotected composite web-core and sandwich panels of more than 30 min (Table 3.7), allows the conclusion that the steel girders fail before the composite-balsa sandwich bridge deck. Unprotected supporting steel members below or above the bridge deck are thus normally more critical than unprotected composite bridge decks. The use of unprotected cellular composite profiles as bridge girders is, however, as critical as using steel girders, since the profiles are normally exposed to three-sided fire and their fire resistance is thus low, see Table 3.7. Even worse are open-section composite profiles, which do not exhibit any notable fire resistance.

In building construction, unprotected cellular profiles, composite web-core or composite-balsa sandwich panels, used as beams or slabs can offer significant fire resistances of 30–60 min if the fire is one-sided. Active or passive fire protection systems can further increase the fire resistance and are required if beams are exposed to three-sided fire. Since columns and load-bearing walls are normally more critical than beams and slabs, because their individual failure may lead to the collapse of larger parts of the structure, the use of composites for such members should be thoroughly assessed. Although web-core walls exposed to one-sided fire may exhibit a significant fire resistance (Table 3.7), applying a fire protection system and designing a redundant structural system are highly recommended.

It should be noted, however, that these conclusions are based on one-dimensional experiments on individual members under standard fire curves. The example of a three-dimensional dynamic building fire simulation demonstrates that factors such as floor plan layout, location of fire, and fresh air supply (in particular), may have significant effects on the temperature responses and fire damage. Testing under standardised fire conditions may thus provide overly conservative fire resistances in some cases. In critical cases, fire resistance evaluations can therefore be supported by three-dimensional dynamic fire simulations.

3.10 DURABILITY

3.10.1 Introduction

Composite structures can be considered durable if structural safety and serviceability are maintained throughout the design service life, which is normally 50 years for buildings and 100 years for bridges [18]. Important factors that may affect durability during these very long periods are (i) unexpected actions, (ii) limited knowledge concerning the long-term effects of environmental conditions, (iii) inappropriate material selection and processing, (iv) ineffective protection systems, (v) inappropriate detailing, and (vi) absence of maintenance.

Unexpected actions may concern sustained gravity loads and elevated temperatures, which cause unforeseen viscoelastic effects, or fatigue-related actions, which may gradually increase over time during such long periods (e.g., bridge traffic loads). Environmental conditions, such as exposure to temperature, moisture (humidity and water), and ultraviolet (UV) radiation, and the combinations thereof, may degrade mechanical and physical properties. Knowledge about such damage mechanisms is still limited in the case of composites [40], and is often based on accelerated laboratory investigations, whose relationship to natural environmental exposure effects is difficult to establish. Long-term studies of structures or components exposed to different types of natural climates, which also include quantified effects on mechanical properties, are still scarce and limited to about 25 years duration.

Material selection and manufacturing process, the latter also regarding curing procedures, should be adapted to the harshness of the environmental conditions. In this respect, it should be considered that bio-based composites, consisting of natural fibres and bio-based polymers – no matter how desirable their implementation is from a sustainability point of view – are not normally able to resist outdoor exposure over the long periods mentioned above, without significant degradation of their mechanical properties, see Section 3.11.4.

If the durability performance of the material itself is still insufficient, protection systems can be applied, e.g., coatings against moisture and UV ingress. Such systems should be effective throughout the whole design service life of the structure, however, and thus normally need a periodic renewal.

Appropriate detailing is particularly important regarding serviceability, fatigue, and environmental protection. Details such as supports and joints should allow structures to deform or move as considered in the design. Stress concentrations should be reduced by appropriate detailing to extend the fatigue life, as discussed in Section 3.8.8. With respect to environmental exposure, stagnant water or water running along surfaces should be prevented by appropriate drainage or drip edges. Generic examples of composite bridge details fulfilling such requirements can be found in [1, 40]. Furthermore, regular inspections and maintenance should be performed in a timely manner to recognise and prevent possible damage, see Section 5.5.

3. Composites – Overcoming Limitations   227

Aspects concerning durability can be found in different sections of the book. This section focuses on irreversible damage and detailing regarding durability. The content is mainly based on findings from long-term surveys of composite structures exposed to different natural climates over periods of up to 25 years. Reversible effects of temperature and moisture are discussed in Section 3.4. Physical ageing is addressed in Section 3.5, since significant interactions occur between curing, physical ageing, and moisture at an early age. Basic information about the durability of fibres and resins is provided in Section 2.2.1.

3.10.2 Long-term surveys of Pontresina Bridge and Eyecatcher Building

3.10.2.1 Overview

The Pontresina Bridge is a two-span truss bridge for pedestrians, which was built by 20 architecture students in two weeks from 17 to 28 November 1997, and installed on 15 December 1997, in Pontresina over the Flaz River. The joints of the pultruded profiles are bolted in one span while they are bonded with an epoxy adhesive in the other span (and complemented by back-up bolts); further details are summarised in Appendix A.2. The bridge is used during winter and removed and stored on the riverbanks during summer due to possible floodwater; the lifting is done by helicopter, see **Fig. 3.140** (left and middle). Periodic inspections of the bridge occurred during the summers of 2005, 2014, and 2022 (after 8, 17, and 25 years of use, respectively), where the two spans were transported to EPFL and mechanical and visual investigations were performed [165, 166].

Occasionally, and mainly in recent years, only the gratings of the bridge were removed during summer to prevent crossing, instead of moving the whole bridge to the banks. The bridge was thus subjected to floodwater in some cases, e.g., in 2003, see Fig. 3.140 (right), without causing any damage. The same occurred on 28 August 2023, but the floodwater level reached an all-time peak and the whole bridge, including the middle concrete pier, were destroyed see **Fig. 3.141**. Fragments of the bridge were found several hundred meters downstream.

The Eyecatcher Building is a mobile building installed in December 1998 for the annual Swissbau building fair, held in Basel from 2 to 6 February 1999. The five-story structure consists of three parallel composite frames and timber floors.

Fig. 3.140 Pontresina Bridge, annual installation during winter (**left**); on riverbanks during summer (**middle**); and during floodwater on 19 June 2003 where only bridge deck gratings were removed (**right**)

Fig. 3.141 Destruction of Pontresina Bridge caused by all-time floodwater on 28 August 2023: flood level, bridge already pulled down, abutment and level of supports below water, and gratings on banks **(left)**; shifted and overturned middle concrete pier **(middle)**; destroyed bridge segments **(right)**

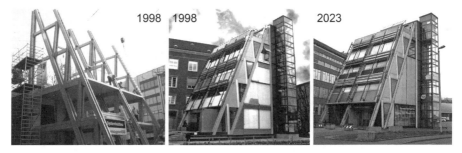

Fig. 3.142 Mobile Eyecatcher Building, construction and 1st location in 1998 **(left and middle)**; location in 2023 **(right)**, both in Basel

The frames are composed of adhesively bonded pultruded built-up sections and detachable bolted joints; further details are given in Appendix A.3. After the first use, the building was dismantled and re-installed at another location in Basel, see **Fig. 3.142**; since then, it has been used as an office building. On-site visual inspections of the building were performed in 2006, 2014, and 2023 (after 8, 16, and 25 years of use, respectively) [166].

Both structures are composed of pultruded profiles manufactured by the same company in 1997 and 1998. The materials used are E-glass fibres and isophthalic polyester, and polyester surface veils were implemented to improve the surface quality. The profiles are unprotected, i.e., no coating was applied. In 1997 and 1998, the prevailing opinion was that protection was not necessary, as was the case for maintenance, which was not considered necessary for these materials.

3.10.2.2 Climates and environmental expositions

Pontresina is located in an alpine, i.e., continental subarctic climate (Dfc, according to the Köppen climate classification), while Basel is situated on the central plateau, in a warm summer continental climate (Dfb). The main characteristics of the two different climates are summarised in **Table 3.8**, while the intensities of sun irradiation,

3. Composites – Overcoming Limitations

Table 3.8 Climate data, annual average values, measuring stations 5.5 km from Pontresina Bridge and 3 km from Eyecatcher Building, based on [166]

Location	Altitude (m)	Climate (Köppen)	Tempera-ture (°C)	Sunshine (hours)	Frost (days)	Snow cover (days)	Relative humidity (%)	Rainfall (mm)
Pontresina	1 790	Dfc	2.0	1 733	235	159	73	713
Basel	240	Dfb	10.5	1 637	64	25	75	842

Table 3.9 Intensity of sun irradiation in Pontresina and Basel, of which approx. 5% is UV-irradiation, "total" indicates direct plus diffuse irradiation, based on [166]

Irradiation direction	Annual total (kWh/m^2)	
	Pontresina	Basel
Direct normal	1 858	1 085
Total on horizontal surface	1 502	1 132
Total on vertical surface, south	1 417	907
Total on vertical surface, north	451	379

during different exposure situations, are listed in **Table 3.9** [166]. Comparing the data underlines the significant differences of the two climates, particularly in terms of sun irradiation and frost days, i.e., combined UV irradiation and freeze-thaw cycles.

The orientation of the Pontresina Bridge is southwest-northeast. However, since the bridge spans are symmetrical and annually moved, and can thus be turned by 180°, their sun exposure randomly changes. The inclined envelope of the Eyecatcher Building is directed southeast.

3.10.2.3 Stiffness and strength degradation

Two types of identical mechanical analyses are performed during each inspection of the Pontresina Bridge: (i) full-scale serviceability limit state (SLS) loading of both spans and (ii) measurements of the tensile strength and tension elastic modulus on coupon specimens taken from transverse I-beams of the bolted span, which can be replaced by new beams, as shown in **Figs. 3.143–144** (left). The full-scale load is applied by brick pallets, always in the sequence shown in Fig. 3.143 (right). Pallets Nos. 1–8 represent a uniformly distributed load of 4 kN/m^2, while Nos. 9 and 10 simulate a 10 kN snow removal vehicle at mid-span.

The load vs mid-span deflection responses of the bolted and bonded spans in 1997 and 2022 (after 25 years) are compared in **Fig. 3.145** (left). Since the brick pallet weight varied slightly during the 25 years, the responses are adjusted to the load of 1997. The changes in the slopes of the responses are caused by the loading sequence. As expected (see Section 2.4.1), the deflections of the bonded span are slightly smaller

Fig. 3.143 Pontresina Bridge, periodic serviceability (SLS) loading **(left)**, and sequence of loading with brick pallets (representing 4 kN/m^2 superposed by a 10 kN vehicle load at mid-span) **(right)**

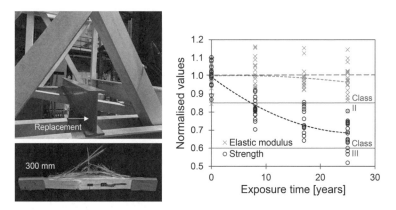

Fig. 3.144 Pontresina Bridge, replaced transverse beam (grey) and splitting failure mode of 300 mm long coupon specimen, [166] **(left)**; normalised tensile strength and tension elastic modulus vs years of exposure, polynomial fits of experimental data and conversion factors for moisture exposure, Class II (0.85) and III (0.60) of CEN/TS 19101 [58] **(right)**

than those of the bolted span, as is also shown in Fig. 3.145 (right). These results and the fact that the 1997 values were measured in a different laboratory allow the conclusion that the overall span stiffnesses did not notably vary during the 25 years. The span/deflection ratios are, in rounded values, $L/800$ for the bonded span and $L/750$ for the bolted span, and thus still much smaller than the limit of $L/500$.

The development of the coupon tensile strength and tension elastic modulus over the 25 years is shown in Fig. 3.144 (right); the experimental results are normalised by the initial values of 1997 and fitted by second-order polynomial functions. The tensile strength continuously decreases but stabilises at approximately 68% (on average) of the initial strength after 25 years. This strength reduction can be associated with an average conversion factor for moisture according to CEN/TS 19101 [1], which originates from irreversible damage in this case (since the specimens were dry), see also Section 3.4.6. The conversion factors for moisture according to [1] for exposure Class II (0.85, outdoor exposure) and Class III (0.60, continuous exposure to water) are also added. As specified for Class II in [1] based on the results above, the 0.85

3. Composites – Overcoming Limitations   231

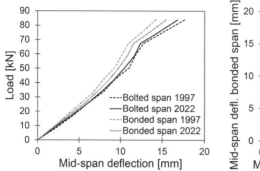 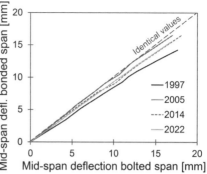

Fig. 3.145 Pontresina Bridge, load vs mid-span deflection of bolted and bonded span in 1997 and 2022 **(left)**; mid-span deflection of bolted vs bonded span in 1997, 2005, 2014, and 2022 (after 25 years) **(right)** (1997 at ETHZ, 2005–22 at EPFL) [58]

value is not applicable for combined UV irradiation and frequent freeze-thaw cycles, which cause more severe damage than normal outdoor exposure. The UV irradiation leads to superficial resin degradation and associated fibre blooming. Through the exposed fibres, moisture can ingress into the material along the more porous fibre-matrix interphases through wicking effects (see below, Sections 3.10.2.6–7); freeze-thaw cycles then cause damage at deeper depths [166]. To maintain a normal outdoor condition conversion factor of 0.85, a protective coating should have been applied after 7–8 years. The elastic modulus, in contrast, does not yet exhibit a notable change, although a slight decreasing trend initiates after 25 years. This result is in agreement with the full-scale experimental results discussed above.

Long-term results concerning the degradation of the mechanical properties are also available for the Kolding Bridge, built in 1997 in Denmark from profiles of the same pultruder [167]. The bridge is in the same climate zone as the Eyecatcher Building. In 2012, an L-section profile was removed from the bolted wind bracing below the bridge deck and three-point bending experiments were performed on coupon specimens. After 15 years of use, the bending strength decreased, on average, by 12%, while the elastic modulus increased by 9% (considering only Type 1B and 2 specimens in [167]). The Class II conversion factor of 0.85 thus seems appropriate in this case, assuming that the strength decrease also levels off.

In 1997, the design of the Pontresina Bridge was performed according to the Design Manual (DM) of the profile manufacturer [168]. A conversion factor for environmental conditions was not considered at that time. The critical verifications were those of the bolted joints between the vertical tube posts and the transverse beams (and not those of the profiles between the joints). The transverse load on the railing and superimposed wind cause a transverse moment on these joints, resulting in bolt forces as visualised in **Fig. 3.146** (left, force couple). According to the DM, two bolts M42 (with 42 mm diameter) would have been necessary per joint, which would have increased the profile sections considerably and reduced the use of their capacity between the joints significantly. Furthermore, as explained above, adhesive joints

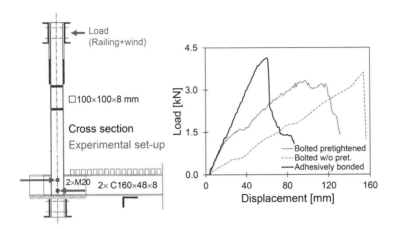

Fig. 3.146 Pontresina Bridge, experimental full-scale investigations in 1997 for design of bolted and adhesively bonded joints: bridge cross section with overlaid experimental set-up and applied forces **(left)**; load vs handrail displacement responses for bolted and bonded joints **(right)** [58]

were used in one of the spans; corresponding design provisions did not yet exist in 1997, however. Due to these two reasons, full-scale joint experiments were conducted by the profile manufacturer, as sketched in Fig. 3.146 (left), to (i) explore whether the bolt diameter could be reduced from 42 to 20 mm (without increasing the bolt number), and thus obtain reasonable profile sections, and (ii) verify the capacity of the bonded joints.

The load vs displacement responses from these experiments are shown in Fig. 3.146 (right). As can be seen, in one experiment, the bolted joint was slightly pretightened. This did not improve the joint resistance, however, while the adhesive joint exhibited the highest resistance. From these experimental results, it was concluded that the bolt diameter could be reduced to 20 mm and the adhesive joint resistance was sufficient. The profiles between the joints, however, were still used to only about 30% of their capacity.

Based on the measured significant strength reduction, the structural safety of the bolted and bonded joints was verified again in 2022, this time according to the new European Technical Specification CEN/TS 19101 [1]. Concerning the bolted joint, the updated ratio of the design values of the joint resistance to the action effects results is $R_d/E_d = 1.10$. This low value is still considered to be tolerable, however, since (i) the strength reduction levels off after 25 years, (ii) the (comparably high) SLS full-scale loading of the bridge does not show any change in the behaviour or local damage after 25 years of use (see above), and (iii) the assumed critical design railing load, i.e., 1.2 kN/m, is unlikely to occur on this bridge. In the adhesively bonded joint, the bending moment from the handrail is transmitted through torsion in the two adhesive layers to the transverse beams. The structural safety condition is still fulfilled without any reduction.

As was done for the Pontresina Bridge, full scale experimental investigations of the adhesively bonded connections of the Eyecatcher Building's built-up sections

3. Composites – Overcoming Limitations

Fig. 3.147 Eyecatcher Building, adhesively bonded built-up sections during manufacturing in 1998, [166] **(left)**; full-scale four-point bending experimental investigations and failure mode inside profile (and not adhesive connection) **(right)**

were carried out in 1998, as shown in **Fig. 3.147**. Failure always occurred inside the profiles and never in the adhesive connections. Since the stresses in the large adhesive connections are small, the new consideration of conversion factors for moisture under outdoor conditions and temperature does not affect the structural safety of the adhesive joints.

These examples demonstrate how the perception and consideration of durability in design has changed since the 1990s and what possible effects it can have on structural safety. Design manuals or handbooks of the time, such as [20, 168], did not yet consider any environmental degradation of mechanical properties in design verifications, except for the reversible effects of below-T_g temperatures in [20]. Experimentally justified conversion factors with a comprehensive background were only recently derived, however, in [40] for the new European Technical Specification, CEN/TS 19101 [1].

3.10.2.4 Adhesively bonded connections

Both structures include adhesively bonded joints, whose age is now 25 years. Since the joint dimensions of the Pontresina Bridge are small and such primary load-bearing joints were used for the first time, additional back-up bolts were added to increase the structural redundancy. In the frames of the Eyecatcher building, adhesive connections were implemented in the built-up sections, as shown in Appendix A.3. Since the bonding surfaces are large and the stresses thus small, and structural redundancy already exists, no back-up bolts were added. The design of both types of joints was based on experimental investigations, as described above.

The full-scale bridge experiments demonstrated that the stiffness of the bonded span is still higher than that of the bolted span. The adhesive joints did thus not lose stiffness over the 25 year period. In 2014, it was concluded, from the existence of small gaps at some adhesive edges, that detachments occurred in some adhesive joints due to transverse impact, see **Fig. 3.148** (left) [166]. Furthermore, small cracks were visible in the fillets of some adhesive layers. In 2022, the bolts were removed in these joints to better examine their status. No detachments could be found in these cases, however; the gaps were just caused by missing adhesive material, and the cracks had not penetrated into the connections. In the Eyecatcher Building, only the adhesive layer edges are visible for monitoring. In 2023, as in the previous inspections, no

 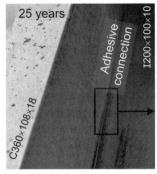 

Fig. 3.148 Adhesively bonded connections after 25 years, Pontresina Bridge hybrid joint (no detachment after removing of bolts) **(left)**; Eyecatcher Building (inclined middle column on 5th floor terrace) **(right)**

signs of degradation and no cracks could be found at these locations, as can be seen in Fig. 3.148 (right). It has thus been concluded that the adhesive joints are still fully functional in both structures, after 25 years.

3.10.2.5 Impact damage

Impact damage occurred only at the Pontresina Bridge and was thus attributed to the annual movements of the structure, which occurred about 40 times during the 25 years. The damage concerned the outstand (cantilevering) flanges of open-section profiles where cracks and material crushing were found in all three inspections. Cracking occurred in the web-flange junctions due to transverse impact on the outstand flanges, as shown in **Fig. 3.149** (left and middle). Such impact was mainly caused by inappropriate lifting and storing of the bridge spans on the banks, or during the installation of the gratings on the transverse beams. Typical crushing damage in a bottom chord open-section flange is furthermore depicted in **Fig. 3.150** (left).

Crack and crushing damage in open-section profiles can be rehabilitated relatively easily, as demonstrated in Figs. 3.149–150 (right). The cavities of open-sections can be filled with high-stiffness polymer filler blocks, which support the damaged flanges. Crushing damage and cracks can furthermore be bridged by overlaid adhesively bonded flat or L-section profiles [165].

The bottom flanges of the bottom chord channel section profiles exhibited significant cracks after 25 years, however (Fig. 3.149, left). The structural safety of these profiles was thus re-examined. In 1997, only 32% of the material strength was used (see above). Taking an average conversion factor of 0.65 into account and assuming that one or two of the four flanges in total per truss girder chord are lost (i.e., disconnected by cracks at the web-flange junctions) increases the use of strength to 60% and 73%, respectively, thus to values still significantly below the full strength. Nevertheless, it was decided to apply a coating to these four bottom chord channel sections in 2023, as was already done in 2015, after the second inspection, on the four top chord channel sections (see below).

3. Composites – Overcoming Limitations   235

Fig. 3.149 Pontresina Bridge, cracks in open-section outstand flanges due to transverse impact, bottom view on bottom chord channel section **(left)**; top view on end transverse beam **(middle)**; rehabilitation of end transverse beam by supporting flanges with high-stiffness polymer filler block, [165] **(right)**

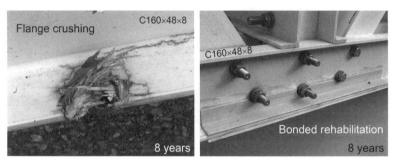

Fig. 3.150 Pontresina Bridge, flange crushing in bottom chord channel section profile due to impact after 8 years **(left)**; rehabilitation by adhesively bonded bridging profiles **(right)** [165]

3.10.2.6 Fibre blooming

As already introduced above (Section 3.10.2.3), strong UV irradiation may degrade the outermost resin layer and uncover the topmost fibres, resulting in a pattern of exposed fibres denominated "fibre blooming". At the beginning of this process, the glossy and smooth surface – normally apparent on pultruded profiles – is altered to matte and slightly rough. The subsequent fibre blooming first concerns the outer mat and/or fabric layers, but subsequently may go deeper to roving layers. To characterise this progress of degradation, four levels were defined in [166], Levels 0–3, as listed in **Table 3.10**, where Level-0 represents the reference surface without any degradation.

The different levels are illustrated with examples selected from the Pontresina Bridge and Eyecatcher Building, as shown in **Figs. 3.151–152**, respectively. Level-1 and -2 were already visible after eight years on the bridge profiles, during the first inspection. Surprisingly, they even appeared on the same surface, separated by a straight line, as shown in Fig. 3.151 (left). After consultation with the profile manufacturer, it was concluded that the surface veil was missing in the Level-2 area, i.e., it shifted away during the pultrusion process. Later field inspections [169] confirmed the important role of surface veils, which effectively hinder the penetration of UV radiation into the material. Level-3 damage was found on the most exposed surfaces of the bridge after 17 years, i.e., on the horizontal top chord flanges, as shown in Fig. 3.151 (right), but this damage was limited to a few smaller regions.

Table 3.10 Fibre blooming Levels 0–3, based on [166]

Level	Description
0	Intact glossy and smooth surface (reference)
1	Matte and slightly rough surface (pattern of black dots possible)
2	Exposed fibres of outer mat/fabric layers
3	Exposed fibres of deeper rovings

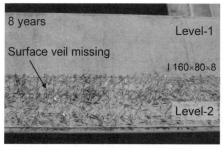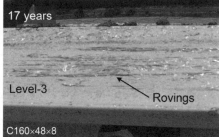

Fig. 3.151 Pontresina Bridge, fibre blooming Level-1 and -2 and effect of missing surface veil on transverse beam top flange after 8 years, [165] **(left)**; Level-3 on top chord channel section flange after 17 years **(right)**

The examples of the Eyecatcher Building depict the significant effect of orientation with respect to the sunlight. The same profile is shown from two slightly different angles in Fig. 3.152, after 25 years of exposure. Depending on the orientation, the levels vary from Level-0 to Level-2. On the less exposed inclined back side of the flanges close to the envelope, the surfaces were still intact and glossy (Level-0), while on the most exposed front side of the flanges, the highest levels (Level-1 and -2) were observed. The vertical web surfaces and inclined back side flange surfaces away from the envelope were in between, exhibiting Level-1.

The development of the four levels of fibre blooming on the exposed surfaces of the Pontresina Bridge and Eyecatcher Building (exterior surfaces only), over 25 years, is compared in **Fig. 3.153**. Level-0 surfaces rapidly decreased in the bridge case from the beginning, while a decrease started after only about five years in the building case; the subsequent rates of decrease were, however, similar in both cases. Level-1 surfaces reached a peak of approximately 50% after about 16 years in the bridge case; the following decrease was caused by the significant transformation from Level-1 into Level-2, which reached about 60% after 25 years. The development of Level-1 started after about five years on the building surfaces and then continuously increased up to 90%, since Level-2 appeared only after about 10 years and remained very small. A very limited amount of Level-3 developed only in the bridge case, up to 17 years when these surfaces were coated (see below). As already explained above, the building did not exhibit any Level-3 after 25 years.

3. Composites – Overcoming Limitations 237

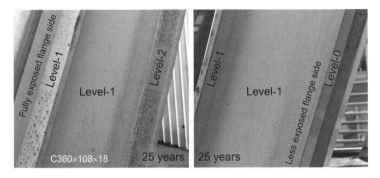

Fig. 3.152 Eyecatcher Building, fibre blooming Level-0 to Level-2 on inclined middle column of fifth floor terrace, seen under two slightly different angles (2× C360×108×18, north-eastern orientation), after 25 years

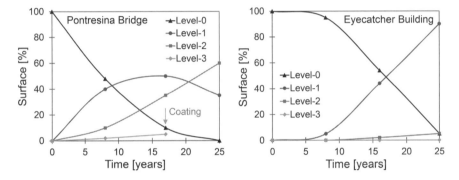

Fig. 3.153 Progress of fibre blooming levels during 25 years for Pontresina Bridge, continental subarctic climate (**left**) and Eyecatcher building, warm summer continental climate (**right**) [58]

This comparison highlights the effects of the different climates and associated UV irradiation to which the two structures were exposed. The average rate of development of fibre blooming Level-2 on the Pontresina bridge surfaces was about 10 times higher (3.1%/year) than the average rate on the surfaces of the Eyecatcher Building (0.3%/year). This difference was caused by an approximately 1.7 times higher UV irradiation in the former case (based on the "direct normal" values of Table 3.9).

The effect of fibre blooming on the mechanical properties was furthermore investigated using the coupon specimens of the Pontresina Bridge, cut from the top and bottom flanges of the transverse beam, which was removed after 25 years (see above, Section 3.10.2.3). Coupons of the top flange exhibited Level-2 fibre blooming on the top surface, while Level-1, without fibre blooming, appeared on top of the bottom flanges. Tensile strength and tension elastic modulus are compared in **Fig. 3.154**, where the measured values are normalised by those from 1997. Fibre blooming caused a reduction in strength (11%) and, to a lesser extent, in modulus (4%) for coupons taken in the middle of the top flanges. The differences are less obvious for coupons cut from the edges of the flanges. The results suggest that fibre blooming can have a

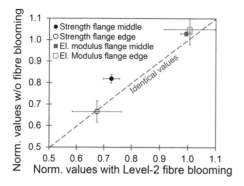

Fig. 3.154 Pontresina Bridge, effect of Level-2 fibre blooming on normalised tensile strength and tension elastic modulus after 25 years (av. ± stdev. values, 3–8 specimens per point) [58]

detrimental effect on tensile strength; however, the database is too small for a robust conclusion. The significant scatter of the mechanical properties shown in Fig. 3.144 (right) may be explained, nevertheless, by the fact that results with and without fibre blooming are comprised.

Based on results of the second inspection, the decision was taken in 2014, after 17 years, to apply a coating to the top chord channel section profiles, which exhibited Level-2 and Level-3 fibre blooming, since the Level-3 strength reduction was most likely to be more severe than that for Level-2. A second reason for this decision was the tactile discomfort for the users of these handrails. The coated top chord channels can be seen in Fig. 3.157 (below).

3.10.2.7 Moisture effects

Moisture ingress into the polymer network can cause physical or chemical degradation of a composite material. The most relevant physical degradation mechanism is plasticisation, which normally is reversible after drying, as has already been discussed in Section 3.4.6. If moisture exposure is sustained, irreversible chemical degradation may occur, such as hydrolysis, i.e., chain scission in the molecular network, or degradation of the fibre-matrix bond. A detailed overview of the moisture-related damage mechanisms that can occur is given in [40].

The most efficient way to prevent moisture damage is to avoid or limit moisture ingress into the composite material. Basically, moisture ingress may occur through diffusion through the matrix or capillarity (also denominated "wicking") along fibres in the fibre-matrix interphases, which exhibit higher porosity than the surrounding matrix. The ingress through diffusion is much slower than that through capillarity, and the latter should thus be prevented [142].

Selecting an appropriate resin, which is subsequently fully cured, can effectively reduce moisture ingress through diffusion. Sealing cut profile and laminate ends, or bolt-borehole surfaces, and implementing surface veils to retain fibres from penetrating the surfaces can avoid ingress through capillarity. Applying protective coatings is the most efficient way to reduce moisture ingress through both diffusion and capillarity (as long as the coating is effective and not damaged). In cases of continuous

3. Composites – Overcoming Limitations   239

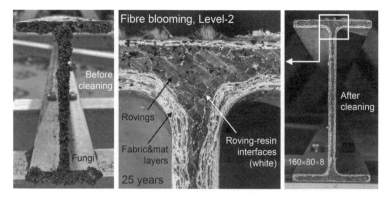

Fig. 3.155 Pontresina Bridge, transverse beam end after 25 years, cut surface with lost sealing and fungi growth **(left)**; white coloured roving-matrix interphases **(middle)**; central roving and outer combined fabric and mat layers after cleaning **(right)**

Fig. 3.156 Eyecatcher Building, moisture migration through unsealed bolt holes during rainy period **(left)** and dry period **(middle and right)**, recognisable by colour changes, inclined middle column on fifth floor terrace, southwestern orientation (note that overlaid darker pattern of dry period is dirt)

exposure to water or high relative humidity, i.e., Class III according to [1], the application of a protective coating is recommended [1, 40].

To visualise capillarity or wicking, two examples are shown in **Figs. 3.155–156**. The first case concerns a transverse beam of the Pontresina Bridge, which lost the protective coating of the cut ends after 25 years. Fungi growth confirms the increased humidity on these surfaces and, after cleaning, a white colouring of the interphases along the central rovings is visible, whose origin and nature has not yet been determined. The second example concerns the bolted connections of the Eyecatcher Building. The initially sealed bolt holes lost their sealing during the demounting and second installation. Since then, the colour around the bolts changes with changing humidity. The colour becomes darker during longer humid periods, while this colour change disappears during extended dry periods. After consultation with the profile manufacturer, this phenomenon was attributed to moisture ingress through the unsealed bolt holes, and, due to the assumed reversibility, was not considered to be a concern regarding structural safety.

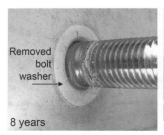

Fig. 3.157 Pontresina Bridge, colour change compared to UV-protected surface below bolt washer, after 8 years **(left)**; differences between white-coated top chord after 18 years and uncoated profiles in 2022 **(right)**

Fig. 3.158 Pontresina Bridge in 1997 **(left)** and after 25 years **(right)**, effect of yellowing due to UV radiation

3.10.2.8 Colour changes

UV radiation may not only cause fibre blooming, as discussed above (Section 3.10.2.6), but also discolouration, which often appears as yellowing [40]. The white colour of the Pontresina Bridge profiles is the natural colour of the resin; this colour was also most welcome since this bridge merges well into the winter landscape. Concerns already existed at that time about a potential yellowing of the bridge. Based on consultation with the manufacturer about the resin type and the presence of surface veils, which hinder the deeper penetration of the UV radiation into the material, the risk was nevertheless taken to expose the unprotected profiles to the environment.

The discolouration was surveyed during the inspections, as can be seen in **Fig. 3.157**. A slight yellowing occurred during the years, if expressed in RAL colours, from RAL9016 (1997) to RAL9012 (2022). The discolouration, however, seems to have stabilised between 17 and 25 years, and does not clash with the landscape, as shown in **Fig. 3.158**. Unlike the Pontresina Bridge, the resin of the Eyecatcher Building profiles was grey-coloured. No notable discolouration was observed during the 25-year period.

3.10.3 Further long-term studies

Further projects realised with the involvement of CCLab and referred to in this work – but not yet as old as the Pontresina Bridge and Eyecatcher Building – are the Novartis

3. Composites – Overcoming Limitations 241

Fig. 3.159 Novartis Building, schematic top view on composite sandwich roof after 12 years exhibiting blistering and cracks along adhesive joints **(left)**; levelling of roof surface during manufacturing **(middle)**; recess edges in face sheet laminates above adhesive joints **(right)**

Building and the Avançon Bridge. They were built in 2006 and 2012, and are 18 and 12 years old in 2024, see Appendices A.4 and A.5, respectively, and Section 4.7.

The first inspection of the composite-PUR web-core sandwich roof of the Novartis Building was performed in 2018, after 12 years in use. The roof was full of dirt at that time since no maintenance had been performed during this period. Two types of damage were found on the top side of the roof, i.e., blistering in the region of the smallest roof thickness and cracks along the module joints, see **Fig. 3.159** (left). Since the face sheets were applied by hand lamination, the outer surfaces required levelling and smoothening (Fig. 3.159, middle); a short-fibre compound was used locally for this purpose. A protective coating was subsequently applied above the whole roof surfaces. Blistering in the coating occurred in regions of extensive levelling and was attributed to an inappropriate quality of the levelling compound. Furthermore, recesses in the face sheet laminates, which were left during the module manufacturing (Fig. 3.159, right), were subsequently filled on-site, after application of the adhesive joints. Cracks appeared in the coating along the edges of these recesses. Since the damage spread over larger parts of the top surface, a new laminate layer was applied on the whole top surface. One conclusion from this experience was that similar members of complex shapes should preferably be manufactured by vacuum infusion to prevent such levelling, as was subsequently planned for similar roofs, see Appendices B.2 and B.3. Furthermore, laminate butt joints should be over-laminated if a periodic inspection and renewal of the coating cannot be ensured.

Inspections of the Avançon Bridge (Appendix A.5) were performed eight and eleven years after construction, the first by the bridge owner and the second by myself. No maintenance was performed during the first eight years. Since the bridge was built with a semi-integral concept, simple elastomer plug expansion joints could be installed at the bridge ends, see **Fig. 3.160** [5]. Due to the steep longitudinal slope of the bridge (8%) and presumable braking forces applied above these joints, they deformed in the downhill direction and thus were replaced, after eight years, by a more rigid construction, composed of steel and intermediate rubber profiles, see **Fig. 3.161** (left and middle). In the second inspection after eleven years, some small cracks and associated debonding were found in the protective coating on the most exposed top surfaces of the kerbs, see Fig. 3.161 (right). These cracks were not noted during the

Fig. 3.160 Avançon Bridge, built in 2012 **(left)** and after 11 years **(middle and right)**

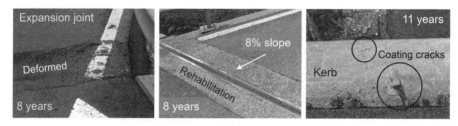

Fig. 3.161 Avançon Bridge, deformed elastomer plug joint due to steep slope and rehabilitation after 8 years **(left and middle)**; cracks in protective coating of most exposed kerb top side after 11 years **(right)**

first inspection, although they were certainly already present. According to the bridge owner, such damage often occurs with this type of coating and is thus not related to the composite structure. No other visually detectable damage appeared, the adhesively bonded deck-to-girder and element joints did not exhibit any visible changes.

3.10.4 Detailing for durability

An appropriate detailing to maintain durability should primarily ensure that the assumptions made in the design, in this respect, correspond to the real on-site exposure and damage mechanisms. If a Class II exposure according to [1] is assumed in design, for instance, i.e., a normal outdoor exposure, locations of stagnant water or water frequently running along surfaces should be prevented by implementing appropriate slopes on surfaces and drainage measures or drip edges. Such constructive measures are simple and, in the long term, normally more effective, reliable, and less expensive than coatings, which can be damaged (see above, Fig. 3.161, right) and require periodic renewal.

An example of how simple appropriate detailing with composites can be conceived is shown in **Fig. 3.162**, where the detailing of a concrete and a composite bridge kerb is compared. The composite example represents the detailing as executed in the Avançon Bridge (Appendix A.5). In both cases, the detailing should (i) protect the deck slab from moisture ingress through the surfacing, (ii) prevent stagnant water on the surfacing, and (iii) prevent water from running along the bottom deck surface to the underlying main girders and their deck connections.

A clear advantage of composites in bridge construction is that a waterproofing layer on deck slabs is not necessary, as it is on concrete slabs to protect the steel

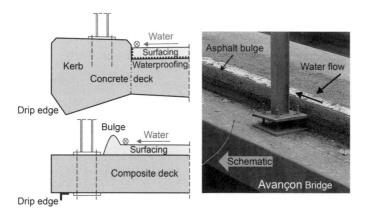

Fig. 3.162 Concrete vs composite bridge kerb, former exhibiting complex shape and requiring waterproofing layer **(top left)**; simple kerb detail of Avançon Bridge **(bottom left and right)**

reinforcement against corrosion. The advantage is even greater since the design service life of the waterproofing layer is much shorter than that of the bridge and it must thus be periodically replaced, which normally causes significant traffic disruptions. To prevent stagnant water on the surfacing, bridge decks normally have a transverse and a longitudinal slope. The transverse slope makes the water flow to one kerb side, where it is collected and then flows along a raised border in the longitudinal direction to a drainage point. This raised border, in the composite case, can be conceived as a simple bulge in the surfacing, while a complicated detail is required in the concrete case, since the waterproofing layer also must be anchored and sealed at this location. To prevent the water from running along the bottom deck surface, a simple L-shaped profile can be bonded to the composite deck to act as a drip edge. To conceive a drip edge in concrete is more complicated, and together with the raised border, a complex shape of the kerb results, while the composite slab section does not need any change to form the kerb.

Without a functional drip edge, the water could flow along the bottom surface down to the adhesive connection between the composite bridge deck and main girders, and penetrate into the adhesive layer, as sketched in **Fig. 3.163** (left). Based on experimental investigations using an epoxy adhesive and applying the Arrhenius extrapolation method, it was shown, however, that the penetration length of moisture into the adhesive layer, and associated length of strength reduction, are small compared to the width of such flange connections, i.e., approximately 50 mm after 100 years compared to a connection width of 400 mm, in the case of the Avançon Bridge, see Fig. 3.163 (right) [60]. Nevertheless, as also recommended in [1], exposed adhesive edges should always be sealed, as also schematically indicated in Fig. 3.163 (left).

A further sensitive detail concerns bolted connections through closed-section profiles, as shown in **Fig. 3.164**. Tightening of the bolts may exceed the intralaminar tensile strength of the profile walls and cause longitudinal cracking on the cell side, which may remain undetected and cause significant long-term damage. This potential

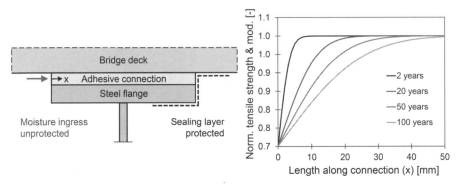

Fig. 3.163 Bridge deck-steel girder adhesive connection, moisture-unprotected and -protected **(left)**; retained normalised tensile strength and tension elastic modulus vs length of moisture ingress into epoxy adhesive layer if unprotected from 2 to 100 years **(right)**, based on [60]

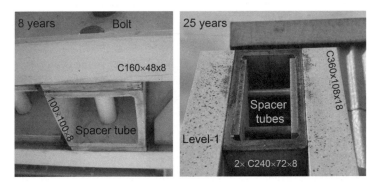

Fig. 3.164 Spacer tubes to prevent cracking of closed-section profiles due to bolt tightening: Pontresina Bridge after 8 years, view on bottom chord, [165] **(left)**; Eyecatcher Building after 25 years, joint of built-up profile on top of terrace **(right)**

damage can be prevented by implementing spacer tubes, against which the tightening forces are applied.

3.10.5 Conclusions

Durability, as discussed in this section, concerns irreversible damage caused by environmental conditions such as moisture, UV irradiation, freeze-thaw cycles, and combinations thereof. Results of long-term exposure to natural weathering demonstrate that climatic conditions have a significant effect on durability. The same unprotected composite profiles, i.e., pultruded by the same manufacturer in 1997 and 1998, exhibit very different damage patterns after being exposed to a continental subarctic or warm summer climate over a period of 25 years. In the former case, a protective coating would need to be applied after a few years to maintain material strength, while in the latter case, based on visual inspection, this seems to still be unnecessary after 25 years.

Detailing is one of the most effective ways to maintain resistance against moisture ingress, since water is effectively removed from the surfaces. Coatings can provide a similar protection; they may become ineffective through cracking and debonding, however, and depending on inspection periods, potential damage can remain undetected for many years. If appropriate maintenance is not performed, i.e., periodic renewal of the coating, the situation may become much more critical.

Applying a coating may not be necessary if only Level-2 fibre blooming occurs, as long as tactile discomfort or aesthetic impairment are not a problem. Deeper Level-3 fibre blooming, however, requires a coating to prevent the degradation of mechanical properties. Considering the design service life of coatings, which is normally in the range of 15–20 years, they should be renewed approximately three to six times during the 50–100-year design service life of composite buildings and bridges; the same applies for steel structures. Depending on the climate and exposure, composites may be left uncoated during the first period of 15–20 years, i.e., the cost of a first coating application may be saved if compared to steel for instance. It is unthinkable today, however, that composite structures can remain uncoated for 50–100 years.

On the material level, the selection of an appropriate resin system is important, but the implementation of surface veils, which effectively provide resistance against moisture ingress and UV irradiation, is as significant. Furthermore, cut profile ends, bolt-borehole surfaces, and exposed adhesive edges should always be sealed.

In contrast to the prevailing opinion in the late 1990s, periodic inspections and appropriate maintenance are required for composite structures to detect possible damage and perform rehabilitation in time. These inspections cannot be performed, however, as part of the same routine as for the examination of steel and concrete structures. Inspectors should be familiar with potential damage patterns in composite materials, adhesive joints, and coatings. Further instructions concerning inspections of composite structures can be found in [170].

3.11 SUSTAINABILITY

3.11.1 Introduction

Composites are often classified as not being sustainable due to their petroleum-based polymeric matrix and the difficulties regarding recycling. Petroleum, in this respect, is considered to be a non-renewable material. Surprisingly, this disqualification is not unusual among people who are involved in matters of sustainability, as I have witnessed on several occasions. When taking a closer and more considered look at the issue, however, it becomes clear that things are not as black and white as they are often presented – due to conclusions drawn prematurely (or to other opaque reasons). Based on my own perspective, I would thus like to emphasise the following points in this respect:

1) Petroleum is a renewable material created by natural processes over millions of years. Today's shortage of the material is caused by our society, which has been able to burn almost all of it, as fuel oil or gasoline, within less than one century.

2) Only an insignificant fraction of the petroleum is used for high-tech materials, including composites. Even if production were to increase tenfold, the fraction

would remain very small. Petroleum would still be an almost inexhaustible resource if used only for high-end applications.

3) Focusing on the petroleum-based matrix, the fact that glass and basalt fibres, which represent up to 60% of the composite volume, originate from quasi-inexhaustible natural resources is overlooked in most cases. Glass fibres, furthermore, can be produced from glass recycling.

4) The design service life of composite buildings and bridges is in the range of 50–100 years, which should be clearly differentiated from that of disposable plastics, which represent a large majority of the non-burnt petroleum-based materials. The embodied energy is conserved during this long period.

5) Linked to the previous point is the matter of recyclability. The shorter the service life and the higher the material quantities, the more significant the question of recycling. It is certainly important that, as an example, millions of only 10-year-old car parts can be recycled. However, if considering an 80-year-old bridge superstructure, and the benefits obtained from this infrastructure during this long period, the question of recycling seems to become less significant.

6) Sustainability indicators (e.g., CO_2 emissions or energy demands) are often compared for a material unit. However, to obtain a meaningful and fair comparison, the indicators should be weighted by the material, component, or system performance. Superior mechanical properties of composites, for instance, may reduce the required material quantity to achieve a desired performance, compared to a solution with conventional materials.

7) To appropriately assess the sustainability of composites, the entire life cycle, and not only the problematic production and end-of-life stages, should be taken into account, as the compensating advantages offered by composites most often occur during the construction and long use stages.

It would be advantageous for the discussion about the sustainability of composites to be more fact-based and better contextualised. Admittedly, the somewhat compromised reputation of composites, in this respect, has also been self-inflicted; a striking example are the millions of tonnes of waste caused by wind turbine rotor blades, after service lives of only about 20 years. A clear end-of-life concept should be developed for the future production of these blades, and the service life significantly increased.

In this section, the basic method of sustainability assessment will be discussed first, with regard to composites. Results of such assessments in bridge construction will then be summarised. Finally, bio-based composites will be briefly addressed, whose fibres or matrix, or both, are based on natural materials.

3.11.2 Sustainability assessment of composites

The term "sustainable development" was introduced as early as 1987, in the Brundtland Report [171], and defined as "development that meets the needs of the present without compromising the ability of future generations to meet their own needs". A three-pillar conception of sustainability has emerged over time, considering social, environmental, and economic aspects [172], which can be characterised and quantified by sets of specific indicators.

In the case of social aspects, indicators in bridge construction, as an example, may cover (i) work-zone safety on site, (ii) users' convenience (including traffic delays), and (iii) noise and dust emissions [173].

Environmental aspects are quantified through life cycle assessment (LCA). Environmental impacts are investigated across the different life cycle stages, from the extraction of raw materials to final disposal, or reuse. The typical stages considered in [174] are (i) product stage (also referred to as cradle-to-gate: raw material supply, transport, manufacturing), (ii) construction stage (or gate-to-completion: transport, installation), (iii) use stage (use, maintenance, repair, replacement, refurbishment, operational energy and water use), (iv) end-of-life stage (or grave: de-construction, transport, waste processing for reuse, recovery and/or recycling, waste disposal), and (v) beyond life cycle (reuse-, recovery-, recycling-potential). LCA follows a systematic approach dividing the analysis into four phases: (i) goal and scope (identifying system boundaries, functional unit, allocation procedures, and targeted environmental impacts), (ii) life cycle inventory, LCI (data collection, e.g., from data inventories), (iii) life cycle impact assessment, LCIA (score calculation), and (iv) interpretation [175]. The assessment can be performed for different impact categories and associated indicators, such as climate change (indicator example: global warming potential, quantified in [kg] CO_2 per functional unit), or resource use (indicator example: embodied energy, in [MJ] per functional unit).

Economic aspects can be quantified through life cycle costing (LCC). LCC calculations consider cash flows across the different stages. By discounting and summing them, the net present cost can be obtained and compared. Costs can be categorised into (i) initial, (ii) operation or energy consumption, (iii) maintenance and repair, (iv) replacement, and (v) end-of-life costs, including social costs (e.g., due to traffic delays), and environmental costs (e.g., energy consumption).

LCA and LCC methods are the subject of much criticism, in general, and in the application to composites in particular, as comprehensively discussed in a review in [176]. The main limitations are based on uncertainties in environmental and long-term cost input data, and, in addition, on the incompleteness of required data for composites. Most of the existing databases and LCA tools do not provide information for composites. If information is available, it mainly concerns the product stage, but the data is often not sufficiently broken down, e.g., to distinguish different manufacturing processes. To take the manufacturing process appropriately into account is important, as the data in **Table 3.11** demonstrates. Pultrusion, for instance, requires less than one third of the energy of vacuum infusion (according to this dataset). Furthermore, data from literature for energy demands and CO_2 emissions exhibit large scatter, as demonstrated by the values for composite constituent materials components, listed in **Table 3.12**; this scatter may significantly affect the results of LCA studies. Data about the stages where composites may present significant advantages, i.e., during construction and use (e.g., maintenance), are not contained in current databases and have to be collected case by case, or assumed based on engineering judgement, which makes such analyses subjective in many cases.

Table 3.11 Energy demand of manufacturing processes [177]

Manufacturing process	Energy demand [MJ/kg]
Pultrusion	3.1
Vacuum infusion	10.2
Prepreg production	40.0
Spray up	14.9
Filament winding	2.7

3.11.3 Results from case studies

A significant number of case studies has been performed on the sustainability of composite bridges, see reviews in [176, 178]. In most cases, bridges or bridge decks made of conventional and composite materials were designed for the same case, and the environmental impacts were then compared. After analysing these studies, the following conclusions can be drawn:

1) Most of the findings are in favour of composite solutions. However, the results are not always fully trustworthy, and comparisons of the results are difficult as the studies do not cover all life cycle stages. The end-of-life stage, in particular, is often missing or only partially considered. One of the reasons for this deficiency is the lack of reliable data.

2) In general, results are based on input data which are scarce and thus combined and somehow arbitrarily selected from different references and databases. Their uncertainty is high and the sensitivity of the results to these selections is often not discussed. An example is the assumption that the bridge closure required for the replacement of a concrete with a composite deck is only 15 hours [173]. The conclusion that the deck replacement is a sustainable solution largely depends on this assumption. Similarly, assumptions concerning required maintenance during the use stage are often subjective and significantly influence the results.

3) The possible economic savings are clearer than the environmental benefits.

4) The energy demands and CO_2 emissions of composites (particularly the resins) are high compared to other materials. However, since the material quantity of composites is often less, these higher values can be compensated over all life cycle stages.

5) Significant savings are always achieved at the construction stage.

6) The effects of traffic delay can be significant; for instance, they can account for (i) up to 75% of the total emissions during construction and maintenance stages of steel-concrete bridges [179], or (ii) up to 90% of the total bridge cost of concrete bridges [180]. Similar values can be expected for composite bridges.

7) Results depend on the assumed lifespan. The longer the lifespan, the more the results are favourable to the use of composites.

3. Composites – Overcoming Limitations

Table 3.12   Energy demand and CO_2 emissions of composite constituent materials components [13]

Raw material	Energy demand [MJ/kg]	CO_2 emissions [kg CO_2/kg]
Carbon fibre	183–1 150	24–31
Glass fibre	29–55	1.4–3.0
Basalt fibre	18	0.4
Epoxy	76–141	4.7–8.6
Polyester	63–128	2.3–7.6
PUR foam	67–75	3.0
Balsa wood	54	-3.2

8) The CO_2 emissions per unit of polymer concrete are more than eight times higher than those of asphalt [173] (regarding surfacing selection, see also Sections 2.6.2.3 and 3.8.8).

Concerning the critical end-of-life stage, two options exist for composite members: (i) disposal as waste or (ii) recycling. In the first option, members are disposed of in landfills or incinerated; in the latter, part of the embodied energy can be recovered during the combustion of the organic material components. Landfilling is increasingly prohibited, however, and incineration restricted due to air pollution. Most composite materials and members nevertheless still go the way of disposal as waste.

Recycling comprises different concepts, which are designated with different terms, such as "downcycling" or "upcycling", "reuse", or "repurposing". These terms partially overlap in meaning, but all are related to reuse, and the concepts can be applied at the material, structural component, or member level. Downcycling, related to composites, mainly signifies the shredding of components and members to small particles, which can be reused as filler materials in other applications. The initial material properties cannot be recovered. The separation of fibres and matrix and their reuse is theoretically possible if thermoplastic resins are used. This method has, however, no practical significance, although it is repeatedly emphasised.

Upcycling and repurposing have similar meanings and are also related to reuse. In this case, the material is not downgraded and keeps its initial or current properties, or members can be refurbished, if necessary, to have a second life for the same application or a different purpose. This type of recycling is currently strongly propagated and significant efforts are being undertaken in research. However, the potential success may be limited since a market for reusable components and members with qualified properties, which would be required, does not yet exist, and it seems unlikely that it will be built up anytime soon. The number of such additional life cycles also seems to be limited, and a third cycle probably not feasible. The end-of-life problematic is thus only delayed by one cycle and remains unsolved.

A further problem lies in the fact that structural members are currently designed for one specific purpose, as is impressively demonstrated by wind turbine rotor blades, whose design is optimised exclusively for transverse airflow and cantilever action and fixation. Their reuse for single-span beam structures, as intended for pedestrian bridges for instance [179], or smaller roofs, is strongly compromised by (i) the different boundary conditions, (ii) inappropriate shape (e.g., distribution of effective depth), and (iii) inappropriate fibre architecture, thus resulting in highly matrix-dominated responses caused by the new load configurations. A more appropriate and sustainable approach for reuse, consisting in the "design for reuse", is proposed and outlined in Section 6.3.

3.11.4 Bio-based composites

Many studies conclude that bio-based composites are more environmentally friendly than petroleum-based composites due to reduced energy demands, a reduction in damaging emissions, and bio-degradability [180]. The findings in favour of bio-based composites are even more pronounced where natural fibres are used instead of glass fibres [181]. However, most of these studies do not take into account long service lives of 50–100 years, as is the case for buildings and bridges, and associated durability under exposure to outdoor environmental conditions. The hydrophilic nature and associated significant water absorption and reduced freeze-thaw resistance, in addition to inferior creep and fatigue resistances of today's bio-based composites [180, 182], are significant obstacles for their use in primary bridge and building load-bearing structures.

Against this background of limited durability, the environmental benefits are also relativised, if conducting performance-based comparisons, as the following example demonstrates. The energy demand to produce glass and flax fibres is 31.7 vs 6.7 MJ/kg, according to [181]; glass fibre production thus requires about five times the energy of flax fibre production. Weighting this ratio by the structural performance, e.g., the elastic modulus, which is 70 GPa for glass and about 60 GPa for flax fibres, results in a reduction of the ratio from five to about four. This advantage of flax fibres has to be further corrected, however, due to their much shorter service life, caused by a reduced durability under outdoor conditions, e.g., for the construction of a pedestrian bridge. When assuming reasonable service lives of 60 and 15 years for glass and flax fibre composite bridges, respectively, the performance-based energy demand becomes equal. In other words, four flax fibre-based bridges have to be built to cover the service life of one glass fibre-based bridge. Flax fibres thus do not offer any relevant energy savings over glass fibres, if referring to the same performance in terms of stiffness and durability.

3.11.5 Conclusions

The most sustainable solution in bridge and building construction is maintaining existing structures. Structural strengthening is thus a very sustainable use of composites.

In new construction, the sustainability of using composites largely depends on the boundary conditions of a specific project and the material selection. A vehicular bridge built in an urban context is very likely a sustainable solution, while building the same bridge in the countryside is almost certainly less sustainable. Composites enable traffic delays to be minimised through prefabrication and rapid installation of lightweight components, which can positively influence the sustainability in urban environments.

A similar comparison can be made in building construction. Adding new storeys onto existing buildings to densify urban areas, favoured by function integration and modular lightweight construction offered by composites (see Section 4.7.3), can outperform conventional material solution in terms of sustainability. The same building units installed without taking advantage of lightweight and rapid installation may not, however, meet sustainability requirements.

Using basalt instead of glass or carbon fibres and developing designs that favour fibre- over matrix-dominated responses, thus reducing the material quantity, can also contribute to reducing energy use and CO_2 emissions.

Reducing the lifetime of bridge and building structures, as discussed in Section 4.2.3, would furthermore allow bio-based composites to be used and thus significant contributions to sustainability to be made.

Significant efforts must also be taken, however, to be able to quantify the advantages of composites in a trustworthy way. Data inventories for composites should be developed to the same level of detail as they are for conventional materials.

As demonstrated in this section, the disqualification of composites as being unsustainable is not justified; composites can be used in a sustainable way if the boundary conditions of the application, the material selection, and the design are appropriate.

3.12   REFERENCES

[1]   *CEN/TS 19101: Design of fibre-polymer composite structures*. European Committee for Standardization, Brussels, 2022.

[2]   S. Yanes-Armas, J. de Castro, and T. Keller, "Long-term design of FRP-PUR web-core sandwich structures in building construction", *Compos. Struct.*, vol. 181, pp. 214–228, 2017, doi: 10.1016/j.compstruct.2017.08.089.

[3]   M. Osei-Antwi, J. De Castro, A. P. Vassilopoulos, and T. Keller, "Shear mechanical characterization of balsa wood as core material of composite sandwich panels", *Constr. Build. Mater.*, vol. 41, pp. 231–238, 2013, doi: 10.1016/j.conbuildmat.2012.11.009.

[4]   M. Osei-Antwi, J. De Castro, A. P. Vassilopoulos, and T. Keller, "Fracture in complex balsa cores of fiber-reinforced polymer sandwich structures", *Constr. Build. Mater.*, vol. 71, pp. 194–201, 2014, doi: 10.1016/j.conbuildmat.2014.08.029.

[5]   T. Keller, J. Rothe, J. De Castro, and M. Osei-Antwi, "GFRP-balsa sandwich bridge deck: Concept, design, and experimental validation", *J. Compos. Constr.*, vol. 18, 2014, doi: 10.1061/(ASCE)CC.1943-5614.0000423.

[6]   C. Wu, N. Vahedi, A. P. Vassilopoulos, and T. Keller, "Mechanical properties of a balsa wood veneer structural sandwich core material", *Constr. Build. Mater.*, vol. 265, 2020, doi: 10.1016/j.conbuildmat.2020.120193.

[7]   U. Meier and M. Farshad, "Connecting high-performance Carbon-Fiber-Reinforced Polymer cables of suspension and cable-stayed bridges through the use of gradient materials", *J. Comput. Mater. Des.*, vol. 3, pp. 379–384, 1996, doi: 10.1007/BF01185676.

[8]   T. Keller, "Ground and rock anchor", Inventor, European patent EP2829661, 2017

[9]   H. Fan, "Structural performance of permanent post-tensioned CFRP ground anchors with strap ends", PhD thesis, EPFL, Lausanne, 2017. [Online]. Available: 10.5075/epfl-thesis-7923.

[10]   A. U. Winistoerfer, "Development of non-laminated advanced composite straps for civil engineering applications", PhD thesis, University of Warwick, 1999. [Online]. Available: https://wrap.warwick.ac.uk/109734/1/WRAP_Theses_Winistoerfer_1999.pdf.

[11]   H. Fan, A. P. Vassilopoulos, and T. Keller, "Pull-out behavior of CFRP ground anchors with two-strap ends", *Compos. Struct.*, vol. 160, pp. 1258–1267, 2017, doi: 10.1016/j.compstruct.2016.10.048.

[12]   H. Fan, A. P. Vassilopoulos, and T. Keller, "Load transfer mechanisms in CFRP ground anchors with multi-strap ends", *Compos. Struct.*, vol. 184, pp. 125–134, 2018, doi: 10.1016/j.compstruct.2017.09.111.

[13]   Original contribution.

[14]   T. Keller, "Reinforcement element for absorbing forces of concrete slabs in the area of support elements", Inventor, European patent EP2236686, 2010

[15]   T. Keller, A. Kenel, and R. Koppitz, "Carbon fiber-reinforced polymer punching reinforcement and strengthening of concrete flat slabs", *ACI Struct. J.*, vol. 110, pp. 919–927, 2013, doi: 10.14359/51686148.

[16]   R. Koppitz, "Effect of deformation history on punching resistance of reinforced concrete slabs", PhD thesis, EPFL, Lausanne, 2015. [Online]. Available: 10.5075/epfl-thesis-6472.

[17]   L. Liu, X. Wang, Z. Wu, and T. Keller, "Effect of fiber architecture on tension-tension fatigue behavior of bolted composite joints", *Eng. Struct.*, vol. 286, 2023, doi: 10.1016/j.engstruct.2023.116089

[18]   *prEN 1990-1: Basis of structural and geotechnical design – Part 1: New structures*. European Committee for Standardization, Brussels, 2023.

[19]   Y. Fu and X. Yao, "A review on manufacturing defects and their detection of fiber reinforced resin matrix composites", *Compos. Part C*, vol. 8, 2022, doi: 10.1016/j.jcomc.2022.100276.

[20]   J. L. Clarke (Ed.), *EUROCOMP design code and handbook: Structural design of polymer composites*. E&FN Spon, 1996.

[21]   M. R. Wisnom, M. Gigliotti, N. Ersoy, M. Campbell, and K. D. Potter, "Mechanisms generating residual stresses and distortion during manufacture of polymer-matrix composite structures", *Compos. Part A Appl. Sci. Manuf.*, vol. 37, pp. 522–529, 2006, doi: 10.1016/j.compositesa.2005.05.019.

[22]   I. Baran, C. C. Tutum, M. W. Nielsen, and J. H. Hattel, "Process induced residual stresses and distortions in pultrusion", *Compos. Part B Eng.*, vol. 51, pp. 148–161, 2013, doi: 10.1016/j.compositesb.2013.03.031.

[23]   *EN 13706-2: Reinforced plastics composites – Specifications for pultruded profiles – Part 2: Methods of test and general requirements*. European Committee for Standardization, Brussels, 2002.

[24]   P. Kulkarni, K. D. Mali, and S. Singh, "An overview of the formation of fibre waviness and its effect on the mechanical performance of fibre reinforced polymer composites", *Compos. Part A Appl. Sci. Manuf.*, vol. 137, 2020, doi: 10.1016/j.compositesa.2020.106013.

[25]   Y. Bai and T. Keller, "Shear failure of pultruded fiber-reinforced polymer composites under axial compression", *J. Compos. Constr.*, vol. 13, pp. 234–242, 2009, doi: 10.1061/(ASCE)CC.1943-5614.0000003.

[26]   M. Poulton and W. Sebastian, "Taxonomy of fibre mat misalignments in pultruded GFRP bridge decks", *Compos. Part A Appl. Sci. Manuf.*, vol. 142, 2021, doi: 10.1016/j.compositesa.2020.106239.

[27]   P. Feng, Y. Wu, and T. Q. Liu, "Non-uniform fiber-resin distributions of pultruded GFRP profiles", *Compos. Part B Eng.*, vol. 231, 2022, doi: 10.1016/j.compositesb.2021.109543.

3. Composites – Overcoming Limitations

[28] G. G. Cintra, D. C. T. Cardoso, J. D. Vieira, and T. Keller, "Experimental investigation on the moment-rotation performance of pultruded FRP web-flange junctions", *Compos. Part B Eng.*, vol. 222, 2021, doi: 10.1016/j.compositesb.2021.109087.

[29] C. M. Fame, J. Ramôa Correia, E. Ghafoori, and C. Wu, "Damage tolerance of adhesively bonded pultruded GFRP double-strap joints", *Compos. Struct.*, vol. 263, 2021, doi: 10.1016/j.compstruct.2021.113625.

[30] C. M. Fame, L. He, L. Tam, and C. Wu, "Fatigue damage tolerance of adhesively bonded pultruded GFRP double-strap joints with adhesion defects", *J. Compos. Constr.*, vol. 27, 2023, doi: 10.1061/jccof2.cceng-4015.

[31] J. Betten, *Creep mechanics*. Springer, Berlin, Heidelberg, New York, 2008.

[32] J. Vincent, *Structural biomaterials*, Third edit. Princeton University Press, Princeton, 2012. [Online]. Available: 10.1515/9781400842780

[33] R. Miranda Guedes, Ed., *Creep and fatigue in polymer matrix composites*. Woodhead Publishing, Cambridge, 2019.

[34] L. Liu, X. Wang, Z. Wu, and T. Keller, "Tension-tension fatigue behavior of ductile adhesively-bonded FRP joints", *Compos. Struct.*, vol. 268, doi: 10.1016/j.compstruct.2021.113925.

[35] G. Eslami, A. V. Movahedi-Rad, and T. Keller, "Viscoelastic adhesive modeling of ductile adhesive-composite joints during cyclic loading", *Int. J. Adhes. Adhes.*, vol. 119, 2022, doi: 10.1016/j.ijadhadh.2022.103241.

[36] K. P. Menard and N. R. Menard, *Dynamic mechanical analysis*, Third edit. CRC Press, 2020.

[37] *ISO 6721-11: Plastics – Determination of dynamic mechanical properties – Part 11: Glass transition temperature*. International Organization for Standardization, Geneva, 2019.

[38] *ASTM D7028-07(2015). Standard test method for glass transition temperature (DMA Tg) of polymer matrix composites by dynamic mechanical analysis (DMA)*. ASTM American Society for Testing and Materials, West Conshohocken, 2015.

[39] O. Moussa, A. P. Vassilopoulos, J. De Castro, and T. Keller, "Time-temperature dependence of thermomechanical recovery of cold-curing structural adhesives", *Int. J. Adhes. Adhes.*, vol. 35, pp. 94–101, 2012, doi: 10.1016/j.ijadhadh.2012.02.005.

[40] J. R. Correia, T. Keller, J. Knippers, J. T. Mottram, C. Paulotto, J. Sena-Cruz, and L. Ascione, *Design of fibre-polymer composite structures: Commentary to European Technical Specification CEN/TS 19101:2022*. CRC Press, London, 2025.

[41] Y. Bai, N. L. Post, J. J. Lesko, and T. Keller, "Experimental investigations on temperature-dependent thermo-physical and mechanical properties of pultruded GFRP composites", *Thermochim. Acta*, vol. 469, pp. 28–35, 2008, doi: 10.1016/j.tca.2008.01.002.

[42] W. Sun, A. P. Vassilopoulos, and T. Keller, "Effect of thermal lag on glass transition temperature of polymers measured by DMA", *Int. J. Adhes. Adhes.*, vol. 52, 2014, doi: 10.1016/j.ijadhadh.2014.03.009.

[43] M. Savvilotidou, A. P. Vassilopoulos, M. Frigione, and T. Keller, "Effects of aging in dry environment on physical and mechanical properties of a cold-curing structural epoxy adhesive for bridge construction", *Constr. Build. Mater.*, vol. 140, pp. 552–561, 2017, doi: 10.1016/j.conbuildmat.2017.02.063.

[44] M. Angelidi, A. P. Vassilopoulos, and T. Keller, "Ductility, recovery and strain rate dependency of an acrylic structural adhesive," *Constr. Build. Mater.*, vol. 140, 2017, doi: 10.1016/j.conbuildmat.2017.02.101.

[45] G. Eslami, A. V. Movahedi-rad, and T. Keller, "Experimental investigation of rotational behavior of pseudo-ductile adhesive angle joints exhibiting variable strain rate", *Compos. Part B*, vol. 266, 2023, doi: 10.1016/j.compositesb.2023.111009.

[46] A. V. Movahedi-Rad, T. Keller, and A. P. Vassilopoulos, "Creep effects on tension-tension fatigue behavior of angle-ply GFRP composite laminates", *Int. J. Fatigue*, vol. 123, pp. 144–156, 2019, doi: 10.1016/j.ijfatigue.2019.02.010.

[47] B. Benmokrane, V. L. Brown, K. Mohamed, A. Nanni, M. Rossini, and C. Shield, "Creep-rupture limit for GFRP bars subjected to sustained loads", *J. Compos. Constr.*, vol. 23, pp. 1–7, 2019, doi: 10.1061/(asce)cc.1943-5614.0000971.

[48] S. Yanes-Armas and T. Keller, "Structural concept and design of a GFRP-polyurethane sandwich roof structure", in *Proceedings of the Eighth International Conference on Fibre-Reinforced Polymer (FRP) Composites in Civil Engineering (CICE2016)*, 2016, pp. 1318–1323.

[49] S. Yanes-Armas, J. de Castro, and T. Keller, "Energy dissipation and recovery in web-flange junctions of pultruded GFRP decks", *Compos. Struct.*, vol. 148, pp. 168–180, 2016, doi: 10.1016/j.compstruct.2016.03.042.

[50] A. V. Movahedi-Rad, T. Keller, and A. P. Vassilopoulos, "Modeling of fatigue behavior based on interaction between time- and cyclic-dependent mechanical properties", *Compos. Part A Appl. Sci. Manuf.*, vol. 124, 2019, doi: 10.1016/j.compositesa.2019.05.037.

[51] A. V. Movahedi-Rad, T. Keller, and A. P. Vassilopoulos, "Interrupted tension-tension fatigue behavior of angle-ply GFRP composite laminates", *Int. J. Fatigue*, vol. 113, pp. 377–388, 2018, doi: 10.1016/j.ijfatigue.2018.05.001.

[52] J. Zhao, K. Mei, and J. Wu, "Long-term mechanical properties of FRP tendon-anchor systems – A review", *Constr. Build. Mater.*, vol. 230, p. 117017, 2020, doi: 10.1016/j.conbuildmat.2019.117017.

[53] T. Keller, "Strengthening of concrete bridges with carbon cables and strips", in *6th International Symposium on FRP Reinforcement for Concrete Structures (FRPRCS-6)*, Singapore, 2003.

[54] By courtesy of Prof. Urs Meier, EMPA, Switzerland, 25 April 2024.

[55] M. Savvilotidou, A. P. Vassilopoulos, M. Frigione, and T. Keller, "Development of physical and mechanical properties of a cold-curing structural adhesive in a wet bridge environment", *Constr. Build. Mater.*, vol. 144, pp. 115–124, 2017, doi: 10.1016/j.conbuildmat.2017.03.145.

[56] O. Moussa, A. P. Vassilopoulos, and T. Keller, "Effects of low-temperature curing on physical behavior of cold-curing epoxy adhesives in bridge construction", *Int. J. Adhes. Adhes.*, vol. 32, pp. 15–22, 2012, doi: 10.1016/j.ijadhadh.2011.09.001.

[57] O. Moussa, A. P. Vassilopoulos, J. De Castro, and T. Keller, "Early-age tensile properties of structural epoxy adhesives subjected to low-temperature curing", *Int. J. Adhes. Adhes.*, vol. 35, pp. 9–16, 2012, doi: 10.1016/j.ijadhadh.2012.01.023.

[58] Unpublished CCLab data.

[59] O. Moussa, "Thermophysical and thermomechanical behavior of cold-curing structuraladhesives in bridge construction", PhD thesis, EPFL, Lausanne, 2011. [Online]. Available: 10.5075/epfl-thesis-5244.

[60] M. Savvilotidou, "Durability and fatigue performance of a typical cold-curing structural adhesive in bridge construction", PhD thesis, EPFL, Lausanne, 2017. [Online]. Available: 10.5075/epfl-thesis-7935.

[61] *EN ISO 11357-2. Plastics – Differential Scanning Calorimetry (DSC) – Part 2: Determination of glass transition temperature and step height.* Association Française de Normalisation (AFNOR) 93571 La Plaine Saint-Denis Cedex, France, 2020.

[62] G. M. Odegard and A. Bandyopadhyay, "Physical aging of epoxy polymers and their composites", *J. Polym. Sci. Part B Polym. Phys.*, vol. 49, pp. 1695–1716, 2011, doi: 10.1002/polb.22384.

3. Composites – Overcoming Limitations

[63] O. Moussa, A. P. Vassilopoulos, and T. Keller, "Experimental DSC-based method to determine glass transition temperature during curing of structural adhesives", *Constr. Build. Mater.*, vol. 28, pp. 263–268, 2012, doi: 10.1016/j.conbuildmat.2011.08.059.

[64] O. Moussa, A. P. Vassilopoulos, J. De Castro, and T. Keller, "Long-term development of thermo-physical and mechanical properties of cold-curing structural adhesives due to post-curing", *J. Appl. Polym. Sci.*, vol. 127, 2013, doi: 10.1002/app.37965.

[65] T. Keller, "Conceptual design of hybrid-FRP and all-FRP structures," in *Advanced Composite Materials in Bridges and Structures (ACMBS-IV)*, Calgary, 2004.

[66] T. Keller and J. De Castro, "System ductility and redundancy of FRP beam structures with ductile adhesive joints", *Compos. Part B Eng.*, vol. 36, pp. 586–596, 2005, doi: 10.1016/j.compositesb.2005.05.001.

[67] A. E. Naaman, M. H. Harajli, and J. K. Wright, "Analysis of ductility in partially prestressed concrete flexural members", *J. – Prestress. Concr. Inst.*, vol. 31, pp. 64–87, 1986, doi: 10.15554/pcij.05011986.64.87.

[68] J. De Castro and T. Keller, "Ductile double-lap joints from brittle GFRP laminates and ductile adhesives, Part I: Experimental investigation", *Compos. Part B Eng.*, vol. 39, pp. 271–281, 2008, doi: 10.1016/j.compositesb.2007.02.015.

[69] M. Garrido, J. R. Correia, and T. Keller, "Effects of elevated temperature on the shear response of PET and PUR foams used in composite sandwich panels", *Constr. Build. Mater.*, vol. 76, pp. 150–157, 2015, doi: 10.1016/j.conbuildmat.2014.11.053.

[70] T. Keller, *Use of fibre reinforced polymers in bridge construction.* Structural Engineering Documents, SED 7, International Association for Bridge and Structural Engineering, IABSE, Zurich, 2003.

[71] T. Keller and H. Gürtler, "In-plane compression and shear performance of FRP bridge decks acting as top chord of bridge girders", *Compos. Struct.*, vol. 72, pp. 151–162, 2006, doi: 10.1016/j.compstruct.2004.11.004.

[72] T. Keller and H. Gürtler, "Quasi-static and fatigue performance of a cellular FRP bridge deck adhesively bonded to steel girders", *Compos. Struct.*, vol. 70, pp. 484–496, 2005, doi: 10.1016/j.compstruct.2004.09.028.

[73] H. W. Gürtler, "Composite action of FRP bridge decks adhesively bonded to steel main girders", PhD thesis, EPFL, Lausanne, 2005. [Online]. Available: 10.5075/epfl-thesis-3135.

[74] L. Liu, X. Wang, Z. Wu, and T. Keller, "Resistance and ductility of FRP composite hybrid joints", *Compos. Struct.*, vol. 255, 2021, doi: 10.1016/j.compstruct.2020.113001.

[75] J. de Castro San Román, "System ductility and redundancy of FRP structures with ductile adhesively-bonded joints", PhD thesis, EPFL, Lausanne, 2005. [Online]. Available: 10.5075/epfl-thesis-3214.

[76] J. E. Gordon, *Structures or why things don't fall down.* Plenum Press, New York, 1978.

[77] T. L. Andersen, *Fracture mechanics – fundamentals and applications*, 4th ed. CRC Press, 2017.

[78] M. Shahverdi, A. P. Vassilopoulos, and T. Keller, "Mixed-mode quasi-static failure criteria for adhesively-bonded pultruded GFRP joints", *Compos. Part A Appl. Sci. Manuf.*, vol. 59, pp. 45–56, 2014, doi: 10.1016/j.compositesa.2013.12.007.

[79] R. El-Hajjar and R. Haj-Ali, "Mode-I fracture toughness testing of thick section FRP composites using the ESE(T) specimen", *Eng. Fract. Mech.*, vol. 72, pp. 631–643, 2005, doi: 10.1016/j.engfracmech.2004.03.013.

[80] L. Almeida-Fernandes, J. R. Correia, and N. Silvestre, "Transverse fracture behavior of pultruded GFRP materials in tension: Effect of Fiber Layup", *J. Compos. Constr.*, vol. 24, no. 4, pp. 1–17, 2020, doi: 10.1061/(asce)cc.1943-5614.0001024.

[81] W. Liu, P. Feng, and J. Huang, "Bilinear softening model and double K fracture criterion for quasi-brittle fracture of pultruded FRP composites", *Compos. Struct.*, vol. 160, pp. 1119–1125, 2017, doi: 10.1016/j.compstruct.2016.10.134.

[82] H. Zhao, X. Liu, W. Zhao, G. Wang, and B. Liu, "An overview of research on FDM 3D printing process of continuous fiber reinforced composites", *J. Phys. Conf. Ser.*, vol. 1213, 2019, doi: 10.1088/1742-6596/1213/5/052037.

[83] Y. Zhang, "Fracture and fatigue of adhesively-bonded fiberreinforced polymer structural joints", PhD thesis, EPFL, Lausanne, 2010. [Online]. Available: 10.5075/epfl-thesis-4662.

[84] M. Shahverdi, "Mixed-mode static and fatigue failure criteria for adhesively-bonded FRP joints", PhD thesis, EPFL, Lausanne, 2013. [Online]. Available: 10.5075/epfl-thesis-5728.

[85] A. Cameselle Molares, "Two-dimensional crack growth in FRP laminates and sandwich panels" PhD thesis, EPFL, Lausanne, 2019. [Online]. Available: 10.5075/epfl-thesis-9421.

[86] C. Wang, "Two-dimensional quasi-static delamination in composite laminates under Mode-I and Mode-II conditions", PhD thesis, EPFL, Lausanne, 2024. [Online]. Available: 10.5075/epfl-thesis-10535.

[87] Y. Zhang, A. P. Vassilopoulos, and T. Keller, "Mode I and II fracture behavior of adhesively-bonded pultruded composite joints", *Eng. Fract. Mech.*, vol. 77, pp. 128–143, 2010, doi: 10.1016/j.engfracmech.2009.09.015.

[88] A. Cameselle-Molares, A. P. Vassilopoulos, J. Renart, A. Turon, and T. Keller, "Numerical simulation of two-dimensional in-plane crack propagation in FRP laminates", *Compos. Struct.*, vol. 200, 2018, doi: 10.1016/j.compstruct.2018.05.136.

[89] G. A. Pappas and J. Botsis, "Towards a geometry independent traction-separation and angle relation due to large scale bridging in DCB configuration", *Compos. Sci. Technol.*, vol. 197, 2020, doi: 10.1016/j.compscitech.2020.108172.

[90] G. G. Cintra, J. D. Vieira, D. C. T. Cardoso, and T. Keller, "Mode I and Mode II fracture behavior in pultruded glass fiber-polymer – Experimental and numerical investigation", *Compos. Part B Eng.*, vol. 266, 2023, doi: 10.1016/j.compositesb.2023.110988.

[91] R. Sarfaraz, A. P. Vassilopoulos, and T. Keller, "Experimental investigation of the fatigue behavior of adhesively-bonded pultruded GFRP joints under different load ratios", *Int. J. Fatigue*, vol. 33, pp. 1451–1460, 2011, doi: 10.1016/j.ijfatigue.2011.05.012.

[92] M. Shahverdi, A. P. Vassilopoulos, and T. Keller, "Modeling effects of asymmetry and fiber bridging on Mode I fracture behavior of bonded pultruded composite joints", *Eng. Fract. Mech.*, vol. 99, 2013, doi: 10.1016/j.engfracmech.2013.02.001.

[93] C. Wang, A. P. Vassilopoulos, and T. Keller, "Experimental investigation of two-dimensional Mode-II delamination in composite laminates", *Compos. Part A Appl. Sci. Manuf.*, vol. 173, 2023, doi: 10.1016/j.compositesa.2023.107666.

[94] M. Shahverdi, A. P. Vassilopoulos, and T. Keller, "A phenomenological analysis of Mode I fracture of adhesively-bonded pultruded GFRP joints", *Eng. Fract. Mech.*, vol. 78, 2011, doi: 10.1016/j.engfracmech.2011.04.007.

[95] C. Wang, A. P. Vassilopoulos, and T. Keller, "Numerical investigation of two-dimensional Mode-II delamination in composite laminates", *Compos. Part A*, vol. 179, 2024, doi: 10.1016/j.compositesa.2024.108012.

[96] A. Cameselle-Molares, A. P. Vassilopoulos, and T. Keller, "Experimental investigation of two-dimensional delamination in GFRP laminates", *Eng. Fract. Mech.*, vol. 203, 2018, doi: 10.1016/j.engfracmech.2018.05.015.

[97] A. Cameselle-Molares, A. P. Vassilopoulos, and T. Keller, "Two-dimensional quasi-static debonding in GFRP/balsa sandwich panels", *Compos. Struct.*, vol. 215, pp. 391–401, 2019, doi: 10.1016/j.compstruct.2019.02.077.

3. Composites – Overcoming Limitations

[98] C. Wang, A. P. Vassilopoulos, and T. Keller, "Numerical modeling of two-dimensional delamination growth in composite laminates with in-plane isotropy", *Eng. Fract. Mech.*, vol. 250, 2021, doi: 10.1016/j.engfracmech.2021.107787.

[99] A. Cameselle-Molares, A. P. Vassilopoulos, J. Renart, A. Turon, and T. Keller, "Numerically-based method for fracture characterization of Mode I-dominated two-dimensional delamination in FRP laminates", *Compos. Struct.*, vol. 214, 2019, doi: 10.1016/j.compstruct.2019.02.014.

[100] A. Cameselle-Molares, A. P. Vassilopoulos, and T. Keller, "Two-dimensional fatigue debonding in GFRP/balsa sandwich panels", *Int. J. Fatigue*, vol. 125, pp. 72–84, 2019, doi: 10.1016/j.ijfatigue.2019.03.032.

[101] M. Shahverdi, A. P. Vassilopoulos, and T. Keller, "Experimental investigation of R-ratio effects on fatigue crack growth of adhesively-bonded pultruded GFRP DCB joints under CA loading", *Compos. Part A Appl. Sci. Manuf.*, vol. 43, 2012, doi: 10.1016/j.compositesa.2011.10.018.

[102] C. Wang, A. P. Vassilopoulos, and T. Keller, "Experimental investigation of two-dimensional Mode-II delamination in composite laminates", *Compos. Part A*, vol. 173, 2023, doi: 10.1016/j.compositesa.2023.107666.

[103] G. G. Cintra, J. D. Vieira, D. C. T. Cardoso, and T. Keller, "Novel multi-crack damage approach for pultruded fiber-polymer web-flange junctions", *Compos. Part B Eng.*, vol. 269, no. July 2023, p. 111102, 2024, doi: 10.1016/j.compositesb.2023.111102.

[104] M. J. Laffan, S. T. Pinho, P. Robinson, and L. Iannucci, "Measurement of the in situ ply fracture toughness associated with mode I fibre tensile failure in FRP. Part I: Data reduction", *Compos. Sci. Technol.*, vol. 70, pp. 606–613, 2010, doi: 10.1016/j.compscitech.2009.12.016.

[105] R. H. Martin, *Delamination failure in a unidirectional curved composite laminate.* Nasa Contractor Report 182018, Hampton, Virginia, 1990.

[106] *prEN 1991-2: Actions on structures – Part 2: Traffic loads on bridges.* European Committee for Standardization, Brussels, 2021.

[107] R. Sarfaraz Khabbaz, "Fatigue life prediction of adhesively-bonded fiber-reinforced polymer structural joints under spectrum loading patterns", PhD thesis, EPFL, Lausanne, 2012. [Online]. Available: 10.5075/epfl-thesis-5573.

[108] A. Movahedirad, "Load history effects in fibre-reinforced polymer composite materials", PhD thesis, EPFL, Lausanne, 2019. [Online]. Available: 10.5075/epfl-thesis-9384.

[109] A. V. Movahedi-Rad, T. Keller, and A. P. Vassilopoulos, "Fatigue damage in angle-ply GFRP laminates under tension-tension fatigue", *Int. J. Fatigue*, vol. 109, 2018, doi: 10.1016/j.ijfatigue.2017.12.015.

[110] *EN 1993-1-2: Design of steel structures – Part 1–9: Fatigue.* European Committee for Standardization, Brussels, 2005.

[111] T. Keller, T. Tirelli, and A. Zhou, "Tensile fatigue performance of pultruded glass fiber reinforced polymer profiles", *Compos. Struct.*, vol. 68, 2005, doi: 10.1016/j.compstruct.2004.03.021.

[112] T. Keller and T. Tirelli, "Fatigue behavior of adhesively connected pultruded GFRP profiles", *Compos. Struct.*, vol. 65, 2004, doi: 10.1016/j.compstruct.2003.10.008.

[113] R. Sarfaraz, A. P. Vassilopoulos, and T. Keller, "Experimental investigation and modeling of mean load effect on fatigue behavior of adhesively-bonded pultruded GFRP joints", *Int. J. Fatigue*, vol. 44, pp. 245–252, 2012, doi: 10.1016/j.ijfatigue.2012.04.021.

[114] A. V. Movahedi-Rad, G. Eslami, and T. Keller, "A novel fatigue life prediction methodology based on energy dissipation in viscoelastic materials", *Int. J. Fatigue*, vol. 152, 2021, doi: 10.1016/j.ijfatigue.2021.106457.

[115] *Données concernant le trafic – saisie du poids (WIM) (unpublished data).* Office Fédéral des Routes, Suisse, 2019.

[116] R. Sarfaraz, A. P. Vassilopoulos, and T. Keller, "Block loading fatigue of adhesively bonded pultruded GFRP joints", *Int. J. Fatigue*, vol. 49, 2013, doi: 10.1016/j.ijfatigue.2012.12.006.

[117] R. Sarfaraz, A. P. Vassilopoulos, and T. Keller, "Variable amplitude fatigue of adhesively-bonded pultruded GFRP joints", *Int. J. Fatigue*, vol. 55, 2013, doi: 10.1016/j.ijfatigue.2013.04.024.

[118] Y. Zhang, A. P. Vassilopoulos, and T. Keller, "Environmental effects on fatigue behavior of adhesively-bonded pultruded structural joints", *Compos. Sci. Technol.*, vol. 69, pp. 1022–1028, 2009, doi: 10.1016/j.compscitech.2009.01.024.

[119] R. Sarfaraz, A. P. Vassilopoulos, and T. Keller, "A hybrid S-N formulation for fatigue life modeling of composite materials and structures", *Compos. Part A Appl. Sci. Manuf.*, vol. 43, 2012, doi: 10.1016/j.compositesa.2011.11.008.

[120] M. Shahverdi, A. P. Vassilopoulos, and T. Keller, "Mixed-mode fatigue failure criteria for adhesively-bonded pultruded GFRP joints", *Compos. Part A Appl. Sci. Manuf.*, vol. 54, pp. 46–55, 2013, doi: 10.1016/j.compositesa.2013.06.017.

[121] M. Osei-Antwi, J. de Castro, A. P. Vassilopoulos, and T. Keller, "Analytical modeling of local stresses at Balsa/timber core joints of FRP sandwich structures", *Compos. Struct.*, vol. 116, pp. 501–508, 2014, doi: 10.1016/j.compstruct.2014.05.050.

[122] P. K. Majumdar, J. J. Lesko, T. E. Cousins, and Z. Liu, "Conformable tire patch loading for FRP composite bridge deck", vol. 13, pp. 575–581, 2009, doi: 10.1061/(ASCE)CC.1943-5614.0000033.

[123] W. M. Sebastian, M. Ralph, M. Poulton, and J. Goacher, "Lab and field studies into effectiveness of flat steel plate – rubber pad systems as tyre substitutes for local loading of cellular GFRP bridge decking", *Compos. Part B Eng.*, vol. 125, pp. 100–122, 2017, doi: 10.1016/j.compositesb.2017.05.044.

[124] W. Sebastian, B. Dodds, and C. Benner, "Commentary: Restoring the West Mill GFRP deck road bridge to full capacity", *Proc. Inst. Civ. Eng. Struct. Build.*, vol. 173, pp. 158–160, 2020, doi: 10.1680/jstbu.2020.173.3.158.

[125] L. Liu, J. Maljaars, and T. Keller, "Towards a fatigue load model for composite road bridges," *Under Rev.*.

[126] *EN 1991-1-2: Actions on structures – Part 1–2: General actions – Actions on structures exposed to fire*. European Committee for Standardization, Brussels, 2002.

[127] Y. Bai, T. Vallée, and T. Keller, "Modeling of thermo-physical properties for FRP composites under elevated and high temperature", *Compos. Sci. Technol.*, vol. 67, pp. 3098–3109, 2007, doi: 10.1016/j.compscitech.2007.04.019.

[128] Y. I. Dimitrienko, *Thermomechanics of composites under high temperatures*. Springer, Dordrecht, 1999.

[129] J. R. Correia, Y. Bai, and T. Keller, "A review of the fire behaviour of pultruded GFRP structural profiles for civil engineering applications", *Compos. Struct.*, vol. 127, pp. 267–287, 2015, doi: 10.1016/j.compstruct.2015.03.006.

[130] C. D. Tracy, "Fire endurance of multicellular panels in an FRP building system", PhD thesis, EPFL, Lausanne, 2005. [Online]. Available: 10.5075/epfl-thesis-3235.

[131] Y. Bai, "Material and structural performance of fiber-reinforced polymer composites at elevated and high temperatures," PhD thesis, EPFL, Lausanne, 2009. [Online]. Available: 10.5075/epfl-thesis-4340.

[132] W. Sun, "Temperature effects on material properties and structural response of polymer matrix composites," PhD thesis EPFL, Lausanne, 2015. [Online]. Available: 10.5075/epfl-thesis-6674.

[133] N. Vahedi, "Fire performance evaluation of GFRP-balsa sandwich bridge decks," PhD thesis, EPFL, Lausanne, 2022. [Online]. Available: 10.5075/epfl-thesis-9364.

[134] T. Keller, C. Tracy, and A. Zhou, "Structural response of liquid-cooled GFRP slabs subjected to fire – Part I: Material and post-fire modeling", *Compos. Part A Appl. Sci. Manuf.*, vol. 37, pp. 1286–1295, 2006, doi: 10.1016/j.compositesa.2005.08.006.

3. Composites – Overcoming Limitations

[135] Y. Bai and T. Keller, "Time dependence of material properties of frp composites in fire", *J. Compos. Mater.*, vol. 43, pp. 2469–2484, 2009, doi: 10.1177/0021998309344641.

[136] Y. Bai, T. Keller, and T. Vallée, "Modeling of stiffness of FRP composites under elevated and high temperatures", *Compos. Sci. Technol.*, vol. 68, pp. 3099–3106, 2008, doi: 10.1016/j.compscitech.2008.07.005.

[137] Y. Bai and T. Keller, "Modeling of strength degradation for fiber-reinforced polymer composites in fire", *J. Compos. Mater.*, vol. 43, pp. 2371–2385, 2009, doi: 10.1177/0021998309344642.

[138] N. Vahedi, C. Tiago, A. P. Vassilopoulos, J. R. Correia, and T. Keller, "Thermophysical properties of balsa wood used as core of sandwich composite bridge decks exposed to external fire", *Constr. Build. Mater.*, vol. 329, 2022, doi: 10.1016/j.conbuildmat.2022.127164.

[139] N. Vahedi, J. R. Correia, A. P. Vassilopoulos, and T. Keller, "Effects of core air gaps and steel inserts on thermal response of GFRP-balsa sandwich panels subjected to fire", *Fire Saf. J.*, vol. 134, 2022, doi: 10.1016/j.firesaf.2022.103703.

[140] N. Vahedi, J. R. Correia, A. P. Vassilopoulos, and T. Keller, "Effects of core air gaps and steel inserts on thermomechanical response of GFRP-balsa sandwich panels subjected to fire", *Compos. Struct.*, vol. 313, 2023, doi: 10.1016/j.compstruct.2023.116924.

[141] Y. Bai, T. Vallée, and T. Keller, "Modeling of thermal responses for FRP composites under elevated and high temperatures", *Compos. Sci. Technol.*, vol. 68, pp. 47–56, 2008, doi: 10.1016/j.compscitech.2007.05.039.

[142] N. L. Post, F. Riebel, A. Zhou, T. Keller, S. W. Case, and J. J. Lesko, "Investigation of 3D moisture diffusion coefficients and damage in a pultruded E-glass/Polyester structural composite", *J. Compos. Mater.*, vol. 43, pp. 75–96, 2009, doi: 10.1177/0021998308098152.

[143] L. Tranvan, V. Legrand, and F. Jacquemin, "Thermal decomposition kinetics of balsa wood: Kinetics and degradation mechanisms comparison between dry and moisturized materials", *Polym. Degrad. Stab.*, vol. 110, pp. 208–215, 2014, doi: 10.1016/j.polymdegradstab.2014.09.004.

[144] N. Vahedi, C. Wu, A. P. Vassilopoulos, and T. Keller, "Thermomechanical characterization of a balsa-wood-veneer structural sandwich core material at elevated temperatures", *Constr. Build. Mater.*, vol. 230, 2020, doi: 10.1016/j.conbuildmat.2019.117037.

[145] L. Yang and J. L. Thomason, "The thermal behaviour of glass fibre investigated by thermomechanical analysis", *J. Mater. Sci.*, vol. 48, pp. 5768–5775, 2013, doi: 10.1007/s10853-013-7369-7.

[146] I. C. Rosa, J. P. Firmo, and J. R. Correia, "Experimental study of the tensile behaviour of GFRP reinforcing bars at elevated temperatures", *Constr. Build. Mater.*, vol. 324, 2022, doi: 10.1016/j.conbuildmat.2022.126676.

[147] *EN 1993-1-2: Design of steel structures – Part 1–2: General rules – Structural fire design*. European Committee for Standardization, Brussels, 2005.

[148] F. T. Wallenberger, P. A. Bingham (Editors), *Fiberglass and glass technology*. Springer, New York, 2010.

[149] W. Sun, A. P. Vassilopoulos, and T. Keller, "Effect of temperature on kinking failure mode of non-slender glass fiber-reinforced polymer specimens", *Compos. Struct.*, vol. 133, pp. 178–190, 2015, doi: 10.1016/j.compstruct.2015.07.054.

[150] Y. Bai and T. Keller, "Modeling of mechanical response of FRP composites in fire", *Compos. Part A Appl. Sci. Manuf.*, vol. 40, pp. 731–738, 2009, doi: 10.1016/j.compositesa.2009.03.003.

[151] W. Sun, A. P. Vassilopoulos, and T. Keller, "Experimental investigation of kink initiation and kink band formation in unidirectional glass fiber-reinforced polymer specimens", *Compos. Struct.*, vol. 130, pp. 9–17, 2015, doi: 10.1016/j.compstruct.2015.04.028.

[152] T. Goodrich, N. Nawaz, S. Feih, B. Y. Lattimer, and A. P. Mouritz, "High-temperature mechanical properties and thermal recovery of balsa wood", *J. Wood Sci.*, vol. 56, pp. 437–443, 2010, doi: 10.1007/s10086-010-1125-2.

[153] *EN 1363-1. Fire resistance tests – Part 1: general requirements*. European Committee for Standardization, Brussels, 2020.

[154] C. Ludwig, J. Knippers, E. Hugi, and K. Wakili, "Thermal and thermo-mechanical investigation of polyester based composite beams", in *Proceedings of the fourth international conference on the response of composite materials to fire*, Newcastle upon Tyne, 2008.

[155] T. Morgado, J. R. Correia, N. Silvestre, and F. A. Branco, "Experimental study on the fire resistance of GFRP pultruded tubular beams", *Compos. Part B Eng.*, vol. 139, 2017, pp. 106–116, 2018, doi: 10.1016/j.compositesb.2017.11.036.

[156] T. Keller, C. Tracy, and E. Hugi, "Fire endurance of loaded and liquid-cooled GFRP slabs for construction", *Compos. Part A Appl. Sci. Manuf.*, vol. 37, pp. 1055–1067, 2006, doi: 10.1016/j.compositesa.2005.03.030.

[157] L. Zhang, Y. Dai, Y. Bai, W. Chen, and J. Ye, "Fire performance of loaded fibre reinforced polymer multicellular composite structures with fire-resistant panels", *Constr. Build. Mater.*, vol. 296, p. 123733, 2021, doi: 10.1016/j.conbuildmat.2021.123733.

[158] M. Proença, M. Garrido, J. R. Correia, and M. G. Gomes, "Fire resistance behaviour of GFRP-polyurethane composite sandwich panels for building floors", *Compos. Part B Eng.*, vol. 224, 2021, doi: 10.1016/j.compositesb.2021.109171.

[159] T. Morgado, J. R. Correia, A. Moreira, F. A. Branco, and C. Tiago, "Experimental study on the fire resistance of GFRP pultruded tubular columns", *Compos. Part B Eng.*, vol. 69, pp. 201–211, 2015, doi: 10.1016/j.compositesb.2014.10.005.

[160] Y. Bai, E. Hugi, C. Ludwig, and T. Keller, "Fire performance of water-cooled GFRP columns. I: Fire endurance investigation", *J. Compos. Constr.*, vol. 15, pp. 404–412, 2011, doi: 10.1061/(ASCE)CC.1943-5614.0000160.

[161] T. Keller, C. Tracy, and A. Zhou, "Structural response of liquid-cooled GFRP slabs subjected to fire – Part II: Thermo-chemical and thermo-mechanical modeling", *Compos. Part A Appl. Sci. Manuf.*, vol. 37, pp. 1296–1308, 2006, doi: 10.1016/j.compositesa.2005.08.007.

[162] Y. Bai and T. Keller, "Fire performance of water-cooled GFRP columns. II: Postfire investigation", *J. Compos. Constr.*, vol. 15, pp. 413–421, 2011, doi: 10.1061/(ASCE)CC.1943-5614.0000191.

[163] Y. Bai and T. Keller, "Modeling of post-fire stiffness of E-glass fiber-reinforced polyester composites", *Compos. Part A Appl. Sci. Manuf.*, vol. 38, pp. 413–421, 2007, doi: 10.1016/j.compositesa.2007.06.010.

[164] C. Ding, Y. Bai, F. Azhari, and T. Keller, "Fire dynamic responses of fibre reinforced polymer composite buildings", *J. Compos. Constr.*, vol. 28, 2024, doi: 10.1061/JCCOF2.CCENG-4504.

[165] T. Keller, Y. Bai, and T. Vallée, "Long-term performance of a glass fiber-reinforced polymer truss bridge", *J. Compos. Constr.*, vol. 11, pp. 99–108, 2007, doi: 10.1061/(ASCE)1090-0268(2007)11:1(99).

[166] T. Keller, N. A. Theodorou, A. P. Vassilopoulos, and J. De Castro, "Effect of natural weathering on durability of pultruded glass fiber-reinforced bridge and building structures", *J. Compos. Constr.*, vol. 20, 2016, doi: 10.1061/(ASCE)CC.1943-5614.0000589.

[167] Ramboll, *Generaleftersyn samt undersogelese af egenskaber for glasfibermateriale*. Bro Nr. 22439 Gl. Strandvey, 2013.

[168] *Fiberline design manual*, 1st ed. Fiberline Composites A/S, Denmark, 1995.

[169] A. Castelo, J. R. Correia, S. Cabral-Fonseca, and J. de Brito, "In-service performance of fiber-reinforced polymer constructions used in water and sewage treatment plants", *J. Perform. Constr. Facil.*, vol. 34, pp. 1–19, 2020, doi: 10.1061/(asce)cf.1943-5509.0001449.

[170] A. Castelo, J. R. Correia, S. Cabral-Fonseca, and J. de Brito, "Inspection, diagnosis and rehabilitation system for all-fibre-reinforced polymer constructions", *Constr. Build. Mater.*, vol. 253, 2020, doi: 10.1016/j.conbuildmat.2020.119160.

3. Composites – Overcoming Limitations

[171] United Nations, *Our common future. Report of the World Commission on Environment and Development*. Oxford University Press, 1987.

[172] B. Purvis, Y. Mao, and D. Robinson, "Three pillars of sustainability: in search of conceptual origins", *Sustain. Sci.*, vol. 14, pp. 681–695, 2019, doi: 10.1007/s11625-018-0627-5.

[173] V. Mara, R. Haghani, and P. Harryson, "Bridge decks of fibre reinforced polymer (FRP): A sustainable solution", *Constr. Build. Mater.*, vol. 50, pp. 190–199, 2014, doi: 10.1016/j.conbuildmat.2013.09.036.

[174] *EN 15804: Sustainability of construction works. Environmental product declarations. Core rules for the product category of construction products*. European Committee for Standardization, Brussels, 2022.

[175] *ISO 14040: Environmental management – Life cycle assessment – Principles and framework*. International Organization for Standardization, Geneva, 2006.

[176] T. Jena and S. Kaewunruen, "Life cycle sustainability assessments of an innovative FRP composite footbridge", *Sustain.*, vol. 13, pp. 1–20, 2021, doi: 10.3390/su132313000.

[177] Y. S. Song, J. R. Youn, and T. G. Gutowski, "Life cycle energy analysis of fiber-reinforced composites", *Compos. Part A Appl. Sci. Manuf.*, vol. 40, pp. 1257–1265, 2009, doi: 10.1016/j.compositesa.2009.05.020.

[178] R. Jain and L. Lee, *Fiber reinforced polymer (FRP) composites for infrastructure applications: Focusing on innovation, technology implementation and sustainability*. Springer, 2012. doi: 10.1007/978-94-007-2357-3.

[179] A. André, J. Kullberg, D. Nygren, C. Mattsson, G. Nedev, and R. Haghani, "Re-use of wind turbine blade for construction and infrastructure applications", *IOP Conf. Ser. Mater. Sci. Eng.*, vol. 942, 2020, doi: 10.1088/1757-899X/942/1/012015.

[180] B. P. Chang, A. K. Mohanty, and M. Misra, "Studies on durability of sustainable biobased composites: a review", *RSC Adv.*, vol. 10, pp. 17955–17999, 2020, doi: 10.1039/c9ra09554c.

[181] S. V. Joshi, L. T. Drzal, A. K. Mohanty, and S. Arora, "Are natural fiber composites environmentally superior to glass fiber reinforced composites?", *Compos. Part A Appl. Sci. Manuf.*, vol. 35, pp. 371–376, 2004, doi: 10.1016/j.compositesa.2003.09.016.

[182] A. T. Shahid, M. Hofmann, M. Garrido, R. J. Correia, and C. R. Rosa, "Freeze-thaw durability of basalt fibre reinforced bio-based unsaturarated polyester composite", *Materials*, vol. 16, 2023, doi: 10.3390/ma16155411.

CHAPTER 4

COMPOSITES – EXPLORING OPPORTUNITIES

4.1   INTRODUCTION

The previous chapter dealt with the limitations of composites, together with measures to overcome these limitations. The opportunities created by composites are the subject of this chapter.

Opportunities offered by the unique properties of composites enable design options in structural engineering and architecture that are yet to be explored. These options are either not possible or are difficult to achieve using conventional materials. An overview of the unique properties of composites and how they can be related to different design options is given in **Table 4.1**. The fields of opportunities concerned by these design options in terms of architecture, sustainability, and economy are also indicated. In this respect, energy and structural engineering aspects are covered by sustainability and economy, respectively, since the former mainly concerns energy saving and harvesting, while the latter has an impact on cost in most cases.

Unique properties of composites related to structural engineering and architecture mainly concern physical and mechanical material properties and system properties. System properties are related to the nature of the material and its associated processing into characteristic and material-tailored member shapes, such as sandwich panels.

Unique physical properties of composites include low density and thermal conductivity (if also considering core materials, and except carbon fibres), and degree of transparency in the visible and infrared light spectra in the case of glass fibres; the degree of transparency depends on the optical refractive index ratios of fibres and matrix. Unique mechanical properties are based on the high strength and thinness of the fibres, which can be arranged in multidirectional lay-ups of laminates. Such laminates are still thin when compared to cross-laminated timber, for instance (see Fig. 1.7), and can be tailored to different stress fields. Furthermore, the high flexibility of the fibres together with the liquid state of the resin during processing allow such laminates to be curved in one or two directions.

Unique system properties exist at the member level, where laminates can be combined to form homogeneous core or web-core sandwich panels. These in turn allow for the integration of building physics and architectural functions into structural sandwich panel members, thereby offering physical and mechanical properties at the system level. In addition to the physical material property of low density, for instance, further lightness can be achieved by hollow sections at the system level.

Table 4.1 Design options offered by composites, derived from unique material and system properties, and associated fields of opportunities ($\checkmark$ = applicable, - = not applicable) [1]

Design options (and monograph sections)		Unique properties (material & system level)	Fields of opportunities		
			Architecture	Sustainability	Economy
Lessons learnt from nature (Section 4.2)		Multifunctionality	$\checkmark$	$\checkmark$	$\checkmark$
Structural form (4.3)	Language of composites	Fibre thinness and flexibility / Mouldability	$\checkmark$	-	-
	Material-tailored structural form	Low density / Lightness High specific strength and stiffness / Tailorable fibre architecture	$\checkmark$	$\checkmark$	$\checkmark$
	Material-tailored structural system				
Transparency (4.4)		Refractive index ratio[1]	$\checkmark$	-	-
Function integration (4.5)	Thermal insulation	Low thermal conductivity on sandwich panel level	$\checkmark$	$\checkmark$	$\checkmark$
	Thermal activation	Refractive index ratio[2] on sandwich panel level	-	$\checkmark$	$\checkmark$
	Solar cell encapsulation	Refractive index ratio[1] on laminate level	$\checkmark$	$\checkmark$	$\checkmark$
	Colour, light, and texture	Pigmentation Refractive index ratio[1] on laminate level	$\checkmark$	-	-
Hybrid construction (4.6)		Ease of material combinations	-	$\checkmark$	$\checkmark$
Modular construction (4.7)		Low density / Lightness Mouldability Low thermal conductivity on sandwich panel level	$\checkmark$	$\checkmark$	$\checkmark$

[1] Of glass fibres and matrix, in visible light spectrum
[2] In infrared light spectrum

Six design options were derived from the unique material and system properties, as listed in Table 4.1; they are discussed in the following Sections 4.2–4.7. The first section "Lessons learnt from nature" explores opportunities, which may be derived from multifunctional construction concepts developed and optimised by nature over millions of years. Opportunities may relate to architecture, sustainability, and economy.

The second section "Structural form" concerns the specific structural language of composites, which mainly results from the thinness and flexibility of the fibres and thus their mouldability. Additionally, low density, high specific strength and stiffness, and tailorable fibre architecture render lightweight construction possible, which

4. Composites – Exploring opportunities

265

requires novel material-tailored structural forms and systems to be developed. Such optimised forms and systems represent opportunities for new ways of architectural expression, while supporting economy.

The third section "Transparency" focuses on the possible different degrees of transparency of load-bearing composite laminates in the visible spectrum of light, which opens up new opportunities in architecture.

The fourth section "Function integration" explores opportunities based on the integration of functions (other than the structural one) into structural sandwich panel members or laminates. The low thermal conductivity of core materials allows the merging of load-bearing structure and building envelope. Water circuits can be integrated into sandwich cores to provide hydronic thermal activation, i.e., to heat and cool buildings and provide fire resistance. The thermal transparency of composites, obtained through an optimisation of the refractive index ratio in the infrared spectrum, can increase the efficiency of such systems. The possible transparency of laminates in the visible spectrum of light may enable the encapsulation of solar cells and thus energy harvesting. Finally, colour can be added through pigmentation, and light and surface texture can be integrated. All these design options provide opportunities in all aspects, i.e., architecture, sustainability, and economy.

The sixth section "Hybrid construction" discusses how composites can be combined with other structural materials to fully benefit from the excellent mechanical properties and not be economically penalised by certain weaknesses, e.g., lacking material ductility or a significant matrix-dominated behaviour. Since opportunities are primarily seen in hybrid construction, rather than measures to overcome limitations of the materials, this design option is included in the present Chapter 4.

The last section, "Modular construction", demonstrates how large-scale, lightweight and function-integrated sandwich construction may offer opportunities in modular construction and thus architecture, by also supporting sustainability and economy.

Specific conclusions are included at the end of each section, while overall conclusions about the opportunities of composites are drawn in Section 6.2.

4.2 LESSONS LEARNT FROM NATURE

4.2.1 Introduction

The structures of nature have undergone millions of years of optimisation processes. One of the targets of optimisation is the minimisation of energy use and hence material consumption, amongst other things. The complex structural forms and design principles specific to nature are results of these minimisation processes [2]. In contrast to engineering structures, complex forms in nature result from growth and do not entail any extra cost for manufacturing. The question thus arises as to what extent such "ideal" natural structural forms and design principles can serve as models for engineering structures, as was the case in the example shown in **Fig. 4.1**, where the complex form and modular structure of diatoms inspired the conception of a pedestrian bridge. Responses to this question are provided in the following, but also in Section 4.3, where ideal structural forms are further discussed.

 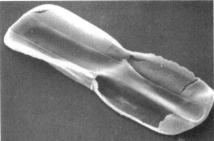

Fig. 4.1 Model of TSCB Bridge (see Appendix B.4) **(left)**, inspired by diatoms (genus Donkinia, length ~100 μm) **(right)**

As designers use composites in engineering structures, so too does nature use composites as structural materials. Carbon, found in all forms of life, is one of the basic elements of nature and is arranged in polymer chains. This is also the case for fibre-polymer composite resins and some types of fibres, such as carbon or aramid fibres. A similar question to the one above thus arises, i.e., whether new insights can be derived from how nature uses composite materials in its structures, or to what extent the composite structural concepts of nature can inspire those of composite engineering structures.

The composition and material properties of natural composites and how they are used are summarised hereinafter, followed by a discussion on the design principles of nature and to what degree they correspond to those of engineering structures. Since the natural design principles correspond approximately to today's principles of sustainability, the potential of composites to also respect the latter is further discussed. Finally, some selected structural principles of nature and their applicability to composites are examined.

The content of this section was part of a Master's course in "Complex structures" for architecture and civil engineering students offered at EPFL from 2010–2020.

4.2.2 Materials of natural structures

Materials in nature are either of an inorganic nature (such as stone and metal) or organic nature (e.g., natural fibres or wood); this section discusses only the latter due to their close relationship to fibre-polymer composites. The basic components of many organic materials that fulfil structural functions are biopolymer fibrils, such as chitin fibrils in animal exoskeletons, cellulose fibrils in plants, and silk fibrils in spider webs, which represent three of the most abundant biopolymer fibrils in nature [3]. Their polymer chains comprise alternating amorphous and crystalline segments, which are connected by hydrogen bonds and van der Waals forces. Chitin fibrils of exoskeletons are embedded in a mineral-protein matrix, while the cellulose fibrils of plants are embedded in a matrix of amorphous hemicellulose and lignin to form the cell walls. The cellulose fibrils are helically wound in specific layers of the cell walls, while the mechanical properties can vary from fibre- to matrix-dominated, depending on the fibril angle [3].

4. Composites – Exploring opportunities

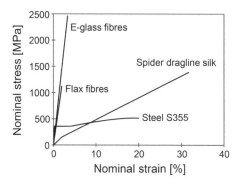

Fig. 4.2 Nominal stress vs strain curves of spider dragline silk at 20°C and 50% RH [4], flax fibres [5], and comparison to E-glass fibres [6] and structural steel S355, derived from [7]

Biopolymer fibrils may exhibit outstanding mechanical properties, as shown in **Fig. 4.2**, where silk fibril and flax fibre properties (containing ~70 wt% cellulose fibrils) are compared to structural steel (grade S355) and E-glass fibres. Spider dragline silk clearly outperforms steel with regard to strength and strain at failure, and flax fibres exhibit a similar stiffness to that of E-glass fibres. All these natural materials are lightweight, and their specific properties (related to density) are thus high (see Fig. 1.10 for flax fibres).

Similar to fibre-polymer composites, natural organic materials thus offer a high performance and in many cases are composed of fibres embedded in a matrix, with the former normally oriented in the loading direction and fulfilling the primary load-bearing function.

4.2.3 Design principles of nature

Nature transmits forces in tension and compression, either directly along the fibres, or by aligning the directions of basic building blocks and those of the principal stresses, as in the famous example of the femur's trabecular structure [8]. In many cases, the flow of these forces is visualised and can thus be understood or "read" quite easily.

While the design principles in structural engineering are mainly based on structural safety considerations (avoiding collapse being the most important principle), serviceability (limiting deformations), durability (over the long design service life), aesthetics (considered from case to case), sustainability (principle of increasing importance), and economics (through optimised construction processes, amongst other things), see Section 5.1, the design principles of constructions in nature are significantly different. As derived from [9] and supplemented in the following, they mainly include the following components:

1) Prevention of loading through large deformations

Structures in nature attempt to prevent loading as much as possible through large deformations, as demonstrated in **Fig. 4.3**, taking the examples of a tree and a pole or

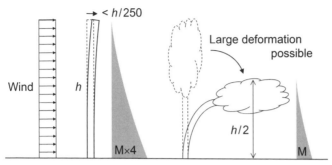

Fig. 4.3 Wind action on man-made and natural structures (M = bending moment per meter depth) [1]

tower subjected to lateral wind. The bending moment to be transmitted by the roots of the tree into the ground is considerably reduced by the large bending deformation, while the pole must resist the entire wind force due to the prescribed limited serviceability deformations at the top. Interestingly, the bending of the tree is facilitated by an internal self-stress state [10]. Similarly, spider webs can store a significant amount of elastic energy thanks to the large deformations possible, as demonstrated in Fig. 4.2. These very efficient design principles of nature are unfortunately not applicable in engineering structures owing to the above-mentioned serviceability restrictions (with some exceptions, such as seismic dampers, which are based on a certain deformation capacity).

2) Multifunctionality instead of monofunctionality

In current construction techniques, each element normally has its own function, and elements are added to each other linearly. A building wall, for instance, normally consists of four layers: weather protection, vapour barrier, thermal insulation, and load-bearing structure (from the exterior to interior). Structures of nature, however, instead of being monofunctional, additive, and linear are multifunctional, integrated, and cross-linked.

As discussed in Section 4.5, and in contrast to conventional construction materials, composites – thanks to their physical properties – also allow for multifunctional construction. Elements can fulfil several functions and be integrated, such as the building envelope and load-bearing structure, which can be merged into a single-layer structural envelope.

Basic structural concepts cannot be directly copied from nature, as demonstrated above, but cross sections of plant stems, for instance, can inspire technical sections and lead to interesting solutions, as depicted in **Fig. 4.4**. The stem envelope is multifunctional and structurally reinforced by concentrated vascular bundles; a similar concept was used for the floor plan of the Dock Tower project, see Appendix B.1. The composite envelope of the circular section fulfils the building physics and structural functions; five additional outer elevator and staircase shafts are merged with the envelope and contribute to the structural function. The large effective depth of 27 m (i.e.,

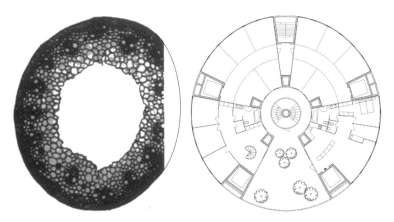

Fig. 4.4 Cross section of stem of Lolium perenne (ø 2–3 mm) **(left)** and composite Dock Tower (ø 27 m, see Appendix B.1) **(right)**

the section diameter) significantly reduces the stresses in the envelope, in contrast to conventional towers, where the load-bearing structure is normally in the centre with much reduced effective depth and thus much higher stresses. The comparably low stresses allow the integration of translucent and transparent envelope parts between the opaque shafts, see Appendix B.1.

3) Optimisation of entire system instead of individual elements

In additive and linear technical systems, each element is normally optimised according to its specific function. In nature, however, the overall multifunctional system is optimised, and compromises are thus required regarding the individual functions. Similarly, if composites are used in multifunctional applications, the entire system should be optimised and not just individual subsystems.

4) Energy saving instead of energy waste

Energy saving or minimisation is an advantage in survival and thus a basic principle of nature. Composites, thanks to their lightweight nature and suitability for prefabricated modular construction, may offer significant advantages in this respect, see Section 4.7.

5) Use of solar energy

Nature uses solar energy directly or indirectly, e.g., directly to warm objects (cold-blooded animals, for instance), or indirectly by using solar-induced wind movements for cooling purposes, or by converting light energy into chemical energy through photosynthesis. As demonstrated in Section 4.4, composites allow for the encapsulation of solar cells into load-bearing laminates thanks to their possible transparency and thus may add an energy harvesting function to enhance the multifunctionality of building elements.

6) Temporal limitation instead of unnecessary durability

Nature limits durability to the natural lifetime of its structures and thus further optimises overall energy consumption. In engineering structures, however, the design service life of structures is normally prescribed in codes; the typical design service life of buildings and bridges is 50 and 100 years, respectively [11]. However, as today's reality demonstrates in numerous cases, building and bridges need heavy transformation or upgrading after only a few (two-three) decades, since they become obsolete or no longer fulfil the requirements of our rapidly developing society. The question thus arises, from the economic and sustainability points of view, whether these obviously exaggerated prescribed lifetimes and the associated durability are still appropriate or whether they should not be reduced to reasonable durations and levels. In this respect, composites may offer interesting solutions based on their potential for lightweight modular construction, employing reusable building blocks of adaptable durability. Furthermore, reduced lifetimes would allow bio-based composites to be used and thus significantly contribute to the sustainability of composites, see Section 3.11.4.

7) Complete recycling instead of waste accumulation

Nature completely recycles its structures and materials. In the framework of sustainability, this principle is also gaining importance in construction. Complete material recycling (i.e., reuse of separated material components), however, still presents a challenge for composites, see Section 3.11.3, (as it also does for other construction materials, such as steel-reinforced concrete). However, if engineering structures are designed in a modular way, they can be disassembled at the end of the life and reused at the structural member level, if designed accordingly (see Sections 3.11.3 and 6.3). Composites are particularly suitable for this type of recycling at the member or module level.

8) Development according to the trial-and-error process

Development in nature is driven by the principles of evolution and occurs without any precise aim, basically following a trial-and-error process. Obviously, this type of process is not applicable in the design of engineering structures, mainly for economic reasons.

4.2.4 Structural design concepts of nature

4.2.4.1 Tensegrity structures

A structural principle based on tension and compression, where the former is continuous (materialised by cables) and the latter discontinuous (consisting of rigid bars), is known as tensegrity (tensional integrity), as developed by Snelson since the 1950s [12], see **Fig. 4.5** (left). This principle is widespread in nature, applied from the atomic scale up to the scale of the universe, as summarised in **Table 4.2**. Since the stiffness of pure tensegrity structures with "floating" compression bars is limited, however, and normally does not allow serviceability criteria to be met, compression is also made continuous. The compression bars are thus joined together, as done by Buckminster Fuller, and as is also the case in the human bone system.

4. Composites – Exploring opportunities

Fig. 4.5 Tensegrity structure designed by Snelson, Easy-K, Gibbs Farm, New Zealand, 1970 (aluminium tubes and stainless-steel cables, 32 m length) **(left)**; TCy (Tensegrity Cylinder) structure, Geneva, 2022 (suspended cylinder, carbon fibre tubes, and stainless-steel cables, 6.4 m height, see Appendix A.7) **(right)**

Inherent to these minimal tensegrity structures is their extremely lightweight nature. Using lightweight composite materials of high specific strength and stiffness is thus in line with this structural principle, as demonstrated, for instance, in the TCy tensegrity structure, shown in Fig. 4.5 (right). The compression bars, which circumscribe the space of this suspended sculpture in continuous and crossed spirals from top to bottom, are composed of lightweight carbon fibre composite tubes of high strength and stiffness, see Appendix A.7.

4.2.4.2 Convertible structures

Structures of an organic nature grow and may be convertible, such as flowers that open during the day and close at night, or the human muscle/tendon-bone system. Since the movement of elements is normally hindered by friction in the hinges and friction depends on the element weights, lightweight composites are particularly appropriate for use in convertible structures.

Engineering structures, however, cannot grow yet, but natural growth can inspire engineering structures, as shown in **Fig. 4.6**: the concept of the first all-composite house was derived from a snail shell and allows for the sequential adding of spaces [13, 14].

Table 4.2 Tensegrity concepts applied in various scales [1]

Structure and scale	Compression (discontinuous)	Tension (continuous)
Atom	Nucleus, electrons	Electrostatic forces
Human	Bones*	Tendons and muscles
Tensegrity	Bars	Cables
Universe	Suns, planets	Gravitational forces

* Continuous

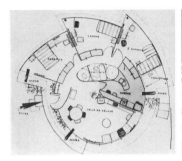 

Fig. 4.6 First composite house, 1955/56, inspired by growth of snail shell

Fig. 4.7 Kinetic composite envelope of (permanent) Thematic Pavilion, Expo 2012, Yeosu, South Korea

A further example of using composites in convertible structures is the kinematic envelope shown in **Fig. 4.7**. Actuators impose elastic bending combined with a lateral rotation to the 108 louvers made of glass fibre-polymer composites, thus enabling the opening and closing of the envelope and creating an animated pattern [15].

4.2.4.3 Fractal structures

The geometry of nature cannot be described by Euclidian theory as appearances and shapes follow fractal laws [16]. Fractal geometry is characterised by self-similar patterns, appearing in cascades across the scales and exhibiting a mixture of order and disorder or randomness, as demonstrated by the example of the Koch curve, shown in **Fig. 4.8**. This curve cannot be characterised by its length since the length is infinite. However, the rate of growth of the length allows the quantification of the fractal nature, which is expressed by the fractal dimension. The higher this fractal dimension, the more significant the fractal aspect of the geometry. In the case of the Koch curve, the fractal dimension, D, is [17]:

$$D = \log N / \log n = \log 4 / \log 3 = 1.26 > 1.0 \tag{4.1}$$

where N is the number of segments, and n is the reduction factor. The straight line in this example has $D = \log 3 / \log 3 = 1.0$ and thus does not exhibit any fractal dimension.

Fractal structures of nature result from natural growth and are usually associated with beauty [18]. In many cases, structural aspects govern the fractal composition,

4. Composites – Exploring opportunities

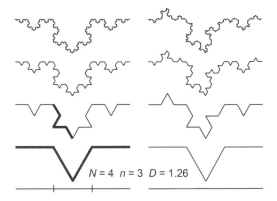

Fig. 4.8 Fractal Koch curve in ordered configuration (**left**) and considering randomness (**right**), derived from [17]

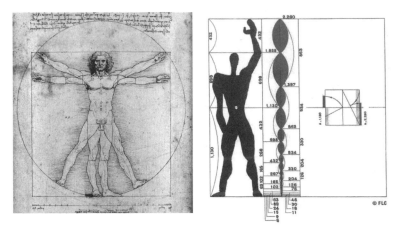

Fig. 4.9 Fractal proportion systems in architecture: the Vitruvian Man by Leonardo da Vinci (**left**) and the Modulor by Le Corbusier (**right**)

examples are the ramified (multifunctional) structures of tree branches and leaves or the wings of insects. In architecture and design, systems of proportions, such as Leonardo da Vinci's Vitruvian Man or Le Corbusier's Modulor, also have inherent elements of fractal geometry, see **Fig. 4.9**. Buildings designed by renowned architects, such as Frank Lloyd Wright or Alvar Aalto, who were consciously inspired by nature, exhibit significant fractal dimensions [17] – and are perceived as beautiful. In these cases, however, the fractal patterns can, but do not necessarily, result from structural considerations. In architecture, spatial arrangements of volumes, texture, or the interaction of light and shadow, may also contribute to significant fractal dimensions.

Composite materials are basically flexible and can be manufactured in complex shapes, closely tailored to global and local structural or multifunctional requirements, including varying degrees of transparency and texture (see Sections 4.3–4.5). They may thus provide textural patterns over various scales, and in this way integrate significant fractal dimensions and associated aesthetic expression.

4.2.5 Conclusions

Although many natural structures are composed of composite materials, a direct copying or mimicking of structural concepts is normally not possible since some of the design principles of nature are clearly different from those of engineering structures, particularly that of large deformability. Several other principles do conform, however, to today's sustainability requirements for engineering structures, and their application is thus desirable. As discussed above, composites exhibit the potential to comply with most of those principles and may do so significantly better than conventional construction materials, such as reinforced concrete or structural steel.

4.3 STRUCTURAL FORM

4.3.1 Introduction

How to use composites is the subject of Chapter 4 and of this section on structural form. As the fibres of composites are flexible and of high tensile strength, these fibres are most efficiently used when they are under pure tension, in members such as cables, bars, and strips, as shown in **Fig. 4.10**, where the fibres are stressed in their axial direction, and the matrix has no relevant structural function (except in the strap ends). This fact is also the reason that composites in civil engineering were first successful in strengthening existing structures with thin unidirectional carbon fibre strips, stressed under tension.

However, it is not possible to design load-bearing structures consisting solely of tension members. In order to satisfy the principle of equilibrium, tension cannot exist without its counterpart compression, as demonstrated in **Fig. 4.11**, which shows the simplest ways how a force (P) can be redirected by three additional forces to create space, i.e., by two tension forces and one compression force, or by one tension and two compression forces. Using composites in compression is much less efficient than in tension, however, since the structural behaviour changes from fibre-dominated (in tension) to matrix-dominated (in compression); the compression properties thus depend on the much lower matrix strength and stiffness. The response to the question of how to use composites efficiently is thus not trivial and requires further reflection.

How materials may influence architectural and structural concepts has been studied for almost 20 years at EPFL in my teaching unit entitled "Architecture and Structures", within which master's students of architecture and civil engineering collaborated in the design of lightweight pavilions comprising three pre-defined spaces, i.e., a reception, exhibition, and projection space. Students were required to use timber, aluminium, or composites as the construction material, with the aim of exploring and visualising the best way to employ the materials' characteristics in their architectural, structural, and constructive designs.

Fig. 4.10 Composite ground/rock anchor with unidirectional carbon fibres, self-anchored strap ends (specimen length 1380 mm)

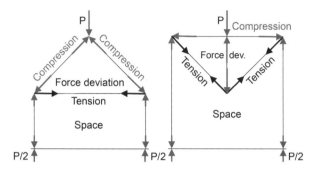

Fig. 4.11 Basic force deviation through combined tension and compression to create space [1]

Some examples of these pavilions can be seen in **Fig. 4.12**. Significant differences between timber on the one hand, and aluminium and composite solutions on the other, could be identified in most cases, while the distinction between aluminium and composite pavilions was not always evident. The differences resulted mainly from the fact that timber components are linear (except for agglomerated panels), while the selected basic components of aluminium and composites are similar, i.e., two-dimensional thin sheets or laminates. Composite profiles were not considered since they were rated as not respecting the nature of the material, and extruded aluminium profile sections were too small to be applied in large-scale structures. Unlike brittle composites, ductile aluminium sheets can be folded easily, this being the reason why folded large-scale members prevailed in the aluminium pavilions. Since it is impossible to fold fibres without causing damage, in most cases large-scale composite members were curved, often in two directions, which is also much easier to produce in composites than in aluminium. A further complication of aluminium is that its thermal conductivity is high compared to (glass fibre) composites; it is thus much more complicated to prevent thermal bridges.

These results demonstrate that multiple aspects must be considered for a design to respect and consider the nature of the material. Opting for a design that is driven by the intention to optimise the use of the materials' properties may also allow for the creation of economical and sustainable solutions, as will be demonstrated in this section.

Below, the typical formal and technical language of composite materials used in bridge and building construction will be discussed, before concentrating on so-called

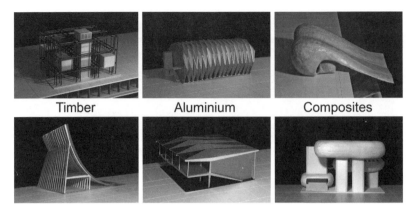

Fig. 4.12 Lightweight pavilions with identical space programmes in timber, aluminium, and composite construction, emerging from collaboration of EPFL master's students in architecture and civil engineering [19]

material-tailored structural forms and how they develop, and then focusing on material-tailored structural systems, as a basis for the structural design.

4.3.2 Language of composites

Each material has its own characteristic "language", where the term is understood as the formal and technical appearance or shape. Steel, for instance, is in most cases used in linear profiles of typical open or closed sections (the I-section being the most representative), which are bolted or welded at the joints. This characteristic language is basically the result of a material-tailored use of steel, as discussed in the next Section 4.3.3.

The language of fibre-polymer composites is mainly based on the high flexibility and very small diameter of the fibres, and the liquid nature of the polymer matrix during processing. Thanks to their flexibility, the fibres can easily adapt to complex or freeform shapes. However, the fibres cannot be folded, the corners are thus always slightly rounded; in this respect, anisotropic carbon fibres are much more sensitive to small radii of curvature than isotropic glass fibres. Furthermore, the small fibre diameter allows a multilayered ply architecture, which can easily be tailored to complex stress fields by combining different fibre directions in still thin-walled members. The polymer matrix finally "freezes" the complex form.

One of the examples that best illustrates the language of composites is the chair developed by Verner Panton in the 1960s, shown in **Fig. 4.13**. The glass fibre-polyester construction is thin-walled, in a complex double-curved shape, which also stabilises the thin laminates, and all the corners are rounded. The structural system is typically form-active, as discussed in Section 4.3.4.

The language of composites is also depicted in **Fig. 4.14**, in a more technical sense in this case. The support construction of a web-core sandwich roof on a load-bearing glass wall envelope is shown [20]. The thin composite laminates are combined with the compact foam structure in a way that is typical, i.e., the glass fibre-polyester face

4. Composites – Exploring opportunities

Fig. 4.13 Panton Chair made of glass fibre-polyester composites, 1967/68

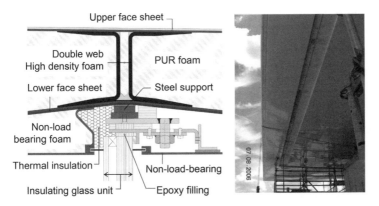

Fig. 4.14 Web-core sandwich roof of Novartis Building, support detail on glass envelope, based on [20] **(left)**; roof with prepared supports before installation of glass panels **(right)**

sheets enclose the core foam. To introduce the support reaction forces into the sandwich panel member, double-webs are implemented above the supports by additional thin laminates, which enclose a high-strength foam and are connected to the face sheets and transverse webs (not shown) by lap shear connections. The laminate ends are tapered according to the stresses to be transmitted, which is easily done by staggering the ply lengths; the laminate deviations are not angular but curved. The necessary spaces and recesses for laminates and their connections can simply be cut into the foam by CNC milling, as shown in **Fig. 4.15**.

4.3.3 Material-tailored structural form

4.3.3.1 Introduction

Examining how fibre-polymer composites have been used in bridge and building construction since the mid-1990s reveals that one-dimensional (1D) profiles were

Fig. 4.15 CNC milling of foam blocks (0.96×0.96×0.5 m³) of sandwich roof of Novartis Building **(left)**; foam block with cut-outs for anchoring against roof uplift, ready for lamination **(right)**, based on [20]

Fig. 4.16 Composite form development: material-tailored sandwich form of 1st phase (Futuro House, diameter 8 m, Matti Suuronen, 1968) **(top left)**; 3rd phase (Novartis Building, module length 18.5 m, 2006, Appendix A.4) **(top right)**; transition from 2nd phase material substitution profile-truss form (Pontresina Bridge, 2×12.5 m span, 1997, A.2) **(bottom left)** to 3rd phase material-tailored frame-sandwich form (TSCB Bridge, 18.0 m span, B.4) **(bottom right)**

implemented in most cases, manufactured by the pultrusion process and exhibiting the same open or closed section shapes as steel profiles [21]. The connections of these composite profiles were normally bolted, as is also done in steel construction. Examples of this structural type are the Pontresina Bridge, composed of glass fibre composite profiles, shown in **Fig. 4.16** (bottom left) (although adhesive joints were used in one of the two spans, see Appendix A.2), and the Eyecatcher Building (see Appendix A.3). Pultruded profiles were also adhesively bonded together to form bridge deck slabs, see Section 2.3.3, resulting in structural forms that mimic those of steel orthotropic decks.

4. Composites – Exploring opportunities

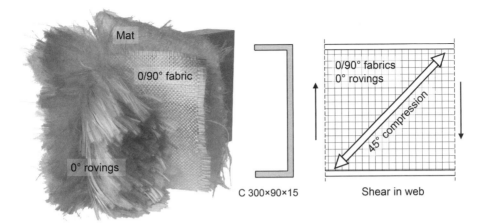

Fig. 4.17 Fibre architecture of pultruded C-section glass fibre composite profile, cross-sectional dimensions 300×90×15 mm, consisting mainly of 0° rovings and only one outside woven 0/90° fabric and combined mat (specimen taken during pultrusion of Eyecatcher beam profiles, see Appendix A.3) **(left and middle)**; web with 0° rovings and 0/90° fabrics and diagonal compression force from shear causing matrix-dominated response **(right)**

In general, the fibre architecture of such pultruded profiles is mainly unidirectional (UD), as shown in **Fig. 4.17** (left), rendering them strongly anisotropic (in contrast to isotropic steel profiles) and producing all the associated disadvantages known from timber construction. The excellent fibre properties cannot be used fully since the much weaker matrix-dominated behaviour, transverse to the fibres, typically governs the design, particularly under shear (Fig. 4.17, right) and in bolted joint regions (see Section 2.4.2).

On the other hand, there are cases of small-scale composite house constructions dating back to the 1950s and 60s, which were composed of two-dimensional (2D) curved and multifunctional sandwich panels, integrating structural and building physics functions, see Section 2.7.2. They were built in prefabricated modules, as also shown in Fig. 4.16 (top left), with an appearance that differs greatly from that of their 1D counterparts. Similar sandwich constructions, but of much larger scale, re-appeared from the late 1990s onwards in roof or bridge construction, see Fig. 4.16 (top and bottom right), and Sections 2.7.4 and 2.6.2, respectively.

The question thus arises why, from the mid-1990s onwards, did composite structural forms evolve to mimic steel construction, or, in other words, why did a simple material substitution of steel by composites occur, even though this substitution did not allow for the full use of the outstanding properties of the novel materials. This issue was investigated in one of the first PhD theses developed at the CCLab in 2004 [22], which examined how so-called material-adapted structural forms develop.

The aims of the thesis were to (i) clarify what a "material-adapted structural form" is, (ii) understand how it developed in history, and (iii) identify the factors influencing this development. The responses to these research questions are summarised below. Instead of "material-adapted", the synonymous term "material-tailored" is used hereafter.

4.3.3.2 Material-tailored structural forms and their development

The term "material-tailored structural form" was derived in [22] as being a structural form that (i) considers the nature of the material, i.e., its properties, and characteristics related to processing and detailing, (ii) demonstrates structural and material efficiency, (iii) is the aesthetic expression of the material properties, processing technology, and function, and iv) is economic.

Considering this definition, most of the composite house constructions of the 1950–60s, including the example shown in Fig. 4.16, may be subsumed under "material-tailored structural form", while this attribute does not apply to structural forms exhibiting simple material substitution. In the former case, the use of composites in 2D thin laminates, where the fibre architecture can be tailored to the principal stress pattern, takes the nature of the material into account, in contrast to the invariable and strongly anisotropic fibre architecture of 1D profiles, in the latter case. Double curvature and sandwich construction stabilise the thin laminates and thus maximise structural and material efficiency, while structural and material efficiency of 1D profiles is significantly reduced, as explained above. Through an appealing organic and complex shape, as well as function integration (see Section 4.5.1), the structural envelope may directly form the living space and thus offer the aesthetic expression of material properties, processing technology, and function, in contrast to the 1D profiles, where processing technology and function cannot be derived from the form. Function integration, prefabrication, and modular construction furthermore increase the overall economy, in contrast to the 1D case, where an additional envelope must be added to the structure (in the case of building construction). In the meantime, since the completion of that PhD thesis in 2004, the aspect of environmental impact, i.e., sustainability, has gained great significance and must be added as an attribute of "material-tailored form"; this aspect is discussed in Section 3.11.

A reflection on the evolution of structural form reveals that as new materials appeared over the course of history – such as stone, iron, and concrete – these new materials were always initially used in a material-tailored way, as has also been the case with composites. These early uses were experimental and small in scale, however, due to limitations in material quality and processing technologies. Ferro-cement, for instance, was used in 1848 to build the hull of two small boats, see **Fig. 4.18** (top left), thus implementing a (small-scale) material-tailored reinforced concrete shell structure. The numerous composite house types of the 1950–60s, mentioned above, were furthermore limited to one storey as the material properties did not yet allow for the construction of multi-storey buildings at that time [23].

After this first phase of experimental material-tailored use, a much more significant second phase of material substitution would normally occur. Key reasons for this development were that the material-tailored structural forms at large scale were not yet known or could not yet be manufactured. Furthermore, the phase of material substitution was crucial to demonstrate that the performances of the novel materials were at least on a par with those of the conventional materials, instilling clients' confidence in the novel materials, even though the material properties could not yet be fully utilised. The material substitution phase of composites was also supported by the further development of the pultrusion process, which enabled the production of large quantities of profiles at a reasonable cost.

4. Composites – Exploring opportunities 281

Fig. 4.18 Reinforced concrete form development: material-tailored shell forms of 1st phase (ferro-cement boat, Joseph Lambot 1848) **(top left)**; 3rd phase (motorway service station, Deitingen, Heinz Isler, 1982 **(top right)**; transition from 2nd phase hierarchical skeleton material substitution (Reutlinger furniture factory, Karlsruhe, Hennebique system, 1899) **(bottom left)** to 3rd phase material-tailored two-way mushroom flat slab (grain storage of the Swiss Confederation, Altdorf, Robert Maillart, 1912) **(bottom right)**

Nevertheless, the substitution phase must finally be superseded by a material-tailored use (third phase) to achieve an economic and sustainable use of the novel materials. Accordingly, and staying within the example of reinforced concrete, after a second phase of reinforced concrete structures, which mimicked the 1D hierarchical steel-skeleton construction (Fig. 4.18, bottom left), material-tailored large-scale thin shells and two-way mushroom flat slabs appeared, see Fig. 4.18 (top and bottom right, respectively). Similarly, for composites there was a development from the second phase of mimicking steel construction, back to material-tailored sandwich construction. Due to significant improvements in composite material properties and processing technologies, the achievable spans became much larger, as shown in Fig. 4.16 (top right).

4.3.3.3 Influence factors and form evolution

The initial hypothesis of the above-mentioned PhD thesis [22] was that material properties govern the material-tailored structural form. The thesis results revealed, however, that not only one, but several (namely, nine) factors influence the form. The different factors intervene at three different stages of form development, i.e., at the stage of the ideal, constructible, and implemented form, as shown in **Fig. 4.19**.

The ideal form is influenced by four factors: function, socio-political context, knowledge, and technological thought, while still being independent of the material selection. An illustrative example is shown in **Fig. 4.20**, with the development of airships using plywood materials in the early 1900s. The ideal form was composed of two sets of crossing spirals, which formed the shape of the airship and resisted the

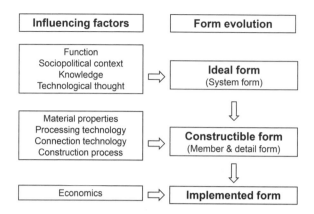

Fig. 4.19 Form influence-interaction model, derived from [22]

Fig. 4.20 Schütte-Lanz Airship made of plywood, 1909–1925, from ideal form (crossed spirals, **left**) to constructible form (crossed spirals composed of Vierendeel bars, **middle**) to implemented form (orthogonal truss and Vierendeel bars, **right**)

internal pressure in an optimal way, independent of the material (wood was used in the model) (Fig. 4.20, left).

The constructible form materialises the ideal form, i.e., member form and details such as joints are designed according to the selected material, see Fig. 4.20 (middle). The influencing factors in this process are the material properties, processing technology of the members, connection technology between the members, and the construction process. In the case of the airship, the spiral wires of the ideal form were transformed into plywood Vierendeel bars, and the joints were adhesively bonded and riveted. Economy finally intervenes in the transformation of the constructible into the implemented, i.e., built form. In the case of the airship, in the end it was more economical to transform the crossed spiral form into an orthogonal form of transverse rings and longitudinal bars (Fig. 4.20, right), although this required more material.

In addition to the material properties, the work in [22] identified that processing technology is particularly important in form development as it allows the ideal form to be materialised, i.e., to transform it into a constructible form that can optimally benefit from the material properties and thus contribute to an economic and sustainable use.

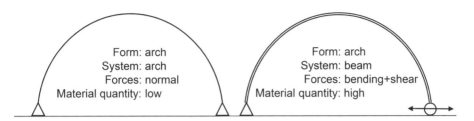

Fig. 4.21 Effect of support conditions on load-bearing mechanism and material quantity: arch system **(left)** vs curved beam system **(right)**

4.3.4 Material-tailored structural system

4.3.4.1 Introduction

In the previous Section 4.3.3, it has been shown how structural forms develop from the ideal to the constructible and finally the implemented form, depending on various influence factors. This section focuses on the ideal form, which is, as previously concluded, not related to material selection and properties. The term "ideal form" can also be associated with the better-known terms "static system" or "structural system"; the latter term is used in the following. The structural system is composed of the axes of the structure or members, complemented by the boundary or support conditions. Analogous to the ideal form, the structural system is independent of the material [24].

Support conditions significantly influence the load-transmission mechanisms of the structure, as illustrated in **Fig. 4.21**. Both structures exhibit the same parabolic shape, the difference being in the support conditions. While in the left case both supports are fixed, i.e., the displacements in the vertical and horizontal directions are blocked (triangle symbols, in-plane rotation is possible), one of the supports is allowed to move horizontally in the right case (circular symbol). This difference fundamentally changes the structural behaviour. In the left case, the load-bearing mechanism is that of an arch, while in the right case, the behaviour is that of a (curved) beam. The internal forces are different (consisting of only axial forces in the left and bending and shear in the right case), which greatly affects the required quantity of material, which is low in the arch and high in the beam case.

The interactions between structural systems and composite materials will be explored below, and the possible effects of the structural system on the required quantity of material will be discussed. Since composite materials are still expensive, a possible reduction in quantity, by selecting an appropriate structural system, may be of interest. Furthermore, prestressed arch and bending-active systems will be discussed, which allow for the construction of complex shapes in an economical way.

4.3.4.2 Form- and section-active structural systems

Structural systems can be classified according to the mechanisms by which forces are redirected from their application points to the support locations. A classification into form-, vector-, section-, and surface-active structural systems is established in [24], which is particularly useful for evaluating the appropriateness of the different types of structural systems for their use with composites.

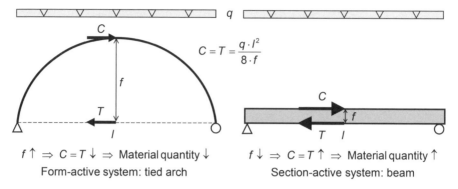

Fig. 4.22 Typology of structural systems: form-active **(left)** vs section-active **(right)** (C = compression, T = tension, q = load, l = span, f = arch rise or lever arm) [1]

Fig. 4.23 Form-active structural system (double-curved composite laminates for cavern interior panelling, engineer Heinz Isler) **(left)** vs section-active system (composite frame of Eyecatcher Building) **(right)**

The difference between form- and section-active systems is visualised in **Fig. 4.22**. Form-active systems, such as arches, shells, or membranes, are basically flexible, i.e., they do not exhibit any significant bending stiffness. The redirection of forces occurs through the particular and appropriate structural form, which causes only internal normal forces (i.e., axial or in-plane compression or tension). In contrast, section-active systems, such as beams and frames, are rigid through their cross section, and the force redirection occurs through the mobilisation of internal bending moments and shear forces. Apart from the stiffness, the difference between form- and section-active systems lies in the rise or lever arm (f). The much larger rise of form-active systems results in much smaller internal forces and thus far fewer material quantities, in comparison to section-active systems, where the lever arm (or effective depth) is short, and internal forces and required material quantities are comparably high, see Fig. 4.22. This difference is also illustrated in **Fig. 4.23**, where a thin-walled composite structure composed of (partially) double-curved laminates is compared to a composite frame structure composed of pultruded profiles – the difference in material quantity is significant.

Composites are usually processed into relatively thin laminates with low or limited bending stiffness. If such laminates are intended to be directly used as structural members, these members must obtain their stiffness from an appropriate structural form or system. As explained above, form-active structural systems, exhibiting single or double curvature, can fulfil this requirement.

Laminates can also be combined to form cross sections of profiles, e.g., flanges and webs of I-sections, thus forming members of the section-active category of structural systems. As explained above and shown in Fig. 4.22, these members nevertheless require a large quantity of material. In this respect, combining laminates with core materials in sandwich panel members is a much more economical way of designing section-active systems, where the laminates only form the face sheets and the core is composed of less expensive and lower quality materials, such as foam or balsa wood, see Fig. 1.11. Since these materials fill the whole core (and are not limited to singular webs as in profiles), the shear stresses are normally low and can be borne.

Profiles and sandwich panel members can, however, be combined on a much larger scale into form-active systems, thus still offering the advantage of an overall reduced material quantity. In this case, the members are not subjected to bending, but to axial or membrane forces. Large-scale shell structures can be designed, for example, so that the shell is not formed by laminates directly, but by profiles (in the case of grid-shells) or sandwich panel members.

Form-active structural systems composed of lightweight materials may, however, exhibit problems of form stability, as illustrated in **Fig. 4.24**, using the simplest case of a form-active system: a flexible cable fixed at both supports. If the permanent load (g) is much higher than a moving variable load (q), the structural form is maintained, and the system remains stable. On the other hand, if the permanent load is much smaller than the variable load due to a lightweight material, the flexible structure will adapt to the moving load and continuously change form, i.e., the system becomes unstable; form stabilisation measures are thus required to ensure stability. An efficient measure in this case is to add a prestressed cable with inversed curvature, i.e., a counter-cable, whose permanent deviation forces (Δg), caused by the prestress, will compensate for the low permanent load.

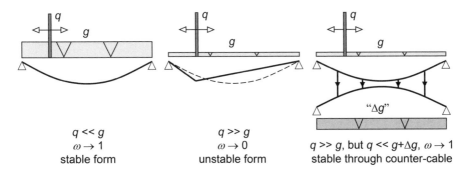

Fig. 4.24 Stabilisation of form-active structure by prestressed counter-cable (ω = self-utilisation ratio, g = permanent load, q = variable load, according to Eq. 4.2) [1]

Fig. 4.25 Form-active double-curved membrane structure (Star-wave Tent over gardens in Thowal, Saudia Arabia, PVC-coated PES (polyester) fabric with fluorine top-coat, 1990)

An exemplary form-active structure of minimal weight is shown in **Fig. 4.25**, i.e., a prestressed membrane structure, where the bearing directions are upwards-curved to the pylon tops, and the stabilisation directions downwards-curved to the floor anchoring points.

The problem of stability of form-active lightweight systems, as described above, can be quantified by using the so-called self-utilisation ratio, ω, defined as follows:

$$\omega = \frac{g}{g+q} = k \cdot \frac{l}{l_{\text{limit}}} \quad \text{with} \quad l_{\text{limit}} = \frac{f_u}{\gamma} \tag{4.2}$$

where k is a factor depending on the structural system, l is the length or span, l_{limit} is the limit length or span, f_u is the material strength, and γ is the specific weight of the material. The limit length can be visualised as a suspended cable or a column, which fails in the material under its self-weight; typical limit lengths are listed in **Table 4.3** for concrete, steel, and composites. The values vary from 1 km for a (non-buckling) lower strength concrete column, up to 150 km for a carbon fibre composite cable.

The resulting self-utilisation ratios are also listed in Table 4.3, where a smaller value indicates the material's greater ability to carry other loads relative to its self-weight. In the case of concrete, the values are high since the self-weight is normally high, and form-active systems are thus appropriate and stable without any further measures, e.g., in the case of a concrete arch. The ω-values of composites are low, however; form-active systems need stabilisation measures in such cases to increase the values, as mentioned above and also shown in Fig. 4.24.

The efficiency of counter-cables to stabilise form-active composite systems is also illustrated in **Fig. 4.26**. Two composite pedestrian bridges are shown where the suspended walkway, in one case, is stabilised by straight back-cables pulling the deck downwards, against the suspension cable forces, whereas, in the other case, the walkway is stabilised by a section-active system, i.e., deep and stiff lateral composite girders. The difference in the required material quantity is obvious. It must be acknowledged, however, that the considerable unsuspended span between the last stay cables and the opposing support (not visible in Fig. 4.26, right) also needs the section-active system.

4. Composites – Exploring opportunities

Table 4.3 Appropriate use of form-active systems [1]

Parameters	Concrete	Steel	Composites
Limit length, l_{limit} [km]	1–2[1]	4–15	15–150[2]
Self-utilisation ratio, ω [-]	0.8–0.5	0.6–0.3	0.4–0.1[2]
Form-active system	appropriate	appropriate from case to case	stabilisation required

[1] Column in compression, buckling excluded
[2] Glass to carbon fibre range

Fig. 4.26 Johnson County Suspension Bridge, USA, 1999 (slender composite deck with lateral back-cables, 128 m span) **(left)**; Kolding Cable-Stayed Bridge, Denmark, 1997 (all-composite, stiff girders, 40 m total length) **(right)**

4.3.4.3 Prestressed arch and bending-active systems

Based on their low weight, high strength, and relatively low bending stiffness, composites are well suited for application in prestressed arch or bending-active systems. In fact, both terms designate the same type of structural system. The designation "prestressed arch" was introduced in 1970 and defined as the shape that results from "buckling a thin strut or plate into an arched curve before attaching it to its supports" [25]. Research on prestressed arches continued and later also included grid-shells. The term "bending-active" appeared at the beginning of the 2010s [26], and research was henceforth conducted under this denomination. Curiously, no references were made to the large knowledge previously gained under the term "prestressed arch". Several of the characteristics of prestressed arches have thus been "re-discovered", such as the fact that the prestress reduces the load-bearing capacity for live loads [27].

Bending-active systems have the advantage that initially straight members can be transported, in a compact way, to the construction site and then installed by just bending the members into the final configuration, which can thus be of complex shape. Materials suitable for this type of structure are, in particular, timber and fibre-polymer composites, as identified in [26]. An early bending-active (prestressed) grid-shell of significant span was built in timber in 1974, as shown

Fig. 4.27 Bending-active (prestressed) grid-shell built of timber, Mannheim Multihalle, Germany, 1974 (spans up to 60 m)

Fig. 4.28 Bending of pultruded glass fibre composite profile (2400×100×10 mm) up to failure

in **Fig. 4.27** [28]. The grid was assembled on the ground, then lifted to its shape, and fixed at the supports before removing the suspension. The bending was thus imposed by the self-weight, and a prestress of the structure remained after removing the suspension. Another option to impose an arch shape to a straight member is to induce second-order deformations (i.e., "buckling", as described above) by shifting one support, as shown in **Fig. 4.28** [29].

Applying sustained prestress to composite arches or grid-shells in building structures, with a design service life of 50 years, may induce viscoelastic responses of the composite material, such as relaxation, creep deformations, and creep rupture. A study performed on the application limits of prestressed composite members for permanent roof structures demonstrates that spans up to about 30 m are nevertheless possible when using carbon fibres. The spans for basalt and glass fibres, however, are reduced to about 20 m and 10 m respectively, due to their higher sensitivity to creep rupture (see Section 3.4.5.2). The investigation also shows that the highest live load capacity can be obtained when members are bent up to the creep rupture stress limit. Moreover, since the prestress reduces the bearing capacity for live loads, possible relaxation may have a beneficial effect since the prestress is reduced [29].

The above-mentioned study was extended by investigating the short-term structural responses of pultruded glass fibre composite profiles under (i) bending the profiles up to failure (Fig. 4.28) and (ii) live load application up to failure (**Fig. 4.29**). In the latter case, the profiles were first bent to a targeted shape (bending degree, or prestress level) and then loaded by symmetric or asymmetric point loads. A bending

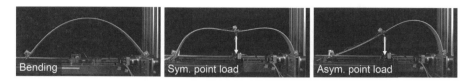

Fig. 4.29 Glass fibre composite profile bent to degree of 55% **(left)**; under subsequent symmetric point load **(middle)**; under asymmetric point load **(right)**

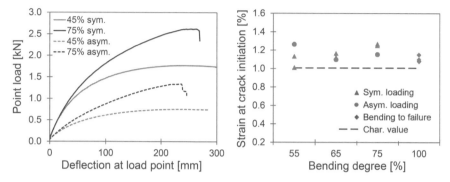

Fig. 4.30 Symmetric and asymmetric point load vs deflection at load point, at bending degrees of 45% and 75% **(left)**; axial strains on top edge at crack initiation vs bending degree and derived characteristic strain value **(right)**, based on data in [27].

degree of 100% is defined to be reached at the tensile stress at the top edge of the cross section when crack initiation occurs – this stress value was obtained from the bending experiments up to failure and associated numerical modelling. A bending degree of, e.g., 55%, thus means bending the profile up to 55% of that tensile stress.

The load vs deflection responses under symmetric and asymmetric point loads are depicted in **Fig. 4.30** (left) for bending degrees of 45% and 75%. As can be seen, the live load bearing capacity increases with higher bending degree. Furthermore, as is normally the case for arches, the capacity is lower under asymmetric than under symmetric loading. Both responses reach a maximum, at which, for bending degrees higher than 45%, cracks initiate and failure occurs. More specifically, transverse cracks initiate first at the edges on the top surface, approximately at the quarter points, where the largest curvatures and thus tensile strains are exhibited, as shown in **Fig. 4.31**. The cracks then propagate over the entire width and start to open and penetrate into the centre of the profile, which is composed of roving bundles. The crack direction then turns and delamination in the plane of the central roving layer initiates and propagates in both profile directions.

The axial strains (measured on the top surface), at which the first cracks initiate, are shown in Fig. 4.30 (right) for different bending degrees and symmetric and asymmetric point loads. The strains are almost independent of the load configuration because they only depend on the curvature. Since the first cracks always initiate close to the maximum load, the characteristic (5% fractile) value of the strain at crack

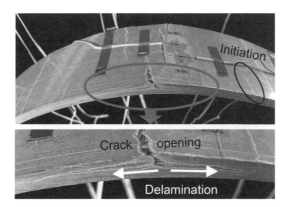

Fig. 4.31 Crack initiation and propagation during progressive creep rupture in bent pultruded glass fibre composite profile (cross section 100×10 mm) [27]

initiation can be applied for the ultimate state verification [27]. Considering that the elastic modulus is 23 GPa, the characteristic strain value corresponds to a characteristic tensile stress at crack initiation of about 230 MPa (which is 96% of the tensile strength indicated by the manufacturer).

A study of the long-term effects of a sustained prestress on creep and relaxation behaviour is currently ongoing at the CCLab. Profiles as used in the previous studies are being bent to different bending degrees (as in Fig. 4.29, left), and reaction forces, axial strains, and vertical deformations are being monitored. First results, after four months duration, show that the arch shapes do not exhibit changes due to creep deformations. The internal forces, however, exhibit a moderate relaxation, as depicted in **Fig. 4.32** for the horizontal reaction force, which may increase up to about 12% after 50 years (independent of bending degree), based on a power function extrapolation (for which the fit exhibits a high coefficient of determination). The most bent profiles (75% and 65%) already exhibit crack initiation during the four months, i.e., initiation of creep rupture, also see Fig. 4.32. Extrapolating the period for crack initiation for the bending degree of 55%, again based on a power function, reveals that crack initiation will occur after about 20 years. Furthermore, limiting the bending degree to 50% can postpone the crack initiation and thus creep rupture to a design service life of 50 years. The investigation of a possible (short-term) live load increase due to (long-term) relaxation will complete this PhD work.

In parallel to gaining scientific results, the design of a 1 000 m span pedestrian bridge in the Swiss Alps is ongoing, see **Fig. 4.33** and Appendix B.5. The suspension bridge spans a deep valley and is composed of 50 bending-active modules of 21.3 m length. Each module is composed of two inclined arches, which are bent from straight profiles at the construction site, after which the entire module is inserted between four cables; the modules are suspended from the two upper cables, and the two lower cables, with counter curvature, stabilise the bridge, according to the principle shown in Fig. 4.24 (right). Cables and arches are made from carbon fibre composites, while the deck is made of glass fibre composites.

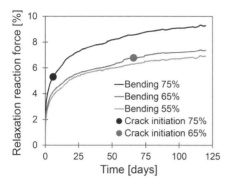

Fig. 4.32 Relaxation of horizontal reaction force vs time (over 4 months), for different bending degrees, and occurrence of crack initiation for two higher bending degrees (research ongoing) [30]

Fig. 4.33 1K-Bridge project: 1 000 m long pedestrian bridge in Lumino, Switzerland, composed of 50 modules of 21.3 m length with two bent arches each, suspended from two cables and stabilised by two counter-cables, cables and arches made of carbon fibre-polymer composites (Appendix B.5)

4.3.5 Conclusions

Each material has its characteristic language in formal appearance and technical detail, provided the material is used in a material-tailored structural form and not, for instance, in a substitution form, mimicking another material. A material-tailored structural form primarily takes the nature of the material into account, in terms of properties, processing, and detailing, and thus maximises the benefits derived from the material's unique properties, which is essential for providing economical and sustainable solutions.

Numerous factors may influence the development of a material-tailored structural form during different design stages, such as ideal, constructible, and implemented forms. Processing technology is one of the most important factors for developing economic and sustainable solutions. Investment in processing technologies is thus an effective way of ensuring that novel materials perform successfully.

The ideal form, in particular, is associated with the static or structural system, whose load-bearing mechanism significantly depends on the boundary or support conditions. Although neither ideal form nor structural system are related to the material, specific types of structural systems may be more or less appropriate for a certain material in terms of required material consumption. For composites, form-active systems are more appropriate in this respect, while section-active systems are less. An exception is the category of section-active sandwich structures, which also minimise the use of composite material. Appropriate structural systems can be designated as material-tailored structural systems, which thus represent a prerequisite for material-tailored structural forms.

4.4   TRANSPARENCY

4.4.1   Introduction

In contrast to conventional structural materials, composites may exhibit certain degrees of transparency in the visible spectrum of light if (i) glass fibres are used, (ii) the refractive indices of fibres and resin match, and (iii) the number of defects in the material is low [31]. As will be demonstrated below, such transparent or translucent composite laminates can still bear loads and may therefore be used in multifunctional members to build skylights of daylit buildings, or translucent beams, or to encapsulate solar cells (Section 4.5.4), thus adding energy harvesting to the structural and building physics functions. The degree of transparency can be graded according to the level of structural loading, with opaque members exhibiting the highest level and fully transparent members the lowest level of loading, as illustrated in the Dock Tower project (see Appendix B.1).

Key metrics regarding transparency are the visible total and diffuse light transmittances through the composite laminate, which need to fulfil specific requirements, as will be discussed below. The scientific background of light transmittance will be summarised, a case study derived, and the building integration of transparent laminates discussed. The theoretical content of this section is based on references [31–34] originating from a PhD thesis [35], completed in 2014 at the CCLab, and further references may be found therein.

4.4.2   Scientific background

The total light transmittance (T_t) of a composite laminate is defined as the ratio of the transmitted to the incident radiant flux of light and is composed of (i) the regular transmittance (T_r), i.e., the light crossing the laminate without changes in direction, and (ii) the diffuse transmittance (T_d), where the crossing light is scattered and changes direction [32]. Based on these two types of transmittances, the degrees of transparency and translucency of a laminate are defined as:

$$T_t = T_r + T_d \quad \text{with} \quad T_r/T_t = \text{ degree of transparency,}$$

and (4.3)

$$T_d/T_t = \text{ degree of translucency}$$

4. Composites – Exploring opportunities

Fig. 4.34 Translucent composite girder under load, 5.70 m span **(left)**; delamination in laminates (parallel grey bands) of corrugated sandwich web under load application point, based on [36] **(right)** (first full-scale experiments at EPFL, in 1998)

Transparency and translucency are thus related to regular and diffuse transmittance, respectively. In an ideal laminate with identical fibre-resin refractive indices and without any defects, the total transmittance is identical to the regular transmittance and is independent of the fibre weight and architecture, i.e., no light is diffused, and the laminate is totally transparent. Slight mismatches in the refractive indices (already in the range of 0.01), defects such as air inclusions (e.g., at fibre crossings), and poor fibre impregnation (due to inappropriate processing and/or curing) decrease the total transmittance and increase the diffuse light fraction, i.e., the translucency.

The same effects may also occur if translucent laminates are overloaded during the design service life and delamination occurs, as demonstrated in **Fig. 4.34** [36] (and Fig. 3.87). Delamination creates new interfaces inside the laminate, which involve multiple reflections and energy losses and thus reduce transparency.

The total light transmittance depends on the amount of light reflected (R) at the two air interfaces of the laminate and the loss of transmittance inside the laminate (L) as follows [31]:

$$T_t = (1-R) \cdot (1-L) \tag{4.4}$$

The total reflectance at the two air interfaces mainly depends on the mismatches of the refractive indices and may be calculated from the refractive indices of the resin (n_r) and air (n_a):

$$R = 2 \cdot [(n_r - n_a)/(n_r + n_a)]^2 \tag{4.5}$$

The loss of transmittance inside the laminate is mainly caused by internal scattering at internal defects and the associated interference and energy reduction. The loss may be assumed to depend on the total fibre weight (w_t), fibre architecture, and manufacturing and curing processes, which all influence the number of defects in the laminate. Assuming a reference fibre weight of 410 g/m^2 and a given loss parameter (c), the loss of transmittance can be expressed as [31]:

$$L = 1 - c^{w_t/410} \tag{4.6}$$

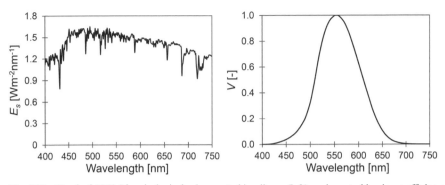

Fig. 4.35 Standard AM1.5 hemispherical solar spectral irradiance **(left)**, and spectral luminous efficiency function for photopic vision **(right)**, both in visible light range, derived from [31]

The total transmittance further depends on the wavelength (λ) of the light. For architectural applications, the visible total transmittance ($T_{t,vis}$) is relevant, i.e., its mean value over the wavelength range of the visible light, which is between approximately 380 and 750 nm. This $T_{t,vis}$-value may be approximated by the transmittance at the 555 nm wavelength ($T_{t,555}$), which is the wavelength where the solar spectral irradiance (E_s) and spectral luminous efficiency function for photopic vision (V) attain their maximum, see **Fig. 4.35**. The latter function represents the sensitivity of the human eye to a specific wavelength [31].

The laminate imperfections and defects that cause energy losses also cause light diffusion. The diffused fraction of the total light transmittance (DF) is identical to the translucency defined above (Eq. 4.3), and may be expressed as [32]:

$$DF = T_d/T_t = 1 - e^{-a \cdot w_{eq}} \quad \text{with} \quad w_{eq} = w_t \cdot (0.67 + V_t) \quad \text{and} \quad V_t = \frac{w_t}{\rho_f \cdot t} \quad (4.7)$$

where the exponential term with a negative sign represents the loss of transparency, a is a second loss parameter, which indicates the rate at which transparency is lost, w_{eq} is the equivalent fibre weight, V_t is the total fibre volume fraction, ρ_f is the glass fibre density (2600 kg/m^3), and t is the laminate thickness. The relevant translucency in the visible spectrum of light (DF_{vis}) may again be approximated by the translucency at 555 nm wavelength (DF_{555}).

For architectural applications, minimum values of total transmittance and translucency are specified as follows [32, 33]:

Daylit buildings: $\quad T_{t,vis} \approx T_{t,555} \geq 0.50 \quad \& \quad DF_{vis} = DF_{555} \geq 0.90$

Solar cell encapsulation: $\quad T_t \geq 0.90$

(4.8)

A high fraction of diffuse light is preferred in daylit buildings to avoid glare from direct sunlight, while the total transmittance should be maximised in the case of solar cell encapsulation. In the latter case, an additional 10% loss in efficiency and thus an additional 10% loss of total transmittance can be permitted for integrated photovoltaic modules with an enhanced aesthetic aspect [31].

In view of a multifunctional use of these transparent/translucent laminates in architectural applications, the question arises as to whether they can also bear a significant load. The design of composites is less a question of strength, but rather a question of the directional stiffness, which depends on the fibre architecture, see Section 5.2.6. To evaluate the directional laminate stiffness, the total fibre volume fraction parameter, previously discussed, is transformed into a directional elastic modulus (E_d) as follows [33]:

$$E_d = E_r + (E_f - E_r) \cdot V_d \quad \text{with} \quad V_{d,UD} = V_t \quad \text{and} \quad V_{d,\text{sym-CP}} = V_t/2 \qquad (4.9)$$

where E_r and E_f are the elastic moduli of resin and fibres, respectively, V_d is the directional fibre volume fraction, $V_{d,UD}$ and $V_{d,\text{sym-CP}}$ are the directional fibre volume fractions, given for a unidirectional (UD) and symmetric (50/50%) cross-ply (CP) laminate, respectively.

4.4.3 Laminate case study

The visible total transmittance and translucency relationships, as derived in the previous Section 4.4.2, are quantified for typical glass fibre-polymer composite laminates, as used in building construction, e.g., in the sandwich roof structure of the Novartis Building (see Appendix A.4). Unidirectional (UD) E-glass fibre fabrics of 410 g/m^2 and an orthophthalic polyester resin are used to compose UD and symmetric (50/50%) cross-ply (CP) laminates of 0.8–3.8 mm thickness, 410–3280 g/m^2 fibre weight (using 1–8 fabric layers), with fibre volume fractions varying from 0.24 to 0.42. Pure resin specimens are also considered. The refractive indices of both glass fibres and resin are measured and derived, and shown in **Fig. 4.36**. The mismatch is small in the range from 500 to 800 nm and increases to about 0.06 at 400 nm. The laminates are manufactured by a hand lay-up process; two different resin catalyst weights and curing temperatures (23 and 18°C) are applied to obtain two different gel times of 15 and 30 min (i.e., time lapse before a significant increase in resin viscosity occurs). Post curing is

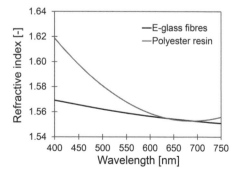

Fig. 4.36 Refractive indices vs wavelength of E-glass fibres and orthophthalic polyester resin, derived from [32]

Table 4.4 Calibrated loss parameters of total and diffuse transmittance models [33]

Laminate	Gel time [min]	c [-]	$a \times 10^{-4}$ [m^2/g]
UD Unidirectional	15	0.952	4.6
	30	0.970	2.1
CP Cross-ply 50/50%	15	0.942	5.7
	30	0.960	2.6

applied to obtain fully cured materials. Spectrophotometry is used to measure the total and diffuse hemispherical light transmittance and total reflectance [31–33].

Based on experimental results, the loss parameters c and a of the total and diffuse transmittance models (Eqs. 4.6–7) are calibrated and the results are summarised in **Table 4.4** for the two different gel times. The measured total reflectance at the two laminate-air interfaces and at 555 nm is $R_{555} = 0.091$. The relevant visible total transmittance ($T_{t,vis}$) may thus be expressed as:

$$T_{t,vis} = 0.909 \cdot c^{w_t/410} = 0.909 \cdot c^{V_t \cdot t \cdot 6.34} \tag{4.10}$$

where w_t is the total fibre weight in [g/m^2], V_t is the total fibre volume fraction [-], and t is the laminate thickness in [mm]. The translucency in the visible spectrum of light may then be expressed accordingly:

$$DF_{vis} = 1 - e^{-a \cdot f_t \cdot t \cdot 2600 \cdot (0.67 + V_t)} \tag{4.11}$$

Based on these calibrated models for these specific laminates, several relationships between the relevant parameters can be derived, as demonstrated in the following Figs. 4.37–43, which also indicate the above-mentioned transmittance limits for daylit buildings and solar cell encapsulation, see Eq. 4.8 [1].

For a certain fibre architecture and curing process, the visible total transmittance, according to Eq. 4.10, is independent of the total fibre volume fraction for a certain fibre weight, as shown in **Fig. 4.37**. An increasing fibre weight makes flawless fibre impregnation more difficult and thus increases the number of defects and decreases the total transmittance. The amount of resin surrounding the fibres of a certain weight, i.e., the fibre volume fraction, does not further affect the transmittance.

The effects of the laminate thickness on the two relationships of visible total transmittance and translucency vs fibre volume fraction are shown in **Fig. 4.38** for CP laminates with shorter gel time. With increasing laminate thickness, the total transmittance decreases. More specifically, if the fibre volume fraction is kept constant and the laminate thickness increases, the fibre weight must increase (see Eq. 4.7). Furthermore, for the same laminate thickness and with increasing fibre volume fraction, the weight must also increase. In both cases, the visible total transmittance thus decreases according to the previous results (Fig. 4.37). The translucency increases in accordance with the decreasing total transmittance, since more light is diffused with increasing fibre volume fraction and laminate thickness. The daylit limits of 50% visible total transmittance and 90% translucency can be met for laminates with $t \approx 6$ mm

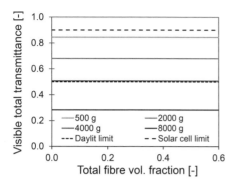

Fig. 4.37 Visible total transmittance vs total fibre volume fraction: effect of fibre fabric weight in [g/m²], for CP laminates and 15 min gel time

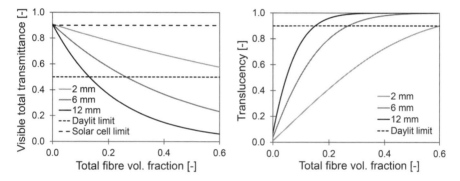

Fig. 4.38 Visible total transmittance (**left**) and translucency (**right**) vs total fibre volume fraction: effects of laminate thickness, both for CP laminates and 15 min gel time

and $V_t \approx 0.25$. The solar cell encapsulation limit of 90% visible total transmittance cannot be reached for this configuration, however, i.e., CP laminates with shorter gel time and 6 mm thickness.

The effects of the fibre architecture (UD or CP) and gel time (15 or 30 min) on the same two relationships of visible total transmittance and translucency vs fibre volume fraction are illustrated in **Fig. 4.39**, for a selected typical laminate thickness of 6 mm. CP laminates exhibit a significantly lower total transmittance and higher translucency than UD laminates. The reasons are mainly the CP fibre crossing points, which are difficult to fully impregnate and small air inclusions may remain, which reduce the transmittance and increase the diffusion. A longer gel time significantly increases the total transmittance and reduces the translucency, as demonstrated in **Fig. 4.40**, since the fibre impregnation is improved and the heat generation during curing and shrinkage of the resin and thus the number of defects is reduced. Fibre architecture and gel time are two parameters that can help to meet the daylit limits. For these 6 mm laminates, the limits are met for fibre volume fractions of approximately 0.25, 0.3, 0.45, and 0.55 if the gel times are CP-15min, UD-15min, CP-30min and

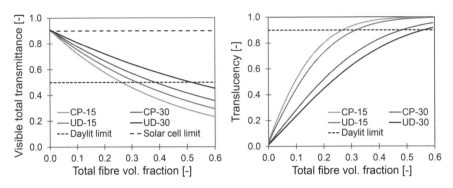

Fig. 4.39 Visible total transmittance **(left)** and translucency **(right)** vs total fibre volume fraction: effects of fibre architecture (CP or UD) and gel time (in [min]), both for 6 mm laminate thickness

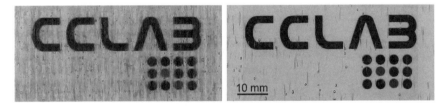

Fig. 4.40 Effect of gel time on light transmittance, UD laminates with text on rear side: 15 min gel time, $V_t = 0.29$, $t = 2.2$ mm: $T_{t,vis} = 0.75$, $DF_{vis} = 0.52$ **(left)**; 30 min gel time, $V_t = 0.24$, $t = 2.6$ mm: $T_{t,vis} = 0.80$, $DF_{vis} = 0.27$ **(right)**, derived from [32]

UD-30min are selected, respectively, see Fig. 4.39. The solar cell encapsulation limit cannot be reached at a laminate thickness of 6 mm.

For daylit applications, the highest and lowest achievable values of visible total transmittance and translucency, within the laminate parameter ranges and as a function of the total fibre volume fraction, are shown in **Fig. 4.41**. The different ranges of fibre architecture (UD or CP), gel time (15 or 30 min), and laminate thickness (2–12 mm) are covered. The highest transmittances and correspondingly lowest translucencies were obtained for the thinnest UD laminate with longest gel time. Accordingly, the lowest transmittances and highest translucencies were obtained for the thickest CP laminate with shortest gel time. The UD laminate cannot meet the daylit limits, while the CP laminate can attain them at a fibre volume fraction of approximately 0.15.

For solar cell encapsulations, the laminate thicknesses above the solar cells are much smaller than for daylit laminates, normally below approximately 2 mm (see example in Section 4.4.4 with 1.5 mm thickness). The highest achievable visible total transmittances are represented in **Fig. 4.42** for laminate thicknesses of 0.5–1.5 mm; they are reached for UD laminates with 30 min gel time. However, the fibre volume fraction required to meet the solar cell encapsulation limit of 0.9 exactly is low. If a

4. Composites – Exploring opportunities

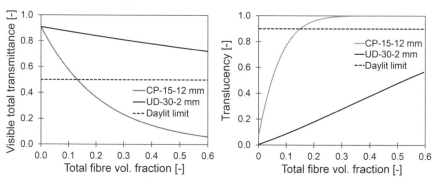

Fig. 4.41 Visible total transmittance **(left)** and translucency **(right)** vs total fibre volume fraction: highest and lowest values for daylit applications, covering different fibre architectures (UD or CP), gel times (15 or 30 min), and laminate thicknesses (2–12 mm)

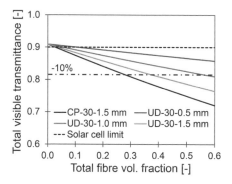

Fig. 4.42 Visible total transmittance vs total fibre volume fraction: highest values for solar cell encapsulation, covering different fibre architectures (UD or CP), and laminate thicknesses (0.5–1.5 mm), for 30 min gel time

10% lower limit of 0.81 is acceptable, reasonable fibre volume fractions of up to 0.4 can be reached and even CP architectures are possible. The maximum transmittance value of 0.909, at 0.0 volume fraction, results from the total reflectance of the laminate, see Eqs. 4.5 and 4.10.

The directional elastic moduli (according to Eq. 4.9) that can be achieved with a typical laminate of 6 mm thickness, which also meet the daylit limits, are shown in **Fig. 4.43**. Elastic moduli of 40–50 GPa can be obtained with a UD laminate and 30 min gel time. However, the elastic moduli of a corresponding 50/50% symmetric CP laminate are much lower, i.e., 15–20 GPa, since half of the fibres are in the transverse direction (see Eq. 4.9) and thus do not contribute to the directional stiffness. The possible UD values are high and the CP values still moderate for glass fibre-polymer composites and thus confirm the feasibility of transparent/translucent load-bearing members that also fulfil the daylit limits.

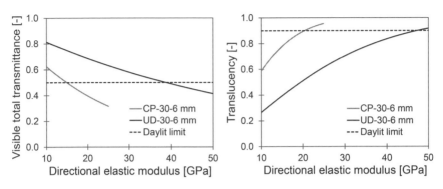

Fig. 4.43 Visible total transmittance **(left)** and translucency **(right)** vs directional elastic modulus: effects of fibre architecture, for 30 min gel time and 6 mm laminate thickness

4.4.4 Building integration

In contrast to conventional transparent zones in building envelopes, in most cases made of glass panels, which are heavy and discontinuous, and do not exhibit a primary load-bearing function, lightweight multifunctional transparent/translucent composite laminates may be directly and continuously integrated into primary composite load-bearing members, without additional joints. Two concepts of such integrations into sandwich roof structures are presented in the following. In the first concept, additional opaque solar cells are encapsulated into the top face sheet around the skylights, while dye-transparent cells may be directly integrated into the skylight in the second concept.

In the first concept, the sandwich structure is composed of glass fibre-polymer composite face sheets and a foam core – to also integrate thermal insulation, see Section 4.5 – as depicted in **Fig. 4.44**. The skylight is created by locally merging the top and bottom face sheets, which can easily be done using a typical vacuum infusion process. Opaque solar cells can be encapsulated into the top face sheet in the opaque sandwich zones.

Taking the composition of the composite sandwich roof structure of the Novartis Building as an example (see Appendix A.4), such a skylight and solar cell integration would exhibit the following characteristics:
- Existing composite face sheets: $t = 6$ mm, $w_t = 4340$ g/m², $V_t = 0.28$,
  mixed CP-UD architecture,
  merging of face sheets: $t = 12$ mm, $w_t = 8680$ g/m², $V_t = 0.28$,
  mixed CP-UD architecture,
  → for 30 min gel time, $c = 0.965$, and $a = 2.35 \times 10^{-4}$ m²/g:
  $T_{t,vis} = 0.43 \approx 0.5$, $DF_{vis} = 0.86 \approx 0.9$.
- Laminate above solar cells: $t = 1.5$ mm, $w_t = 1025$ g/m², $V_t = 0.28$,
  UD architecture,
  → for 30 min gel time, and $c = 0.97$: $T_{t,vis} = 0.84 \approx 0.9$.

The daylit and solar cell limits are thus almost met, the differences in relation to the limits being only 4–14%. If the aesthetic aspect of the solar cell encapsulation

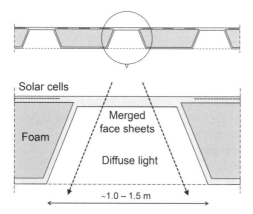

Fig. 4.44 Integration of skylights and opaque solar cells into a composite sandwich roof structure (face sheet thickness not to scale) [1, 33]

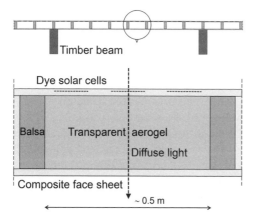

Fig. 4.45 Integration of dye-transparent solar cells into a composite sandwich roof structure (face sheet thickness not to scale) [1, 34]

is considered important and a 10% loss is thus acceptable, the solar cell limit would be met. The directional elastic modulus of the 6 mm face sheets was, based on measurements, 11.2 GPa, and thus quite low since the face sheets were manufactured by hand lay-up.

In the second concept, the whole roof forms a skylight. Translucent sandwich panels are installed on a main load-bearing roof structure, which may be composed, for instance, of timber beams. The sandwich structure comprises translucent glass fibre-polymer composite top and bottom face sheets, which are adhesively bonded to a grid of balsa beams, as shown in **Fig. 4.45**. Dye-transparent solar cells can be encapsulated into the top face sheet. The empty spaces between the balsa beams can be filled with transparent aerogel materials, which provide additional thermal insulation and maintain the translucency.

4.4.5 Conclusions

The quantitative evaluations carried out in the case study above confirm that glass fibre-polymer composite laminates may act simultaneously as load-bearing components of structural members and fulfil requirements for daylit applications and solar cell encapsulation. The achievable visible total, regular, and diffuse light transmittances significantly depend on the fibre architecture and manufacturing and curing processes. Unidirectional architectures may result in the highest total and regular transmittances, and lowest diffuse transmittances, while cross-play laminates normally exhibit the opposite. Hand lay-up manufacturing, as applied in the case study, produces more imperfections and defects than, for instance, vacuum resin infusion would; the diffused light fraction is thus increased. The gel time significantly influences the type of transmittance; longer gel times increase the total and regular transmittance, while shorter gel times increase the diffused light fraction. Furthermore, possible limitations of transparency caused by possibly required (transparent) protective coatings against UV radiation or moisture must also be considered.

Since, depending on the application and in addition to the structural function, opposing requirements may need to be met for the laminates, i.e., a high translucency for daylit applications and a high total transmittance for solar cell encapsulation, the laminate composition, and the manufacturing and curing processes need to be carefully designed. The relationships and values obtained in the case study may be used for preliminary designs in this respect.

The theory and discussion of this section refer to the visible spectrum of light. However, both are also applicable in building physics where thermal transparency, i.e., transparency across the infrared spectrum, is of significance, e.g., in the case of radiant heat exchangers, see Section 4.5.3. The same applies in the field of coloured laminates, where pigments can provide an opaque colour through internal reflections and scattering at the pigment particles, which increase with increasing mismatches in the pigment and resin refractive indices, see Section 4.5.5.

4.5 FUNCTION INTEGRATION

4.5.1 Introduction

Composites are primarily structural materials to be integrated into load-bearing structures, as is the case for reinforced concrete or steel. However, composites exhibit some unique properties, which allow the further integration of non-structural functions into composite members. Such properties are, at the material level, low thermal conductivity (carbon fibres excluded), possible transparency, and fibre flexibility. The use of composites as thin laminates also allows their integration into sandwich panel members with homogenous core or cellular core, i.e., web-core sandwich panels in the latter case, see Section 2.3.3. At such a system level, other functions, in addition to the structural function, can then be integrated into composite members.

Air and water tightness, and low thermal conductivity at the material level, as well as sandwich construction at the system level, allow the integration of building physics functions. The low thermal conductivity of composites and core materials

4. Composites – Exploring opportunities   303

Table 4.5 Possible function integration into multifunctional composite members, based on material or system properties [1]

Design aspects	Integrated functions	Property level	Book Section
Structural	Vertical and horizontal load transmission	Material	4.3
	Serviceability and structural safety	Material	4.3 & 5.1
	Fire resistance (thermal activation)	System	3.9 & 4.5.3
	Durability	Material	3.10
Building physics	Air and water tightness	Material	4.5.1
	Thermal insulation (sandwich construction)	Material and System	4.5.2
	Thermal conditioning (thermal activation)	System	4.5.3
	Energy harvesting (solar cell integration)	Material	4.5.4
	Sound absorption	System	4.5.1
	Fire reaction	Material	3.9
Architectural	Complex shape	Material	4.3.2
	Modular construction	Material	4.7
	Transparency	Material	4.4
	Colour, light, and texture	Material and System	4.5.5

of sandwich panels can provide the necessary thermal insulation for buildings, as listed in **Table 4.5**. Integration of water circuits into the cells of web-core sandwich panels can be used for thermal conditioning, i.e., heating, cooling, and provision of fire resistance. The possible transparency of composite laminates (if glass fibres are used) allows the encapsulation of solar cells into transparent laminates and thus the integration of energy harvesting. Acoustic ceilings for sound absorption can further be integrated into sandwich panels, as demonstrated in the roof structure of the Novartis Building, see Appendix A.4.

The flexibility of fibres and lightweight character of composites permit the design of complex structural shapes, e.g., double-curved or freeform shapes, and development of prefabricated modular construction, see Section 4.7. The possible transparency can serve to design skylights in daylit buildings. Furthermore, composites can be coloured, and light and surface texture can be integrated at the system level. All these options can make composites attractive materials in architecture.

Many of the non-structural functions mentioned above are normally fulfilled by the building envelope, which basically assures thermal insulation and weather protection and integrates a vapour barrier but does not provide any primary load-bearing function. Since conventional structural materials, such as reinforced concrete and steel, have high values of thermal conductivity, the insulating envelope is normally positioned on the exterior side of the building, and the load-bearing structure is thus thermally protected on the interior and not visible from the exterior, as schematically shown in **Fig. 4.46** (left). The load-bearing structure in these cases cannot be used as an architectural feature. Occasionally, the two layers are inverted, in most cases to use the load-bearing structure as an architectural element (Fig. 4.46, middle left). The suspension or fixation of the envelope often causes problems, however, since it is difficult to prevent thermal bridges; comfort and durability problems can occur in such cases,

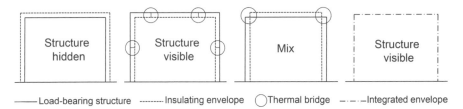

Fig. 4.46 Mutual positioning or merging of load-bearing structure and insulating envelope and effect on architectural expression (from exterior view) [1]

caused by local condensation, for instance. Allowing the interpenetration of structure and envelope (Fig. 4.46, middle right) exacerbates such problems.

The fact that composites can also fulfil the functions normally fulfilled by the envelope opens the possibility of merging the two layers of load-bearing structure and insulating envelope into a single-layer structural envelope, as sketched in Fig. 4.46 (right). The number of building elements can thus be considerably reduced, and thanks to the low weight and prefabrication, large elements can be designed that can be rapidly assembled on the construction site and thus shorten the construction period significantly. Controlled industrial modular prefabrication furthermore helps to increase the building quality and sustainability, see Section 4.7.

In such structural envelopes, the load-bearing structure also becomes an architectural feature, without creating the above-mentioned building physics problem. This reopens the possibility of reverting to architectural concepts and design principles as developed in the era of modernist architecture, which emerged in the first half of the 20th century, see **Fig. 4.47**, where building insulation was not yet a primary concern. The single-layer structural envelope can also exhibit complex shapes and thus directly form living spaces or adapt to the spatial organisation of buildings.

These concepts of merging structure and envelope are not recent since they were already developed in the 1950–60s in the first experimental composite houses, see Sections 2.7.2 and 4.3.3. Function integration, finally, is also a concept found in nature, as demonstrated and discussed in Section 4.2.3.

Fig. 4.47 Load-bearing structure used as architectonical element: Beinecke Rare Book Library, Yale University, Gordon Bunshaft of Skidmore, Owings & Merrill (SOM), 1963, envelope composed of Vierendeel steel truss, cladding in granite, and translucent marble filling-panels, exterior view **(left)** and interior view **(right)**

4. Composites – Exploring opportunities 305

In this section, the aspects of (i) thermal insulation, (ii) thermal activation, (iii) solar cell encapsulation, and (iv) colour, light, and texture, are discussed in detail. Other aspects, such as structural function, complex shape, modular construction, and transparency are covered in other sections, as indicated in Table 4.5. Relevant references can also be found therein.

4.5.2 Thermal insulation

4.5.2.1 Introduction

In Europe today, buildings are responsible for about 40% of the energy consumed and 36% of energy-related greenhouse gas emissions [37, 38]. According to [38], as of 2030, new buildings must be conceived as zero-emission buildings and, by 2050, existing buildings should be transformed into zero-emission buildings.

The energy consumption of buildings is directly related to the thermal insulation performance of the building envelope. In cases where non-insulating structural members penetrate the insulating envelope from the interior to the exterior, such as reinforced concrete balcony slabs, thermal bridges are formed in the insulation layer, which increase energy losses and may have other detrimental effects on indoor comfort [39].

As will be shown in the following, composites normally exhibit low thermal conductivity and are thus advantageous in terms of thermal insulation performance. In cases of reinforced concrete structures comprising thermal bridges, composites can be used to form thermal break components, which integrate structural and insulating functions. Such composite thermal break components, in the case of balcony slabs, for instance, allow the structural continuity of the concrete slab to be maintained through the insulating envelope, and composite materials with low thermal conductivity can significantly reduce energy losses via the thermal bridges [40, 41]. References [39–41] originate from a PhD thesis completed on this topic at the CCLab in 2016 [42]; further references can be found in that document.

4.5.2.2 Thermal conductivity of composite materials

Compared to the most frequently used structural materials today, i.e., steel and reinforced concrete, composites and related sandwich core materials normally offer much lower values of thermal conductivity, comparable to those of another well-established structural material, timber, see **Table 4.6**. The thermal conductivity of aramid and basalt fibres is very low, that of glass fibres is also low, whereas the values of carbon fibres are high and in the same order of magnitude as those of steel. Resin values are also very low, which result in low values for composite laminates, e.g., 0.3–0.6 W/mK in the case of glass fibres of different volume fractions. Typical structural core materials of sandwich panels, such as foam materials or balsa wood, also have very low values of thermal conductivity. The lowest values are offered by non-structural core materials, however, such as aerogel granulate.

The through-thickness thermal conductivity of composite laminates, e.g., sandwich panel face sheets, λ_{fs}, can be estimated by the inverse rule of mixtures as follows [43]:

$$\frac{1}{\lambda_{fs}} = \frac{V_f}{\lambda_f} + \frac{1 - V_f}{\lambda_r} \tag{4.12}$$

Table 4.6 Thermal conductivity of composite and conventional materials at ambient temperature, fibre values in fibre direction [6, 39, 40]

Composite materials				Conventional materials	λ [W/mK]
Fibres and resin	λ [W/mK]	Composites & cores	λ [W/mK]		
Aramid and basalt fibres	0.04	Glass fibre laminate	0.3–0.6	Timber	0.12–0.17
Glass fibres	1.0	Polyurethane foam 50–100 kg/m^3	0.028 -0.033	Steel	50
Carbon fibres	24–105	Balsa wood 100–250 kg/m^3	0.050 -0.085	Stainless steel	17
Polyester resin	0.2	Aerogel granulate	0.013	Reinforced concrete	2.5

where V_f is the fibre volume fraction, and λ_f and λ_r are the thermal conductivities of the fibres and resin, respectively.

In the case of a structural and insulating web-core sandwich panel – i.e., comprising a sandwich core with a parallel or orthogonal system of composite webs integrated into a thermally insulating material, a foam, for instance – the thermal conductivity of the complex core, λ_c, can be derived from the rule of mixtures [6]:

$$\lambda_c = \frac{\lambda_{ins} \cdot A_{ins} + \lambda_w \cdot A_w}{A_{ins} + A_w} \tag{4.13}$$

where λ_{ins} and λ_w are the thermal conductivities of the insulating material and webs, and A_{ins} and A_w are the areas of the insulating material and webs (perpendicular to the through-thickness heat flow), respectively. For a quadratic grid of orthogonal webs, the following can be derived:

$$\lambda_c = \frac{\lambda_{ins} \cdot b_{ins}^2 + \lambda_w \cdot 2 \cdot t_w \cdot (b_{ins} + t_w)}{(b_{ins} + t_w)^2} \tag{4.14}$$

where b_{ins} is the width of the cell-filling insulating material, and t_w is the web thickness.

4.5.2.3 Sandwich structures

Composite sandwich structures are composed of composite face sheets and a usually homogeneous core of structural foam or balsa wood. A structural foam in the core can, however, further serve as thermal insulation and thus, as mentioned above, permit the load-bearing sandwich structure to be merged with the building envelope. In web-core sandwich panel members without a foam core, the core is usually composed of a parallel or orthogonal web system of the same composite material as the face sheets to provide the shear load-bearing capacity. The resulting cell structure can be filled with a non-structural material of much higher insulation capacity than a foam, such as aerogel granulate, to also integrate the insulation function into the structural member.

The thermal insulation performance of a building envelope member is characterised by the thermal transmittance, U, which is calculated according to [44] as follows:

$$U = (R_{si} + 2 \cdot t_{fs}/\lambda_{fs} + t_c/\lambda_c + R_{se})^{-1} \tag{4.15}$$

where R_{si} and R_{se} are the internal and external surface resistances (0.1–0.2 and 0.04 m²K/W, respectively, independent of the materials [44]), t_{fs} and t_c are the face sheet and core thicknesses, and λ_{fs} and λ_c are the face sheet and core thermal conductivities, respectively.

Analysing Eq. 4.15 for typical λ-values of composite sandwich structural materials, as listed in Table 4.6, shows that the contribution of thin face sheets (as defined in [6]) to the thermal transmittance can be neglected, regardless of fibre type. For a typical ratio of core thickness to core web spacing $t_c/(b_{ins}+t_w) \approx 0.5$ (see below), the relationship of U-value vs core thickness can be derived from Eq. 4.15, as illustrated in **Fig. 4.48**, which is valid for structural polyurethane (PUR) foam and non-structural aerogel cores with integrated quadratic-orthogonal glass fibre composite webs of 6 or 12 mm thickness. The reference values of homogeneous cores without webs are also shown, while the case "Aerogel no webs" is theoretical since there is no vertical shear forces capacity.

Typical U-values for opaque envelope elements targeted for zero-emission buildings are in the range of 0.10–0.15 W/(m²K) [39]. According to Fig. 4.48, the thicknesses of, for instance, a sandwich roof with a structural PUR foam core without webs are 270–180 mm in this case. If 6 mm quadratic-orthogonal glass fibre composite webs are added to reinforce the shear capacity, the required thicknesses increase to 340–240 mm. Replacing the PUR foam by a more efficient non-structural aerogel insulation would reduce these high values to 230–170 mm, which, however, are in the same range as those of the structural PUR foam core without webs. The much better insulation capacity of the non-structural aerogel insulation may thus be offset by the necessary addition of structural webs. In contrast to structural foam, however, an aerogel insulation can offer a certain degree of transparency and thus be used in skylight members, for instance, as shown in Fig. 4.45.

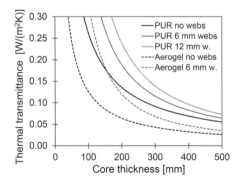

Fig. 4.48 Thermal transmittance vs thickness of sandwich core, consisting of polyurethane (PUR) foam of 50 kg/m³ density or aerogel granulate, with and without quadratic-orthogonal glass fibre composite webs of 6 or 12 mm thickness, for core thickness = 0.5·web spacing, λ_c = 0.028 W/mK, λ_w = 0.40 W/mK, $R_{si}+R_{se}$ = 0.2 m²K/W

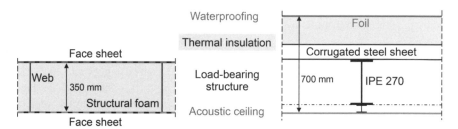

Fig. 4.49 Thickness comparison of function-integrated composite **(left)** and built **(right)** sandwich roof of EPFL Rolex Learning Center (dimensions in [mm]), according to [45]

Taking the example of the sandwich roof of the Novartis Building [20] (see Appendix A.4), with an average core thickness of 450 mm (varying from 50 to 600 mm), and PUR foam reinforced with a quadratic-orthogonal glass fibre composite web system of 925 mm spacing and 6 mm thickness, allows a low average U-value of 0.07 W/(m²K) to be achieved according to Fig. 4.48 (whereby $t_c/(b_{ins}+t_w)$ = (450–2×6)/925 = 0.47 ≅ 0.5, with 6 mm thick face sheet thickness). The glass fibre composite webs thereby increase the thermal conductivity from 0.028 W/mK (PUR) by 17% to an equivalent value of 0.033 W/mK (PUR plus webs).

Comparing a composite sandwich panel thickness of 250 mm, for instance – with U = 0.15 W/(m²K), 10 mm thick face sheets and a PUR foam core with 6 mm thick webs according to Fig. 4.48 – with that of a reinforced concrete slab of the same thickness (250 mm) and an added one-sided PUR foam insulation, results in an equivalent insulation thickness of 200 mm in the latter case and thus a total thickness of 450 mm, compared to the 250 mm of the integrated composite member. The member thickness can therefore be significantly reduced by integrating the thermal insulation into a structural composite envelope.

A similar result was obtained in a study for a composite sandwich roof for the EPFL Rolex Learning Center (see Appendix B.2), as shown in **Fig. 4.49**. The roof thickness was able to be reduced by 50% from 700 to 350 mm in this case by integrating all the functions into a structural composite envelope [45].

4.5.2.4 Zero-emission buildings

According to [37, 38], zero-emission buildings should exhibit a very low energy demand, which should be fully covered by energy from renewable sources generated on-site. As demonstrated above, function-integrated composite sandwich structural envelopes may contribute to significantly reducing the energy demand, in terms of heating energy, due to their possible high thermal insulation performance.

Energy transmission losses principally occur through the opaque and transparent envelope elements, with thermal bridges included in the former. On the other hand, transparent elements can provide solar gains. To evaluate the effects of the transmission losses via thermal bridges, a residential case study building is analysed in the following, as shown in **Fig. 4.50**; details can be found in [39]. Three types of envelopes with different thermal performances are assumed, i.e., a so-called "Minergie"

4. Composites – Exploring opportunities

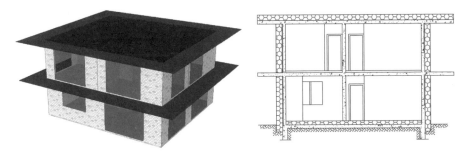

Fig. 4.50 Two-storey case study building with opaque and transparent envelope elements and roof and balcony projections, 7.1 m high, according to [39]

and "Minergie-P" envelope (according to Swiss standards, see [39], the latter being more efficient than the former), and an optimised envelope, comparable to that of a zero-emission building, thus offering the best performance. Thermal bridges exist along the balcony and roof projections of all three envelopes and thermal break components are positioned at these locations accordingly.

The linear thermal transmittances of currently used (conventional) thermal break components, consisting of foam materials and penetrating stainless steel rebars, are in the range of 0.25–0.30 W/mK, for reasonable projections of 1.2–1.5 m [39]. As will be shown later, much lower values of 0.1 W/mK can be achieved if composite materials are used. The case study is thus performed for three values of linear thermal transmittance, i.e., 0.10, 0.15, and 0.30 W/mK. For the nine combinations of envelope and linear thermal transmittance parameters, three metrics are analysed: (i) total transmission losses of the building, which include losses via the transparent (window) elements, (ii) transmission losses via the opaque elements only, and (iii) heating needs, which should be compensated by solar gains in the case of a zero-emission building.

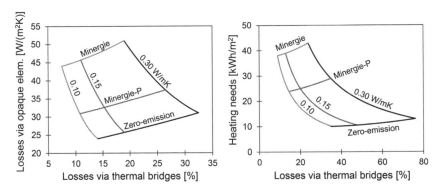

Fig. 4.51 Thermal losses via opaque envelope elements (**left**) and heating needs (**right**) vs proportion of losses via thermal bridges, for different linear thermal transmittances of thermal break components and envelope standards (in Switzerland), derived from [39]

The results demonstrate that the transmission losses via thermal bridges represent a small proportion of the total transmission losses, i.e., a maximum of 13% for the lowest performance thermal break implemented in the best performance envelope. The improvement of the thermal break performance from 0.30 to 0.10 W/mK has a minor effect, and the total transmission losses are reduced by only 6 or 8% for the lowest and best performance envelopes, respectively.

Referring to the losses via the opaque elements, however, changes the results significantly, as shown in Fig. 4.51 (left). The losses via the thermal breaks increase by up to 32%. They are reduced by the best performance thermal break by 12 or 18% for the lowest and best performance envelopes, respectively. The heating needs of the zero-emission envelope are reduced from 76 to 35% by switching from the lowest to the best performance thermal break, respectively (Fig. 4.51, right). The limitation of the linear thermal transmittance of thermal break components is therefore essential.

4.5.2.5 Thermal break components

Thermal break components based on composite materials to reduce the thermal transmittance were developed at the CCLab in several stages. The first designs were hybrid and combined pultruded glass fibre composite components and stainless steel rebars embedded in a layer of closed-cell extruded polystyrene foam. The stainless steel rebars were subsequently replaced by other glass fibre composite components [46, 47].

A later design is shown in **Fig. 4.52** and comprises materials with lower thermal conductivity, i.e., aramid fibre composites and aerogel granulate, see Table 4.6. The concrete slab is interrupted by a highly insulating PVC box, only 50 mm wide, filled with aerogel granulate, which can be adapted to the slab thickness. The box is penetrated on the top by aramid-composite loop elements, which transmit the tension forces across the insulation box, see free-body diagram in **Fig. 4.53** (left). Loop ends are selected to reduce the anchoring length. The bottom compression forces are transmitted via contact with a short rectangular full section glass fibre composite profile – aramid cannot be used in this case due to the limited compression performance. The shear forces are transmitted by a hexagonal aramid-PUR sandwich component; aramid can be used here since the diagonal compression forces are comparatively small. The hexagonal shape was selected to obtain a symmetric configuration, thus making the thermal break insensitive to the way it is placed (from the top) into the already installed concrete reinforcement, as shown in Fig. 4.52 (middle). The sandwich component can be manufactured by an automated aramid UD-tape winding process to produce large quantities of the component. The glass fibre compression element is integrated into the last of the three winding steps, see Fig. 4.52 (right) [1].

The linear thermal transmittance of the thermal break also depends on the U-value of the surrounding opaque element [40]. For $U = 0.10$ (0.20) W/(m^{2}K) of the opaque element and the densest spacing, i.e., 150 mm between loops and two aramid-glass fibre hexagons per loop to obtain significant projections of up to 4.0 m, low values of 0.16 (0.13) W/mK are obtained. If U = 0.15 W/(m^{2}K) and the spacing is tripled, the value decreases to 0.9 W/mK. The linear thermal transmittance values are thus considerably lower compared to those of current designs (see values above).

Fig. 4.52 Thermal break consisting of aramid loops, shear-compression hexagonal sandwich components and aerogel insulation (**left**); thermal break placed from top into previously installed concrete reinforcement (left side: building interior, right side: exterior balcony) (**middle**); tape-wound shear-compression hexagonal sandwich component (**right**) [1, 40]

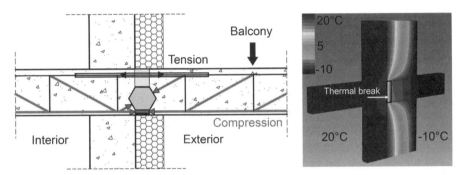

Fig. 4.53 Integration of composite thermal break into envelope insulation layer and free-body diagram (**left**); temperature field in thermal break zone (left side: building interior, right side: exterior balcony) (**right**), according to [40]

The resulting temperature field is also shown in Fig. 4.53 (right), with 20°C and −10°C being the assumed interior and exterior temperatures, respectively, during winter. The effect of the thermal break is clearly visible, the interior-exterior temperature gradient is maintained through the short distance of 50 mm of the thermal insulation. The internal surface temperature (almost 19°C) remains above the dew point temperature, thus avoiding possible interior condensation and maintaining indoor comfort.

With today's rapid development of basalt fibre composites, aramid and glass fibres can be replaced by basalt fibres, which have the same low thermal conductivity as aramid fibres (see Table 4.6), and also exhibit a significantly higher creep rupture strength and better resistance to alkaline concrete pore water (compared to glass fibres) [6].

4.5.3 Thermal activation

4.5.3.1 Introduction

Conventional thermal conditioning systems used in building construction, such as floor heating and ceiling cooling systems, are normally added to the top or bottom of concrete slabs. Both types are hydronic radiant systems and use polyethylene or copper piping with circulating water to supply or remove heat. In thermally activated building systems (TABS), the piping is embedded in the concrete slab.

Such water circuits can also be integrated into the core of composite sandwich slabs, or cells of web-core sandwich slabs can be directly used to circulate water. The possible high thermal/infrared transparency of composites can assist the heat exchange. In addition to the thermal conditioning function, the integrated water circuits can also be used to (i) activate forced slab cooling in the case of fire, and thus maintain structural integrity over the required fire resistance period (see Section 3.9.3), and (ii) cool solar cells encapsulated into laminates of sandwich structures (Section 4.5.4). Structural and building physics functions can therefore be merged into single members.

In the following discussion, a concept of thermal activation of composite sandwich slabs to provide thermal conditioning and fire resistance is presented and the use of composites as radiative heat exchanger is briefly addressed.

4.5.3.2 Composite hydronic radiant systems

A design of a thermally activated composite slab for building construction has been developed at EPFL in several steps [48, 49]; a cross section of the most recent slab design is shown in **Fig. 4.54** (left) [1, 49]. The load-bearing structure consists of a glass fibre-polymer web-core sandwich structure with the cells between the webs filled with a thermally insulating foam, if necessary (e.g., polyurethane foam). Thin water channels of 4 mm thickness are integrated into the top and bottom face panels. The channels are separated above the webs by 40 mm wide carbon fibre-polymer sections. The slab is manufactured by pultrusion of maximum four-cell sections (depending on the available pulling force), which are adhesively bonded together to form the slab. The foam can subsequently be injected into the cells, if necessary. The slab is operated as a Tichelmann system, i.e., with parallel flow in the channels, and water in- and outlets through the columns and connecting beams, as shown in Fig. 4.54 (right). The water flow is laminar at flow rates of between 0.8 and 2.2 cm/s.

The top channels provide floor heating or cooling while the bottom channels serve as ceiling cooling. The small carbon sections have two functions: (i) they increase the possible spans of the slab up to 10 m and (ii), due to their high thermal conductivity, they facilitate a uniform temperature distribution on the top surface and thus increase comfort; without the carbon sections, the temperatures would be lower above the webs.

The performance of the design was compared to that of standard concrete slab radiant systems (an embedded surface system (ESS) for floor heating/cooling and a radiant ceiling panel (RCP) for ceiling cooling). In the heating mode, the water temperatures and heat losses are lower, and the surface temperatures are more uniformly distributed in the composite design. Furthermore, the response time is faster (up to three times), and the overall energy required to operate the system is less. In the cooling mode, the system is slightly less efficient than a standard system; the advantages in the heating mode largely compensate for this difference, however [49].

A 60 min fire resistance can be provided by increasing the water flow rate to about 14 cm/s; the corresponding increase in water temperature is about 25°C. These

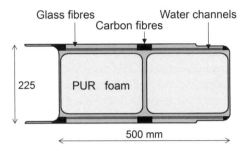 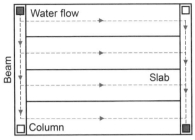

Fig. 4.54 Section through thermally activated fibre-polymer composite building slab **(left)**; slab operated as Tichelmann flow system, in plan view **(right)** [1, 49]

values were derived from fire experiments described in Section 3.9.3. Water channels, integrated into the top face sheets or core, can also be used to cool encapsulated solar cells and thus increase their efficiency, see Section 4.5.4 [49].

4.5.3.3 Thermal transparency

Glass composites may exhibit a high thermal/infrared transparency, i.e., for wavelengths above the visible light of 780 nm. They may thus represent efficient radiative heat exchangers, e.g., in a case as discussed in the previous section, where heat is exchanged between water and environment via the sandwich composite face sheets.

The degree of thermal transparency depends on the mismatches in the refractive indices between glass fibres and resin, and thus on the same rules as those for transparency in the visible light spectrum, as discussed in Section 4.4.2.

4.5.4 Solar cell encapsulation

4.5.4.1 Introduction

Solar (or photovoltaic, PV) cells are in most cases integrated into independent solar panels, which are attached to the building envelope. From an architectural point of view, these solutions are normally not satisfactory. To improve architectural integration, more complex elements were also developed with already integrated cells. These elements are usually part of the building envelope but remain separated from the primary load-bearing structure.

Solar cells must be protected by a transparent material; in most cases this material is glass, which makes the panels heavy, however. Thanks to the possible transparency of lightweight glass fibre-polymer composites (see Section 4.4.2), solar cells can also be encapsulated into composite laminates. Such laminates can then form components of primary structural members, e.g., the face sheets of composite sandwich panels, which can merge building envelope and load-bearing structure (see Section 4.5.1). In addition to the building physics and structural functions, energy harvesting can thus also be integrated into such panels.

Designs of solar cell encapsulation into transparent composite laminates must take three main aspects into consideration: (i) the heat production of the cells during operation, (ii) the structural composite action between cells and laminate, and (iii) the design service life of the cell system and building structure. In the following, different designs of solar cell encapsulation into transparent composite laminates and the limits of such integrations are discussed. The theoretical content of this section is based on references [34, 50], which originated from work done at the CCLab from 2010–2014; further references can be found in those documents.

4.5.4.2 Solar cell types and effects during operation

There are several types of solar cells, with different efficiencies, fields of application, and effects on mechanical behaviour when encapsulated into a composite laminate. Cell types investigated in [34, 50] were (i) polycrystalline silicon cells, (ii) thin-film amorphous silicon cells, and (iii) dye cells; the first two types are opaque, while the third is transparent, see **Fig. 4.55**.

Polycrystalline silicon (c-Si) cells have the highest efficiency (approximately 16%, on average); however, they exhibit high stiffness and brittleness and are thus difficult to encapsulate into laminates without being damaged. They are relatively small, and therefore also require numerous connections to be integrated into the laminate, which further complicates the encapsulation.

Thin-film amorphous silicon (a-Si) cells have lower efficiency (about 6%). However, in contrast to c-Si cells, they are flexible and available in different and larger sizes, so that the number of wire connections is reduced. Thin-film a-Si cells are therefore more suited to encapsulation into composite laminates than c-Si cells. Furthermore, thanks to their flexibility, they can be curved and thus integrated into more complex shapes.

Dye solar cells are transparent and can have different colours, but their efficiency is the lowest of the three types at about 3–4%. They comprise an ionic electrolyte, which needs to be confined between two substrates, normally made of glass. However, research is in progress to replace glass with polymeric materials, which reduces weight considerably and facilitates integration into composite laminates. As demonstrated in Fig. 4.45, due to their transparency, dye solar cells can be directly integrated into transparent areas of multifunctional building members.

Proportional to their efficiency, solar cells produce heat during their operation. As demonstrated in [50], laminates with encapsulated c-Si cells can heat up to the glass transition temperature of composites (80–90°C) on a sunny day (under assumed 700 W/m^2 simulated sunlight radiation on a horizontal laminate surface). The heating effect is amplified if an insulating foam core of a sandwich panel is located below the laminate and cell (as in [50]). In this respect, the glass transition temperature of the foam core should be considered, since often this value is lower than that of the laminate and the foam material may thus soften. Thin-film a-Si cells heat a laminate slightly less, but the temperatures may still approach the onset value of the glass transition temperature [50]. The heating by lower efficiency dye solar cells is not critical, however, and no detrimental effects were observed during experiments in [34].

Fig. 4.55 Encapsulated solar cells: opaque polycrystalline silicon cells into top face sheet of a composite sandwich panel (under mechanical load and simulated light), based on [50] **(left)**; dye-transparent cells into a translucent composite laminate, [34] **(right)**

In water-thermally activated composite sandwich panel members, as described in Section 4.5.3, the member components, affected by the heating of the solar cells, may be cooled by the integrated water circuits. This cooling also increases the cell efficiency, which depends on the operating temperature and decreases with increasing temperature.

4.5.4.3 Composite action between cells and laminates

Solar cells embedded between fibre layers of laminates exhibit the same strain under loading as the surrounding laminate material, i.e., full composite action, if the cell-laminate bond is perfect. Thin-film a-Si and c-Si cells encapsulated into the top face sheet of sandwich panels provide full composite action [50]. The cells are still operating, i.e., no loss of efficiency occurs, under serviceability limit state loads, e.g., during walking on a roof covered by encapsulated cells.

The encapsulation of dye cells into transparent laminates is more challenging [34]. Cells confined within glass substrates and encapsulated into laminates using a silane primer to improve composite action may nevertheless exhibit damage in the glass fibre laminate interfaces during thermal cycle loading. Dummy cells made of much more flexible polymethyl methacrylate (PMMA) substrates do not exhibit any delamination, however. Dye-transparent solar cells should thus only be encapsulated into composite laminates if they are based on polymeric substrates.

4.5.4.4 Design service life

The studies performed in [34, 50] are still of an exploratory nature. The long-term durability of the encapsulated systems has not yet been sufficiently verified. The thin laminate layer above the cells is not able to protect them from moisture during the full design service life. An additional transparent coating of the laminate must be applied to ensure appropriate protection.

Another point to consider concerns the design service life, which for buildings is normally 50 years. Today's design service life of solar cells, however, is shorter, i.e., 30–40 years for c-Si and even lower for a-Si systems, and their efficiency decreases slightly over these periods. Since the encapsulated cells cannot be easily replaced, integration is only useful if the design service life of cells and building members match.

Fig. 4.56 Contrast effect, same central grey square appears lighter or darker depending on frame

4.5.5 Colour, light, and texture

4.5.5.1 Introduction

Colour and light are directly related since colours are inherent to the spectrum of light visible to humans, and both are significant components of architectural design. The human perception of a colour depends on several factors, such as the spectral intensity distribution of the light source, the reflectance of the material surface, and the spectral sensitivity of the eye [51] (also see Section 4.4.2 and Fig. 4.35). This perception is subjective in many cases, as demonstrated by the contrast effect for instance, shown in **Fig. 4.56**; depending on the surrounding frame, the perception of the same shade of grey is different, darker or lighter. Another example is the Bezold-Brücke effect, according to which the perception of a colour shifts depending on the luminance of the object [51].

The appearance of a colour can be described by different attributes. Hue designates the pure colour of a pigment and can also be represented by a specific wavelength; white and black do not have a hue. A tint or shade lightens or darkens a colour by mixing it with white or black, respectively; both affect the colour saturation and colourfulness. Mixing a colour with grey changes its tone [51, 52].

Coloured composite laminates may be produced by directly adding pigments to the resin or applying coloured painting. Thanks to the possible transparency of composite laminates, light can be integrated in architectural concepts, in a natural or artificial way. Using translucent laminates in skylights of roofs, for instance, allows natural light to enter buildings and illuminate spaces, as discussed in Section 4.4.4. Translucent laminates of natural colour or coloured with pigments can also be illuminated by artificial backlights.

The surface finish of a composite laminate represents a significant functional and architectural design component. In addition to coatings and paintings to provide colour (see Section 2.8) and durability (3.10), different types of texture can be produced during manufacturing or post-applied.

4.5.5.2 Pigments

Coloured composite laminates can be manufactured by blending pigments into the resin. The pigments absorb parts of the light and reflect the complementary part; the

4. Composites – Exploring opportunities   317

colour is thus produced by colour subtraction. Depending on the pigment type and quantity, large colour ranges from pale pastels to deep shades can be achieved. Pigments may also be selected not for their colour but for appearance effects, such as fluorescence or phosphorescence; both absorb light, which later is re-emitted [51, 53].

Pigments are available in dry powders, pastes, or liquids. The dispersibility of pigment powders in the resin depends on the particle size, shape, and wettability, and the resin curing process may be altered by the addition of pigments. Inorganic and organic pigments exist but are fundamentally different, the former being metallic compounds, while the latter mainly contain carbon [51, 53].

Inorganic pigments can provide an opaque colour, i.e., they can hide the resin matrix or background colour through internal reflections and scattering at the pigment particles, which increase both with increasing particle size and differences in the pigment and resin refractive indices, see Section 4.4.2. Organic pigments are normally of smaller particle size and thus have less opacifying properties. Inorganic pigments in general have a high thermal and light stability but a limited brightness of colour. Organic pigments in contrast exhibit less resistance to heat and light but can offer brilliant shades [51, 53].

As regards durability, resin and pigment degradation may interact, and together can produce colour shifts (e.g., yellowing) and colour fading if exposed to weathering and daylight, in the latter case particularly due to the ultraviolet component of solar radiation [51, 53, 54].

4.5.5.3 Painting

The simplest way of adding colour is by applying paint to the composite member surface. In most cases, composite members must be protected by coatings against environmental exposure, see Section 3.10. Most of these coatings are polymer-based, and coloured pigments can be added to such coatings, which in this way become multifunctional [51]. Design and maintenance planning should consider that the design service lives of paint and coatings are usually much shorter than those of buildings and bridges (10–20 vs 50–100 years) and they must thus be renewed periodically.

4.5.5.4 Backlighting

Pigments may also be used to colour translucent glass fibre composite laminates and associated backlighting can transform them into expressive elements of architectural concepts, as shown in **Fig. 4.57**; the laminates in this case are not load-bearing, however.

On the other hand, translucent glass fibre composite laminates can remain uncoloured, and colour may be provided by coloured backlights, as shown in **Fig. 4.58**; the 4.5 mm thick laminates of the 12 m high sculpture are load-bearing in this case and have changing red, green, blue, and yellow backlights [55].

4.5.5.5 Texture

The surface of composite laminates can exhibit very different textures, ranging from leaving the fabric structure of carbon fibres visible below a smooth and glossy surface,

Fig. 4.57 Green- and blue-coloured translucent glass fibre composite laminates with white backlights (installed behind the coloured laminates), Zurich main station, 2003

Fig. 4.58 Uncoloured translucent glass fibre composite laminates with LED backlight of different colours, D-Tower, NOX Architecture, Doetinchem, Netherlands, 2004

for instance, to deep reliefs made possible by correspondingly shaping the moulds for manufacturing, as shown in **Fig. 4.59**, or negative mould inlays for smaller-scale textures, see **Fig. 4.60**. Texture can also be post-applied, e.g., by applying sanding to create a non-slip surface on pedestrian bridge decks, see Fig. 2.12 (top and bottom left).

4.5.6 Conclusions

Thanks to the non-structural material and system level properties offered by composites, the integration of non-structural functions into structural composite members is possible. Some of these integrations are already well established, such as thermal insulation integrated into composite sandwich structures, and the integration of colour and light. Other options are still at the research stage, such as hydronic thermal activation and solar cell encapsulation. The possible merging of load-bearing structure and insulating envelope into a single-layer structural envelope may offer significant advantages in terms of construction time, quality, and cost. Floor or roof constructions can also be designed to be much thinner, which may allow higher spaces, or an increased number of storeys, or reduced building heights.

4. Composites – Exploring opportunities   319

Fig. 4.59 Glass fibre composite arches in HR Giger Bar, 2003, Gruyères, Switzerland **(left)**; glass fibre composite envelope, San Francisco Museum of Modern Art expansion, 2015 **(right)**

Fig. 4.60 House of Dior, Seoul, 2015, with diamond-shaped texture: negative mould inlay **(left)**; detail of resulting texture (middle left); installation of envelope element with texture and detail **(right)**

Nevertheless, the integration of functions into multifunctional members has limitations. If the added function has its own components to be integrated into the cells of sandwich panel members, for instance, it should be possible to rehabilitate or exchange those components in case of defects, and access to such components is therefore necessary. Furthermore, the design service life of different components of different integrated functions should be similar; if not, components with shorter design service life must be periodically replaceable – the design of multifunctional members should therefore be conceived in such a way as to allow for such replacements.

4.6 HYBRID CONSTRUCTION

4.6.1 Introduction

The term "hybrid construction" includes hybrid structures and hybrid members, as defined in Section 1.1. Established structural materials offer some individual outstanding properties (which has been the reason for their success), while the remaining properties may be of an average or even lower-than-average level. Concrete, for instance, has a high compressive but very low tensile strength. Steel normally offers excellent ductility, strength, and stiffness, whereas its susceptibility to corrosion and high thermal conductivity may present disadvantages. In general, one material may exhibit an outstanding property where another material is weak and vice versa. The primary aim of hybrid construction is thus to combine different materials in a way that each design aspect can benefit from an outstanding property and is not governed by an average or inferior property. This approach can result in very economical constructions, in contrast to "single material" solutions, where certain design aspects are often unavoidably determined by a lower property level, and material usage and cost must be increased accordingly.

Joints and connections play an important role in hybrid construction, since they must assure the load transmission, i.e., the composite action, between the members or member components of different materials. Adhesive bonding has proven to be an appropriate technique in this respect, since it generally allows different materials and geometries to be easily connected and component tolerances to be absorbed by layer thickness variations (see Section 3.3.3).

4.6.2 Hybrid composite construction

Weaknesses occurring in composites, such as lack of ductility or matrix-dominated responses, may be compensated for by employing hybrid composite construction. There are numerous possibilities for the ways in which composites can be combined and connected with other materials, such as steel, concrete, or timber. Three typical examples demonstrating this wide variety are introduced and discussed below. The examples concern sandwich panels, used as slab bridges (without additional main girders) or bridge decks (on steel girders) with deck-girder adhesive bonds; they were the subject of three PhD theses written at the CCLab [56–58] in 2005–2014. The examples are complemented by a discussion about composite action between member components.

- Example-1: Hybrid composite-balsa sandwich slab bridges and bridge decks
A research project was conducted in parallel with the design and construction of the Avançon Bridge, which is composed of a composite-balsa core sandwich deck, adhesively bonded onto two steel girders (Appendix A.5). The project aimed at further optimising the sandwich panel so that it could also be used for slab bridges (without additional girders) and as bridge deck in new hybrid bridges of larger spans, or for concrete deck replacement [58]. The optimisation was in terms of (i) shear capacity and weight of the balsa core and (ii) in-plane stiffness (and thus composite action). The combination of these composite decks with steel girders can provide the required ductility to the bridge superstructure.

The results of this work are summarised in **Fig. 4.61**. The shear capacity of the sandwich panel core can be increased by inserting a glass or carbon fibre composite arch into the core, and the weight reduced by using a low-density balsa below the arch [59]. The insertion of the composite arch can be done by CNC cutting of the balsa core and subsequent vacuum infusion of the panel. By applying this concept, slab bridges with spans of up to almost 19 m can be designed [60]. To increase the in-plane stiffness of the panel along the bridge span, high-density timber (e.g., Douglas fir) inserts and (if necessary) supplementary steel plate inserts (protected by the face sheets) can

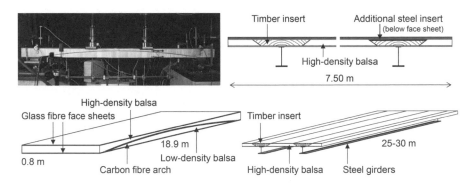

Fig. 4.61 *Example-1*: Hybrid concepts of composite-balsa slab bridges and bridge decks: carbon fibre composite arch between upper high-density and lower low-density balsa core in experimental beam **(top left)**; slab bridge with integrated carbon fibre arch of 18.9 m span and 0.8 m thickness **(bottom left)**; high-density timber and supplementary steel inserts in balsa core of bridge decks above main girders to increase composite action **(top right)**; hybrid deck-girder bridge with spans of up to 30 m, or 25 m in case of concrete deck replacement **(bottom right)**, [1, 59, 60]

be incorporated into the balsa core, above the girders and across the effective width. This stiffness increase makes spans of hybrid deck-girder bridges of up to 30 m possible or allows the replacement of concrete decks by composite decks up to bridge spans of 25 m [60]. Such replacements are normally not possible with lower in-plane stiffness decks, such as sandwich decks with only homogeneous balsa core or web-core pultruded decks, since the decks cannot reach the stiffness of the concrete decks (as would be necessary without reinforcing the steel girders) [1].

- Example-2: Hybrid composite-concrete sandwich bridge deck

A different approach to designing a hybrid sandwich bridge deck was pursued in this second example, as shown in **Fig. 4.62**. Composites are used in the most effective way, i.e., in fibre-dominated tension, in panels placed in the bottom of the cross section, while the top compression layer consists of normal or high-performance concrete, thus avoiding a matrix-dominated use of composites in compression. Lightweight concrete (LC) is selected as the core material, which can be connected (wet-in-wet) to the top concrete layer. To connect the LC to the composite panels (which also serve as formwork), T-upstands are added to the panel sheet to provide mechanical interlock; the panel is composed of pultruded sections. The LC-composite connection can be reinforced by adding an adhesive bond in the interfaces. The manufacturing of this bond is easy, the wet LC can be poured onto the still wet (epoxy) adhesive, and both can cure simultaneously. This hybrid composite-concrete bridge deck is "steel-free" and does not need sealing if dense high-performance concrete is used in the top layer [1].

Also shown in Fig. 4.62 are results of three-point bending experiments (selected to have spans of constant shear). The beams with additional adhesive bond exhibited high ultimate loads and stiffness, but almost brittle behaviour through brittle shear failure in the LC core. The beams with only mechanical interlock failed at much lower

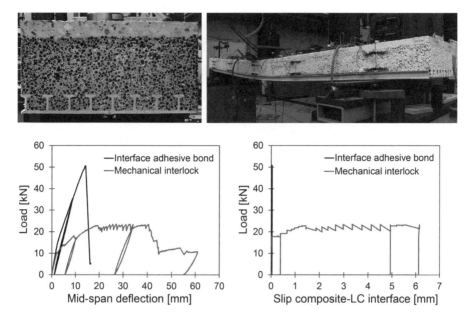

Fig. 4.62 *Example-2*: Hybrid composite-concrete sandwich panel with cross section (400×200 mm) composed of glass fibre-polyester composite sheet (5 mm thick) with T-upstands (35 mm, to also be used as formwork), lightweight concrete (LC, 130 mm thick) and normal or high-performance concrete (30 mm) [1] **(top left)**; beam with mechanical interlock at failure (three-point bending, 3.0 m span) **(top right)**; load vs mid-span deflection responses **(bottom left)**; composite-LC interface slip, for composite-LC interface with and without epoxy adhesive bond **(bottom right)**, based on data in [61]

loads, through progressive shear failure; they exhibited significant pseudo-ductility, however, provided by this progressive failure and friction during slip in the LC-composite interface [61].

Since the pure mechanical interlock may be sensitive to fatigue, an application of this option in bridge construction seems inappropriate. In building construction, however, valuable and ductile solutions may be achieved. The adhesively bonded option is appropriate in bridge construction, where a combination with steel girders can compensate for the lack of ductility.

- Example-3: Hybrid composite web-core deck-steel girder bridges

The subject of one of the first research projects at the CCLab was adhesively bonded connections between pultruded web-core sandwich bridge decks and steel girders. The work started in 2001 and the results, published in [62–64], were the basis for the application of this connection technique in several hybrid-composite bridges in Europe. Previously, the decks were connected in a less material-tailored way, e.g., by steel shear studs, grouted into cavities cut into the decks [21]. The research was based on full-scale four-point bending static and fatigue experiments on hybrid girders composed of two commercial pultruded bridge decks, adhesively bonded onto the

4. Composites – Exploring opportunities

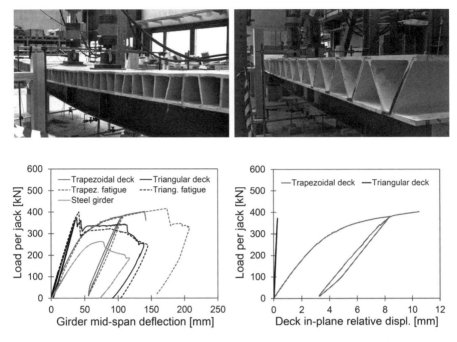

Fig. 4.63 *Example-3*: Hybrid adhesively bonded composite deck-steel girders with trapezoidal or triangular deck section (under four-point bending, 7.5 m span): trapezoidal deck girder at failure with waviness of face panels due to Vierendeel bending **(top left)**; triangular deck girder at failure with straight face panels due to truss axial forces **(top right)**; load vs mid-span deflections, effect of fatigue (10 million cycles), comparison to pure steel girder **(bottom left)**; effect of deck geometry on deck in-plane relative displacements (internal composite action of face panels) **(bottom right)**, based on data in [62, 63]

top flanges of steel girders, see **Fig. 4.63** (top) and Section 2.6.2; ten million fatigue cycles were applied in the fatigue experiments.

The main difference between the two composite bridge decks was the shape of the cellular cross sections, i.e., either trapezoidal or triangular, which led to very different load vs mid-span deflection responses, as shown in Fig. 4.63 (bottom left). While the triangular deck girders exhibited a rather brittle behaviour, the trapezoidal deck girders showed a significant pseudo-ductile response. This pseudo-ductility could be attributed mainly to the trapezoidal deck shape, which caused transverse Vierendeel bending in the sections. The high bending moments in the web-face panel joints produced progressive local failures, which resulted in differential in-plane displacements between upper and lower face panels, i.e., a reduction in composite action inside the deck, as shown in Fig. 4.63 (bottom right). The triangular deck, however, exhibited a truss-like behaviour, i.e., mainly axial forces were induced in the sections. The in-plane composite action was only slightly reduced, which resulted in almost negligible differential face panel displacements. More details of these effects of the cellular shape are discussed in Section 3.6.3 (*Example-1*).

In the trapezoidal deck girders, the entire steel beams yielded at deck failure (the upper parts in compression and the lower parts in tension); while in the triangular deck girders only the bottom parts yielded (in tension). The yielding in this latter case could not fundamentally change the almost brittle behaviour of the hybrid girder. Also shown in Fig. 4.63 are the response of the pure steel girder, and accordingly, the significant increase in bending capacity provided by the two bridge decks in this hybrid configuration [62, 63]. This contribution of the deck decreases, however, if the span increases, and may become insignificant for longer spans, unless the deck stiffness is increased, as discussed in *Example-1* above.

The girder responses further demonstrated that the preceding fatigue loading had no notable effect on the static behaviour. The adhesive bond itself (6–10 mm thickness), resisted the fatigue loading without any signs of damage; full composite action between decks and girders was maintained up to ultimate failure [62, 63], see also next Section 4.6.3.

More critical than the load-bearing behaviour of the adhesive connection in the longitudinal direction is its behaviour in the transverse direction, particularly in multi-girder bridges or bridges with large overhangs, see **Fig. 4.64** (left). Depending on the torsional stiffness of the girders and the position of the load, uplift forces and associated through-thickness tensile stresses may occur in adhesive connections. If the torsional stiffness is small, e.g., in the case of slender open-section steel girders (top left), the top flanges of the girders just follow the rotation of the deck, without resistance, and uplift forces may occur far away from the load application location. In the case of girders with high torsional stiffness, e.g., concrete girders (bottom left), those next to the load location resist the deck deformation and transverse bending moments may apply to the adhesive connections, which may result in significant through-thickness tensile stresses at one of the connection edges. This latter case was experimentally and numerically investigated under static and fatigue loading (using compact and stiff transverse steel girders), as shown in Fig. 4.64 (top right). No stiffness degradation

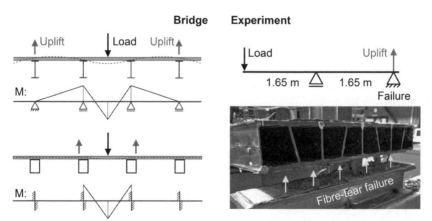

Fig. 4.64 Potential uplift forces in adhesive composite deck-steel girder connections of multi-girder bridges as a function of girders' torsional stiffness **(left)**; experimental set-up with (partial) fibre-tear failure in adhesive connection of trapezoidal deck subjected to uplift forces **(right)**, based on data in [65, 66]

4. Composites – Exploring opportunities   325

was observed after ten million fatigue cycles; the post-fatigue static failure load, however, decreased markedly. This decrease was attributed to the failure mode, which was
mixed (right bottom), in contrast to a deep fibre-tear failure at higher failure loads of
static experiments without preceding fatigue. The differences in failure modes could
be traced back to differences in surface preparation, and not to effects of fatigue loading [65, 66].

4.6.3   Composite action

In the previous section, two types of composite action were presented, i.e., composite
action between the upper and lower face panels of web-core sandwich bridge decks,
and composite action provided by adhesive connections between composite bridge
decks and steel girders. Both types of composite action are further discussed in the
following, using a factor of composite action, γ, derived from [67], which is first
applied to the sandwich panel case as follows (derived from [60, 64]):

$$\gamma = \left(1 + \frac{\pi^2 \cdot E_1 \cdot t_1}{K \cdot l^2}\right)^{-1} \tag{4.16}$$

where E_1 is the elastic longitudinal deck system modulus, t_1 is the thickness of the
upper face panel, K is the in-plane shear stiffness of the deck system, and l is the girder
span. A resulting factor of $\gamma = 1.0$ or 0.0 signifies full or no composite action, respectively. Full composite action is understood, in this respect, as exhibiting a plain strain
distribution through the depth of the cross section.

The corresponding values for the trapezoidal (triangular) deck sections introduced in *Example-3* of the previous Section 4.6.2 are [64]: $E_1 = 11\,700$ (16 200) MPa,
$t_1 = 18.0$ (15.6) mm, $K = 0.026$ (0.209) MPa/mm, respectively. Furthermore, sandwich panels with balsa and balsa/timber core, as presented in *Example-1*, are added
to the investigation; the values in this case are [60]: $E_1 = 40\,000$ MPa, $t_1 = 25.0$ mm,
$K = 1.160$ and 3.364 MPa/mm for balsa and balsa/spruce, respectively. In all these
cases, it is assumed that the effective widths of the lower face panels/sheets and
widths of the upper panels/sheets are identical.

As can be seen in Eq. 4.16, the factor of composite action depends on the
span of the hybrid girders; the corresponding relationships for the different deck-
girder systems, for spans up to 25 m, are shown in **Fig. 4.65** (left). The curves
demonstrate that the degree of composite action increases rapidly with increasing
span and subsequently levels off. High values of $\gamma > 0.9$ are obtained for spans
$l > 10$ m, except in the case of the trapezoidal web-core deck, for the reasons
explained in *Example-3*. The highest values are obtained for the sandwich deck
with mixed balsa/spruce core, which thus allows the largest spans of this bridge
type, up to 30 m, see *Example-1*.

A similar analysis can be performed for the composite action provided by the
adhesive bond between the composite bridge decks and steel girders, as follows
(derived from [60, 64]):

$$\gamma = \left(1 + \frac{\pi^2 \cdot E_1 \cdot A_1}{K \cdot b \cdot l^2}\right)^{-1} \tag{4.17}$$

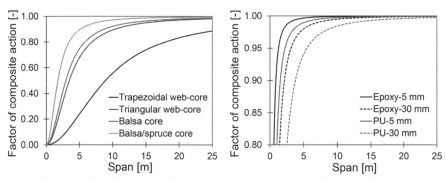

Fig. 4.65 Factor of in-plane composite action (g) between face panels/sheets of different composite bridge decks bonded onto steel girders (**left**); and of epoxy and PU adhesive connections with different thicknesses between composite bridge decks and steel girders, as a function of bridge span (**right**), derived from [60, 64] ($\gamma = 1.0/0.0$ = full/no composite action)

where A_1 is the area of the lower face panel/sheet and b is the width of the adhesive bond on the top steel flange. The analysis is performed for the triangular deck girders with $A_1 = 23\,400$ mm^2 and $b = 196$ mm, and based on four types of adhesive material-thickness combinations, i.e., an epoxy adhesive (shear modulus: $G = 1\,110$ MPa), with layers of 5 mm and 30 mm thickness ($K = 222.0, 37.0$ MPa/mm, respectively), and a polyurethane (PU) adhesive ($G = 360$ MPa), with the same 10 and 30 mm layer thicknesses ($K = 72.0, 12.0$ MPa/mm, respectively). The values are derived from [64].

The resulting degrees of composite action, as a function of the span, are shown in Fig. 4.65 (right). All cases provide high degrees of composite action ($\gamma > 0.95$) for spans larger than 6 m. Even a 30 mm layer of softer polyurethane adhesive produces almost full composite action, if the adhesive remains in its elastic range. The shear stresses in such adhesive layers are, however, normally small since the bonded areas are large, and possible local stress peaks do not affect the composite action. In the girders discussed in *Example-3*, the shear stresses at failure varied between 1.9 and 2.4 MPa, while they were only 0.4 MPa at the serviceability limit state [62, 63].

4.6.4 Conclusions

The efficiency of hybrid composite structures has been discussed and demonstrated. Examples were shown in which composites can be combined with other materials to increase the load-bearing capacity, or to compensate for either limited material ductility or less efficient matrix-dominated behaviour. Hybrid composite solutions, in this respect, may outperform all-composite ones in many cases, since it cannot be excluded, in the latter case, that some dimensions are based on matrix-dominated properties and must thus be increased, or redundancy must be created by additional components to achieve pseudo-ductility.

It has also been demonstrated that adhesive bonding between the different materials is an appropriate connection technique, providing a high degree of composite action and being able to resist fatigue loading in bridge deck applications.

4.7 MODULAR CONSTRUCTION

4.7.1 Introduction

In modular construction, prefabricated units, i.e., modules, are manufactured off-site and subsequently assembled on-site. The advantages of modular construction are normally (i) short on-site construction times, (ii) high quality, thanks to controlled factory processes and environments, (iii) function-tailored modules that reduce material consumption and waste and can be conceived for future disassembly and reuse, and (iv) in buildings, a better life cycle performance with regard to energy and greenhouse gas emissions. The required transportation of the modules, limited module sizes due to transportation, and limited design customisation and flexibility constitute some of the disadvantages [68–70].

In the following, the potential additional advantages arising from the use of composites in modular construction are discussed, and potential fields of application are explored.

4.7.2 Modular composite construction

Interesting characteristics of composites regarding modular construction are the lightweight nature and potential for function integration. Their combination makes composites appropriate for the prefabrication of large-scale lightweight modules. The low weight allows for the easy transport and lifting of the modules on the construction site. Thanks to the function integration possible in building construction, load-bearing structure and envelope can be merged and the number of modules can thus be minimised, and the modules kept compact, see Section 4.5.1.

Function integration is achieved through sandwich construction in most cases, as shown in **Figs. 4.66–67**. In both cases, segments directly form the buildings' outer

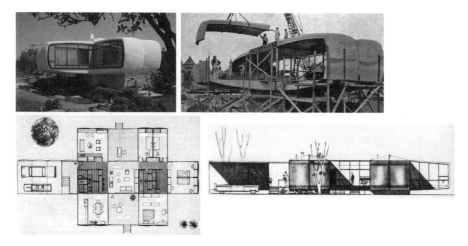

Fig. 4.66 Monsanto "House of the Future" at Disneyland, Anaheim, California, 1957 **(top left)**; composite segments during construction **(top right)**; alternate plan and elevation **(bottom)**

Fig. 4.67 Large-scale modular and lightweight composite sandwich roof segments of Novartis Building during prefabrication **(top left)**; transportation **(top middle)**; installation **(top right)**; local adhesive application to join segments **(bottom left)**; detail adhesive connection on web **(bottom middle)**; segments pulled together with remaining recess in top face sheet **(bottom right)**

and inner skins and interior spaces; the structural and building physics requirements are fulfilled and no other layers need to be added. The floor plan variations of the Monsanto House also demonstrate that a certain flexibility in the configuration of spaces is possible, if the modules are well designed. Similarly, the four large-scale roof segments of the Novartis Building all had different curvatures and an acoustic ceiling was integrated into two of them.

An important role is played by the joints between the modules. They may need to fulfil different requirements such as structural continuity, watertightness, or compensation for module tolerances, or combinations thereof. Joint types that can fulfil all these requirements are adhesive joints, as shown in Fig. 4.67 for instance. In this case, they were applied only locally along the segment edges, before the adjoining segments were pulled together and the recesses in the face sheets closed by bonded laminates. If joints should be detachable for a potential replacement or reuse of modules, bolted joints are more appropriate than adhesive joints; the sealing of such joints is more difficult, however (see Section 3.10.2.7).

4.7.3 Fields of application

The possible fields of application of modular composite construction are various, with three of them seeming particularly appropriate, however, as discussed below, i.e., (i) temporary housing systems, (ii) densification of cities in building construction, and (iii) superstructures in bridge construction.

Prefabricated housing systems are used in remote regions, disaster zones, emergency situations, on construction sites, or for temporary shelters. In these applications, composite systems may provide better performances at a competitive cost compared to conventional solutions in terms of lightness, ease of transportation

4. Composites – Exploring opportunities

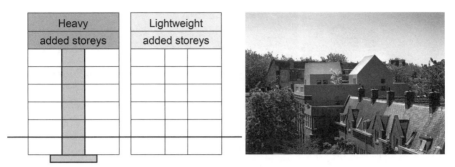

Fig. 4.68 Densification of cities: effect of weight if adding new storeys to existing buildings **(left)**; example of Didden Village, MVRDV Architects, Rotterdam, 2002–06 (timber construction with blue polyurethane coating on large-scale steel frame) **(right)**

and assembly/disassembly, thermal performance, durability, and low maintenance cost [71]; an example of a composite housing system, the ClickHouse, is presented in [71].

The fast-growing population in today's cities demands their densification. Due to the usually already very dense arrangement of buildings in the ground plan, an extension of space is possible only in the vertical direction, e.g., by adding new storeys to existing buildings. Since the load-bearing capacity of existing buildings is normally limited, the added storeys should be as lightweight as possible, or else, the existing structural members must be strengthened, including, in the worst case, the foundations, see **Fig. 4.68** (left). The use of lightweight materials, such as timber or composites, is thus advantageous. An example of a timber elevation is shown in Fig. 4.68 (right). Since the possible supports of the new structure are mainly constituted by the existing exterior envelope walls in this case, a heavy large-scale steel frame construction had to be inserted between the lightweight timber construction and the existing building to bridge the large spans. Using composites, solutions could certainly be found to prevent such heavy transition elements. Function integration, as described above, may offer further advantages of composites over timber, since the modules can be designed to be more compact, and thus lighter, larger and less numerous.

The third interest is bridge construction, where bridge decks are a preferred field of application of composites, see Section 2.6.2. Bridge decks are lightweight and can be prefabricated in large modules, as shown in **Fig. 4.69**. On site, they can be adhesively bonded together and onto steel girders, for instance, and the whole bridge superstructure can then be lifted into place, thus minimising traffic interruptions.

Adhesive joints, as shown in Fig. 4.69, are not detachable, however. A detachable solution is sketched in **Fig. 4.70**, where a steel plate is adhesively bonded to the deck's underside. The deck is then mechanically connected to the steel girder's top flange by bolts. The possible detachment allows the deck to be replaced or reused. In the latter

Fig. 4.69 Modular vehicular bridge composed of prefabricated composite-balsa sandwich deck modules adhesively bonded onto two steel main girders: during sandwich lay-up **(top left)**; after vacuum infusion **(top middle)**; lifting on construction site **(top right)**; removal of peel-ply fabrics to uncover bonding surface **(bottom left)**; application of adhesive on steel top flange **(bottom middle)**; lifting of bridge superstructure into place **(bottom right)**, Avançon Bridge, 2006, see Appendix A.5

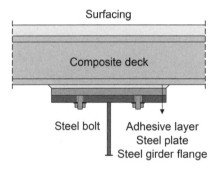

Fig. 4.70 Modular bridge concept with composite deck panels detachable from steel girders, transverse section [1]

case, if properly designed (see Section 6.3), new steel plates can be bonded to the deck to adapt to new transverse spans.

4.7.4 Conclusions

Due to their lightweight nature and possibility of function integration, composites may offer excellent opportunities for modular construction and thus facilitate the further spread of this economic and sustainable construction method in bridge and building construction. Composites also allow the optimal design of module dimensions and shapes, including the corresponding joints, thus minimising module numbers and associated transportation efforts. Progress in numerical controlled manufacturing processes can increase freedom in the design of individually shaped modules.

4.8 REFERENCES

[1] Original contribution.

[2] F. Otto and B. Rasch, *Finding form – towards an architecture of the minimal*. Edition Axel Menges, 1995.

[3] S. Ling, D. L. Kaplan, and M. J. Buehler, "Nanofibrils in nature and materials engineering", *Nat. Rev. Mater.*, vol. 3, pp. 1–15, 2018, doi: 10.1038/natrevmats.2018.16.

[4] G. R. Plaza, G. V. Guinea, J. Pérez-Rigueiro, and M. Elices, "Thermo-hygro-mechanical behavior of spider dragline silk: Glassy and rubbery states", *J. Polym. Sci. Part B Polym. Phys.*, vol. 44, pp. 994–999, 2006, doi: 10.1002/polb.20751.

[5] S. Amroune, A. Belaadi, M. Bourchak, A. Makhlouf, and H. Satha, "Statistical and experimental analysis of the mechanical properties of flax fibers", *J. Nat. Fibers*, vol. 19, pp. 1387–1401, 2022, doi: 10.1080/15440478.2020.1775751.

[6] *CEN/TS 19101: Design of fibre-polymer composite structures*. European Committee for Standardization, Brussels, 2022.

[7] *EN 1993–1–1: Design of steel structures – Part 1–1: General rules and rules for buildings*. European Committee for Standardization, Brussels, 2005.

[8] D. Thompson, *On growth and form*. Cambridge University Press, 1916.

[9] W. Nachtigall, *Vorbild Natur: Bionik-Design für funktionelles Gestalten*. Berlin, Heidelberg, New York, Springer, 1977.

[10] J. E. Gordon, *Structures: Or why things don't fall down*. Dacapo, 1981.

[11] *prEN 1990-1: Basis of structural and geotechnical design – Part 1: New structures*. European Committee for Standardization, Brussels, 2023.

[12] C. Sultan, *Tensegrity: 60 Years of art, science, and engineering*, 1st ed., vol. 43. Elsevier Inc., 2009, doi: 10.1016/S0065-2156(09)43002-3.

[13] A. Schwabe and H. Saechtling, *Bauen mit Kunststoffen*. Berlin, Ullstein AG, 1959.

[14] A. Quarmby, *The plastics architect*. Pall Mall Press, London, 1974.

[15] J. Knippers, H. Jungjohann, F. Scheible, and M. Oppe, "Bio-inspired kinetic GFRP-façade for the thematic pavilion of the EXPO 2012 in Yeosu", in *International Symposium of Shell and Spatial Structures (IASS 2012)*, 2012, vol. 90, pp. 341–347.

[16] B. B. Mandelbrot, *The fractal geometry of nature*. W.H. Freeman, San Francisco, 1982.

[17] C. Bovill, *Fractal geometry in architecture and design*. Birkhäuser, 1996.

[18] P. Prusinkiewicz and A. Lindenmayer, *The algorithmic beauty of plants*. Springer, New York, 1990.

[19] T. Keller and P. Cagna, "La construction légère, construire sur un arteplage d'Expo.02. Unité d'enseignement Architecture et structures", EPFL-CCLab, 2003.

[20] T. Keller, C. Haas, and T. Vallée, "Structural concept, design, and experimental verification of a glass fiber-reinforced polymer sandwich roof structure", *J. Compos. Constr.*, vol. 12, pp. 454–468, 2008, doi: 10.1061/(ASCE)1090-0268(2008)12:4(454).

[21] T. Keller, *Use of fibre reinforced polymers in bridge construction*. Structural Engineering Documents, SED no. 7. International Association for Bridge and Structural Engineering, IABSE, Zurich, 2003.

[22] S. Dooley, "The development of material-adapted structural form", PhD thesis, EPFL, Lausanne, 2004. [Online]. Available: 10.5075/epfl-thesis-2986.

[23] P. Voigt, "Die Pionierphase des Bauens mit glasfaserverstärkten Kunststoffen (GFK) 1942 bis 1980", PhD thesis, Bauhaus-Universität Weimar, 2007. [Online]. Available: https://e-pub.uni-weimar.de/opus4/frontdoor/index/index/docId/821.

[24] H. Engel, *Structure systems*. Hatje Cantz, Berlin, 2009.

[25] J. V. Huddleston, "Behavior of a steep prestressed arch made from a buckled strut", *J. Appl. Mech.*, vol. 37, pp. 984–994, 1970, doi: 10.1115/1.3408728.

[26] J. Lienhard, "Bending-active structures", PhD thesis, University of Stuttgart, Germany, 2014. [Online]. Available: https://elib.uni-stuttgart.de/handle/11682/124

[27] T. Habibi, L. Rhode-Barbarigos, and T. Keller, "Short-term behavior of glass fiber-polymer composite bending-active elastica beam under service load application", *Compos. Struct.*, vol. 337, 2024, doi: 10.1016/j.compstruct.2024.118080.

[28] O. Frei, B. Burkhardt, and M. Bächer, *Multihalle Mannheim*. Stuttgart: Institut für leichte Flächentragwerke (IL): IL 13, 1978.

[29] T. Habibi, L. Rhode-Barbarigos, and T. Keller, "Fiber-polymer composites for permanent large-scale bending-active elastica beams", *Compos. Struct.*, vol. 294, 2022, doi: 10.1016/j.compstruct.2022.115809.

[30] Unpublished CCLab data.

[31] C. Pascual, J. De Castro, A. Schueler, A. P. Vassilopoulos, and T. Keller, "Total light transmittance of glass fiber-reinforced polymer laminates for multifunctional load-bearing structures", *J. Compos. Mater.*, vol. 48, pp. 3591–3604, 2014, doi: 10.1177/0021998313511653.

[32] C. Pascual, J. De Castro, A. Kostro, A. Schueler, A. P. Vassilopoulos, and T. Keller, "Diffuse light transmittance of glass fiber-reinforced polymer laminates for multifunctional load-bearing structures", *J. Compos. Mater.*, vol. 48, pp. 3621–3636, 2014, doi: 10.1177/0021998313511655.

[33] C. Pascual, J. De Castro, A. Kostro, A. Schueler, A. P. Vassilopoulos, and T. Keller, "Optically-derived mechanical properties of glass fiber-reinforced polymer laminates for multifunctional load-bearing structures", *J. Compos. Mater.*, vol. 49, 2015, doi: 10.1177/0021998314567696.

[34] C. Pascual, J. De Castro, A. Schueler, and T. Keller, "Integration of dye solar cells in load-bearing translucent glass fiber-reinforced polymer laminates", *J. Compos. Mater.*, vol. 51, pp. 939–953, 2017, doi: 10.1177/0021998316656393.

[35] C. Pascual Agullo, "Translucent load-bearing GFRP envelopes for daylighting and solar cell integration in building construction", PhD thesis, EPFL, Lausanne, 2014. [Online]. Available: 10.5075/epfl-thesis-6405.

[36] T. Keller, J. De Castro, and M. Schollmayer, "Adhesively bonded and translucent glass fiber reinforced polymer sandwich girders", *J. Compos. Constr.*, vol. 8, pp. 461–470, 2004, doi: 10.1061/(ASCE)1090-0268(2004)8:5(461).

[37] European Commission, "Directive 2010/31/EU of the European parliament and of the council of May 2010 on the energy performance of buildings (recast), 18.06.2010", Brussels, 2010.

[38] European Commission, "Proposal for a directive of the European parliament and of the council on the energy performance of buildings (recast),15.12.2021", Brussels, 2021.

[39] K. Goulouti, J. De Castro, A. P. Vassilopoulos, and T. Keller, "Thermal performance evaluation of fiber-reinforced polymer thermal breaks for balcony connections", *Energy Build.*, vol. 70, pp. 365–371, 2014, doi: 10.1016/j.enbuild.2013.11.070.

4. Composites – Exploring opportunities 333

[40] K. Goulouti, J. de Castro, and T. Keller, "Aramid/glass fiber-reinforced thermal break – thermal and structural performance", *Compos. Struct.*, vol. 136, pp. 113–123, 2016, doi: 10.1016/j.compstruct.2015.10.001.

[41] K. Goulouti, J. de Castro, and T. Keller, "Aramid/glass fiber-reinforced thermal break – Structural system performance", *Compos. Struct.*, vol. 152, pp. 455–463, 2016, doi: 10.1016/j.compstruct.2016.05.038.

[42] K. Goulouti, "Thermal and structural performance of a new fiber-reinforced polymer thermal break for energy-efficient constructions", PhD thesis, EPFL, Lausanne, 2016. [Online]. Available: 10.5075/epfl-thesis-7097.

[43] Y. Bai, T. Vallée, and T. Keller, "Modeling of thermal responses for FRP composites under elevated and high temperatures", *Compos. Sci. Technol.*, vol. 68, pp. 47–56, 2008, doi: 10.1016/j.compscitech.2007.05.039.

[44] *EN ISO 6946:2017: Building components and building elements – Thermal resistance and thermal transmittance – Calculation methods.* European Committee for Standardization, Brussels, 2017.

[45] Ernst Basler+Partner Zurich, EPFL-CCLab, and Scobalit Winterthur, "EPFL Learning Center, feasability study for fiber reinforced polymer (FRP) roof", 2007, unpublished.

[46] T. Keller, F. Riebel, and A. Zhou, "Multifunctional hybrid GFRP/steel joint for concrete slab structures", *J. Compos. Constr.*, vol. 10, pp. 550–560, 2006, doi: 10.1061/(ASCE)1090-0268(2006)10:6(550).

[47] T. Keller, F. Riebel, and A. Zhou, "Multifunctional all-GFRP joint for concrete slab structures", *Constr. Build. Mater.*, vol. 21, pp. 1206–1217, 2007, doi: 10.1016/j.conbuildmat.2006.06.003.

[48] T. Keller, C. Tracy, and E. Hugi, "Fire endurance of loaded and liquid-cooled GFRP slabs for construction", *Compos. Part A Appl. Sci. Manuf.*, vol. 37, pp. 1055–1067, 2006, doi: 10.1016/j.compositesa.2005.03.030.

[49] D. Khovalyg, A. Mudry, and T. Keller, "Prefabricated thermally-activated fiber-polymer composite building slab P-TACS: Toward the multifunctional and pre-fabricated structural elements in buildings", *J. Build. Phys.*, 2023, doi: 10.1177/17442591221150257.

[50] T. Keller, A. P. Vassilopoulos, and B. D. Manshadi, "Thermomechanical behavior of multifunctional GFRP sandwich structures with encapsulated photovoltaic cells", *J. Compos. Constr.*, vol. 14, pp. 470–478, 2010, doi: 10.1061/(ASCE)CC.1943-5614.0000101.

[51] G. De With, *Polymer coatings, a guide to chemistry, characterization, and selected applications.* Wiley-VCH, Weinheim, 2018.

[52] J. Itten, *The art of color.* John Wiley & Sons, 1997.

[53] Y.-K. Lee, "Colorants for composites", in *Wiley Encyclopedia of Composites, 2nd Edition*, John Wiley & Sons, 2012.

[54] J. W. Chin, "Durability of composites exposed to ultraviolet radiation", in *Durability of composites for civil structural applications*, Woodhead Publishing, Cambridge, 2007, doi: 10.1533/9781845693565.1.80

[55] S. Engelsmann, V. Spalding, and S. Peters, *Plastics in architecture and construction.* Birkhäuser, Basel, 2010.

[56] H. W. Gürtler, "Composite action of FRP bridge decks adhesively bonded to steel main girders", PhD thesis, EPFL, Lausanne, 2005. [Online]. Available: 10.5075/epfl-thesis-3135.

[57] E. Schaumann, "Hybrid FRP-lightweight concrete sandwich system for engineering structures", PhD thesis, EPFL, Lausanne, 2008. [Online]. Available: 10.5075/epfl-thesis-4123.

[58] M. Osei-Antwi, "Structural performance of complex core systems for FRP-balsa composite sandwich bridge decks", PhD thesis, EPFL, Lausanne, 2014. [Online]. Available: 10.5075/epfl-thesis-6192.

[59] M. Osei-Antwi, J. De Castro, A. P. Vassilopoulos, and T. Keller, "FRP-balsa composite sandwich bridge deck with complex core assembly", *J. Compos. Constr.*, vol. 17, 2013, doi: 10.1061/(ASCE) CC.1943-5614.0000435.

[60] M. Osei-Antwi, J. De Castro, A. P. Vassilopoulos, and T. Keller, "Structural limits of FRP-balsa sandwich decks in bridge construction", *Compos. Part B Eng.*, vol. 63, pp. 77–84, 2014, doi: 10.1016/j.compositesb.2014.03.027.

[61] T. Keller, E. Schaumann, and T. Vallée, "Flexural behavior of a hybrid FRP and lightweight concrete sandwich bridge deck", *Compos. Part A Appl. Sci. Manuf.*, vol. 38, pp. 879–889, 2007, doi: 10.1016/j.compositesa.2006.07.007.

[62] T. Keller and H. Gürtler, "Quasi-static and fatigue performance of a cellular FRP bridge deck adhesively bonded to steel girders", *Compos. Struct.*, vol. 70, pp. 484–496, 2005, doi: 10.1016/j.compstruct.2004.09.028.

[63] T. Keller and H. Gürtler, "Composite action and adhesive bond between fiber-reinforced polymer bridge decks and main girders", *J. Compos. Constr.*, vol. 9, pp. 360–368, 2005, doi: 10.1061/ (ASCE)1090-0268(2005)9:4(360).

[64] T. Keller and H. Gürtler, "Design of hybrid bridge girders with adhesively bonded and compositely acting FRP deck", *Compos. Struct.*, vol. 74, pp. 202–212, 2006, doi: 10.1016/j.compstruct.2005.04.028.

[65] T. Keller and M. Schollmayer, "Through-thickness performance of adhesive joints between FRP bridge decks and steel girders", *Compos. Struct.*, vol. 87, pp. 232–241, 2009, doi: 10.1016/j.compstruct.2008.01.007.

[66] M. Schollmayer and T. Keller, "Modeling of through-thickness stress state in adhesive joints connecting pultruded FRP bridge decks and steel girders", *Compos. Struct.*, vol. 90, pp. 67–75, 2009, doi: 10.1016/j.compstruct.2009.02.007.

[67] *prEN1995-1-1:2021(E): Design of timber structures – Common rules and rules for buildings – Part 1-1: General. Consolidated Draft*. European Committee for Standardization, Brussels, 2021.

[68] J. Molavi and D. L. Barral, "A construction procurement method to achieve sustainability in modular construction", *Procedia Eng.*, vol. 145, pp. 1362–1369, 2016, doi: 10.1016/j.proeng.2016.04.201.

[69] M. Kamali and K. Hewage, "Life cycle performance of modular buildings: A critical review", *Renew. Sustain. Energy Rev.*, vol. 62, pp. 1171–1183, 2016, doi: 10.1016/j.rser.2016.05.031.

[70] H. Pervez, Y. Ali, and A. Petrillo, "A quantitative assessment of greenhouse gas (GHG) emissions from conventional and modular construction: A case of developing country", *J. Clean. Prod.*, vol. 294, p. 126210, 2021, doi: 10.1016/j.jclepro.2021.126210.

[71] D. Martins, M. Proença, J. R. Correia, J. Gonilha, M. Arruda, and N. Silvestre, "Development of a novel beam-to-column connection system for pultruded GFRP tubular profiles", *Compos. Struct.*, vol. 171, pp. 263–276, 2017, doi: 10.1016/j.compstruct.2017.03.049.

CHAPTER 5

COMPOSITES – HOW TO DESIGN

5.1   INTRODUCTION

The design of composite structures is mainly performed in two phases, namely, the conceptual design and detailed design phases, and is thus in line with the design process for conventional materials, see **Fig. 5.1**. During the conceptual design phase, the basic project idea and concept emerge and are developed within the design team. Decisions crucial to the later robustness, durability, fatigue resistance (where applicable), aesthetics, sustainability, and economy of the project are made in this phase. They concern the member composition and potential function integration, material selection and associated protection systems, structural system and inherent redundancy, concepts of joints and details, manufacturing and installation, and preliminary dimensioning – the latter is to obtain the material quantity for a cost estimate. While the quality of the conceptual design largely depends on the technical competence, creative skills, and experience in composite construction of the design team, the detailed design mainly comprises the meticulous technical verifications of ultimate and serviceability limit states (ULS and SLS verifications), according to standards, such as the European Technical Specification, CEN/TS 19101 [1]. The conceptual design is a highly iterative process while the detailed design process is much more linear.

The detailed design format of composites is not so different from that of conventional materials, as shown in **Fig. 5.2** for the basic ULS and SLS verifications, where E_d is the design value of the effect of actions (internal forces), R_k is the characteristic

Fig. 5.1   Design considerations for composite structures with focus on conceptual design

Design of composite structures	
Conceptual design	Project idea and concept
Detailed design	Member composition & function integration
Verification of limit states	Material selection & protection systems
Execution specification	Structural system and redundancy
Maintenance plan	Concepts of joints and details
Rehabilitation information	Manufacturing & installation concepts
End of life or reuse information	Preliminary dimensioning & cost estimate

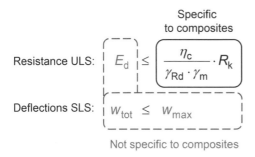

Fig. 5.2 Parameters specific and not specific to composites in basic ULS and SLS design verifications, as defined in [1]

value of the resistance, γ_{Rd} and γ_m are the partial factors (> 1.0) for the uncertainty in the resistance models and material properties, respectively, η_c is the conversion factor (≤ 1.0) for the effects of moisture and temperature, w_{tot} basically involves combinations of the initial and long-term deflections due to permanent and variable actions, and w_{max} is the maximum permissible deflection [1]. In the case of creep rupture, fatigue, and adhesive joints, a single partial factor γ_M is introduced, which accounts for both the uncertainty in material properties and resistance models [1].

The quantities framed with dashed lines in Fig. 5.2 do not need any deeper knowledge about the mechanics of composites, only stiffness values and creep coefficients are required for the SLS verifications, and shear deformations should be considered. Where specific knowledge about composites is required, the quantities are framed with a continuous line. Standards, such as CEN/TS 19101 [1], specify how these quantities can be obtained. The characteristic value of the resistance (R_k) is normally based on characteristic values of material properties and a resistance model. Estimates of the material properties can be obtained from [1] for the preliminary dimensioning, while values for the detailed design are normally derived from tests or obtained from manufacturers. Resistance models, e.g., for wrinkling failure of sandwich panel face sheets, are also provided in [1]. The two partial factors basically depend on the member or joint type and failure mode (γ_{Rd}), and the coefficient of variation of the material properties (γ_m). The conversion factor for moisture and temperature depends on moisture exposure classes and on service and glass transition temperatures and whether the response is fibre- or matrix-dominated [1, 2].

In addition to the conceptual and detailed design phases, the project execution, information about maintenance and potential rehabilitation during the use phase, as well as the planning of the end-of-life stage or possible reuse, are also part of the design.

In this chapter, the conceptual design is first discussed in depth, due to its importance, as emphasised above. Subsequently, regarding the detailed design, fatigue and adhesive joint resistances are focused on, which are among the most difficult subjects in the design of composites. The remaining design verifications are defined and thoroughly presented in [1, 2], and design examples are given in [3]; they will therefore not be further discussed. Finally, execution, maintenance, and rehabilitation are addressed. The end-of-life stage and reuse are discussed in Section 3.11.3 and are not further mentioned. The conclusions of this chapter are drawn in Section 6.2.4 of Chapter 6.

5.2 CONCEPTUAL DESIGN

5.2.1 Member composition and function integration

Depending on the type of construction – e.g., a bridge or a building – and the corresponding bases for planning, e.g., a two-lane vehicular bridge embedded in a specific topography or the type and space programme of a building, initial sketches of these constructions are produced – for instance, for a three-span girder bridge or a four-storey residential building.

Assuming that composites offer significant advantages, due to reasons such as light weight, rapid construction, etc., and are thus considered a valuable option, the member types and their composition can be developed, e.g., a composition of linear and/or two-dimensional members, made entirely of composites or in hybrid construction. A redundant composition is usually preferred, see Section 5.2.3, and in the case of a building, function integration can be taken into account. This process of sketching the construction and member composition is highly iterative and takes into consideration further knowledge about materials, manufacturing, and potential fire protection (also see Section 4.3.3 regarding ideal, constructible, and implemented forms). If the principal designer is inexperienced in composite construction, a team should be formed that can combine the required knowledge and experience. To fully benefit from function integration in building constructions, for instance, architect, structural engineer, building physicist, and manufacturer should already closely cooperate in this early project phase.

5.2.2 Material selection and protection systems

Material selection primarily concerns the selection of the fibre, resin, and core material types, and of the adhesive type in the case of adhesive connections, but also the basic fibre architecture. It also includes protection systems that may be required to maintain the targeted material properties over the whole design service life or fulfil fire reaction or fire resistance requirements.

Fibre, resin, core material, and adhesive type selection are based on the required mechanical performance, mutual compatibility, and resistance against environmental conditions, in particular moisture and UV radiation. The time and temperature dependence of the properties should also be considered, including effects of creep on deformations, creep rupture on strength, and glass transition temperature on strength and stiffness, as discussed in Section 3.4, also see [1, 2].

The fibre architecture is ideally designed to obtain, as much as possible, a fibre-dominated structural response (see Section 3.2.3) and should comprise (i) interspersed mat layers to generate fibre bridging for increased fracture resistance (Section 3.7.2), (ii) surface veils to prevent early fibre blooming due to UV impact (Section 3.10.2), and (iii) if possible, peel-ply fabrics on surfaces to be adhesively bonded

(Section 2.4.3). Furthermore, laminates are preferably conceived as symmetric and balanced (Section 2.3.1).

If mechanical properties experience excessive reduction over the service life due to the impact of environmental conditions, protection systems can be applied, e.g., coatings. Such systems are strongly recommended (i) in the case of permanent exposure to moisture (Class III in [1]), (ii) on bore hole walls of bolted connections, (iii) across edges of adhesive connections, or (iv) on any kind of cut surfaces where fibre ends are exposed, in order to prevent moisture uptake through wicking effects (Section 3.10.2). Sandwich panel face sheets should also protect the core and face sheet-core interfaces from moisture ingress, via an appropriate resistance to moisture diffusion. Furthermore, epoxy resins are particularly sensitive to UV radiation and require corresponding protection.

However, since protection systems can be damaged and generally require periodic replacement, their efficiency may not always be ensured (Section 3.10.3). In many cases, it is more efficient to protect the materials through appropriate constructive detailing, see Section 5.2.4. Further protection may be required against fire; active or passive protection systems can be installed in this case (Section 3.9.1). Active systems, e.g., sprinklers, allow the composite material to remain visible and easily controllable, while passive systems hide the material.

5.2.3 Structural system and redundancy

As a basis of the preliminary dimensioning, and later detailed structural verifications, a structural (static) system is fitted to the member composition, which serves for the determination of the internal forces in the members. The assumed boundary conditions of the structural system, and the possible real displacements and rotations at the member supports, should align. Joints between members should be considered in the system if constructed as hinges or semi-rigid joints.

To provide system ductility in the case of a lack of material ductility, the member composition and structural system can be conceived as redundant (or, in other terms, statically indeterminate or fail-safe) to allow progressive failure and associated energy dissipation at the structural system level (Section 3.6.2). Similarly, failure of adhesive joints should not lead to the collapse of the entire structure (Section 5.3.1). A redundant vertical load-bearing system is also recommended for buildings to better resist fire (Section 3.9.5). Redundancy generally increases structural safety and robustness. If taken into consideration from the beginning, redundancy is not necessarily related to a higher material quantity.

5.2.4 Concepts of joints and details

Since joints between members are always critical locations at which stress concentrations may occur due to geometry and/or material discontinuity, and thus premature failure can initiate, e.g., through fatigue, their number should be minimised when possible. The light weight of composites and appropriate manufacturing methods allow for the production of large-scale members free of joints (the size of which is normally only limited by transportation), thus allowing for a decrease in the number of on-site joints. Joints should furthermore not be designed compact, but extend over a certain distance, to allow smooth transitions of stiffness due to geometry or material changes, and thus reduce associated stress concentrations. To achieve and ensure a high joint efficiency (Section 3.6.3), it is important to develop the joint concept during the conceptual design phase. Whether joints should be selected as bolted, bonded, or hybrid can be decided based on the evaluation presented in Section 2.4.1.

Similarly to joints, points of concentrated load introduction into members, e.g., at the supports or in the case of crash barrier fixations on bridge decks, are also critical locations for detrimental stress concentrations. Core materials of sandwich panels or webs of profiles may need reinforcements at such locations to prevent indentation or web crippling, respectively. Significant through-thickness tensile stresses should also be avoided to prevent delamination.

Appropriate detailing is also crucial to ensure durability. Locations of detrimental stagnant water can be avoided by conceiving appropriate slopes of surfaces and drip edges. In general, effective dewatering systems should be installed, e.g., on roofs or bridge decks.

Furthermore, dark surface colours, if exposed to direct sun radiation, may heat the composite material above the glass transition temperature. This is particularly critical for materials which often exhibit a lower glass transition temperature, such as adhesives or foam cores (Section 2.2). Colour selection should thus be done in this context.

5.2.5 Manufacturing and installation

Member type selection and composition are directly related to possible manufacturing and cost. As discussed in Section 2.5, each manufacturing method – such as hand lay-up, pultrusion, or vacuum infusion – has its advantages, but also its limitations. Taking the example of a beam, for instance: an off-the-shelf pultruded profile can be selected, or the beam can be manufactured custom-made by vacuum infusion. In the first case, the fibre architecture is constant along the length and cannot be adapted to the stresses, e.g., in zones of joints or load introduction, while an optimisation of the architecture along the length is possible in the second case. Consequently, the material is fully used only in some sections in the first case and the material consumption is thus high, while in the second case the material use is high on the whole length and the consumption thus low. The second case also allows for the visualisation of the flow of forces through adaptations of the shape. However, the cost amount of the pultruded beam most likely will be lower than that of the vacuum-infused beam, due to the off-the-shelf production, although the material consumption is higher. In addition to such structural and cost aspects, imperfections and associated tolerances also vary between and within the different manufacturing methods (Section 3.3.2).

A further important aspect of manufacturing is the curing process. A properly high curing degree increases the mechanical properties, glass transition temperature, and resistance to moisture diffusion (Section 3.5.3), and should thus be targeted, i.e., curing under elevated temperature or post curing is preferable. However, on the construction site, applying elevated temperature is often difficult and the mechanical properties and glass transition temperature, e.g., of adhesive joints, may thus not yet be fully developed at the beginning, particularly during winter. Such effects should be considered when defining construction stages and inauguration dates (Section 3.5.3).

5.2.6 Preliminary dimensioning and cost estimate

The aim of the preliminary dimensioning is to estimate the maximum dimensions of the members, with an accuracy of about ±20%, as a basis to establish a robust cost estimate. The critical verifications in this respect are normally (i) the global SLS deflection verification under variable actions and (ii) local ULS verifications of joints and sections of load introduction (including supports). Based on the structural system and the actions and their combinations, the action effects (internal forces or stresses) and deflections can be determined, which are required for these verifications. At this stage, it is not normally necessary to refer to finite element modelling, as appropriate beam statics are often sufficient.

The member dimensions result from an iterative comparison of the obtained deflection and action effect values with deflection limits (in SLS), and resistance values obtained from resistance models (in ULS), according to Fig. 5.2. Both include material properties – stiffness values for SLS and strength values for ULS – estimates of such properties can be obtained from [1]. Rules of thumb, e.g., for the structural span-to-depth ratio, as they exist for reinforced concrete or steel structures (≈ 20 for beams and ≈ 30 for bidirectional slabs) are not always applicable for composite structures since this ratio depends on the fibre type (e.g., glass or carbon) and the type of response, i.e., fibre- or matrix-dominated. An example is given in Section 2.7.4, where the ratio is 29.8 for a unidirectional carbon fibre roof (vs ≈ 20 for steel) and 20.2 for a bidirectional glass fibre roof (vs ≈ 30 for concrete).

For symmetric and balanced adhesive joints, a preliminary dimensioning may be performed based on a characteristic value of an average shear strength of $\tau_{av,k} = 4$ MPa, assumed to be uniformly distributed along the entire bonding length, independent of the adhesive type and occurrence of out-of-plane tensile (peeling) stresses. This value can be derived from the data shown in **Fig. 5.3** (where the characteristic value of the 70 datapoints is even 18% higher, i.e., 4.7 MPa). The different trends of the epoxy and acrylic adhesive curves confirm the differences in the effective stress distributions, shown in Fig. 2.15 (left), which are based on the same dataset. While the average shear strength is maintained in the acrylic-bonded joints for overlap lengths >100 mm, due to an almost uniform stress distribution (as assumed in this calculation), the average strength values rapidly decrease for the epoxy-bonded joints due to a highly nonlinear stress distribution (see Fig. 2.15, left).

In case of significant permanent actions, time-dependent effects – namely creep deformations (SLS) and creep rupture (ULS) – should be considered. Corresponding creep coefficients and rupture values can also be found in [1], the former depend on the fibre architecture (i.e., a fibre- or matrix-dominated response) and the latter on the fibre type (with glass and carbon fibres exhibiting the lowest and highest rupture values, respectively, also see Fig. 3.26, right). Furthermore, the temperature in the material should always be 20°C lower than the glass transition temperature, and the

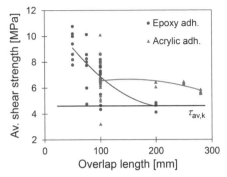

Fig. 5.3 Average experimental shear strength (τ_{av}) vs overlap length with polynomial fits, and derived characteristic value ($\tau_{av,k}$) for preliminary design, obtained from data in [4, 5] for double-lap joints with pultruded glass fibre composite adherends and epoxy and acrylic adhesives ($\tau_{av} = P_u/A$, with P_u = experimental joint failure load, A = total bonding area)

latter should not fall below 60°C [1] (with the exception of rubbery state elastomer adhesives (Fig. 3.18, bottom, with negative T_g-values) and certain polymeric foams, in which cases such high values are difficult to achieve, see Section 2.2.2).

5.3 DETAILED FATIGUE AND ADHESIVE JOINT RESISTANCE VERIFICATIONS

5.3.1 Verification concept

Fatigue and adhesive joint resistances are both closely related to the occurrence of stress concentrations. The latter can be prevented or minimised by an appropriate detailing, see Sections 3.8.8 and 5.2.4. However, stress concentrations can also occur at locations of defects due to deficiencies in manufacturing, and they may remain undetected in such cases. This is particularly the case for adhesive joints if they are applied on-site.

Since a local fatigue or adhesive joint failure can thus not be completely excluded, the structural design should prevent the entire structural collapse in these two cases, more specifically, the load-bearing structure should comprise a certain redundancy or, in other words, have a fail-safe design. Accordingly, the verifications of fatigue and adhesive joint resistances can be similarly conceived, as adopted in [1], where fail-safe (or redundancy) behaviour is supported or imposed by (i) adjusting the partial factors (γ_M) for the fatigue resistance, see **Table 5.1**, and (ii) declaring the fail-safe condition mandatory for adhesive joints. In the latter case, adhesive joint failure is considered as an accidental design situation, according to [6], i.e., the structure is subjected to an exceptional event, which includes local failure, and partial and combination factors of actions can be adapted (i.e., reduced) accordingly.

Furthermore, in both cases, the partial factors for the resistances depend on periodic inspection and maintenance, and the accessibility of the detail or joint, to detect potential damage at an early stage. For adhesive joints, in addition, the partial factor

342   Composites in Structural Engineering and Architecture

Table 5.1 Partial factors (γ_M) for fatigue and adhesive joint resistances, according to [1, 2] (fail-safe condition is mandatory for adhesive joints)

Inspection and access	Fatigue resistance		Adhesive joint resistance	
	Fail-safe	Non-fail-safe	Factory application	On-site application
Periodic inspection and maintenance, detail or joint accessible	1.5	2.0	1.5	2.0
Periodic inspection and maintenance, limited accessibility	2.0	2.5	1.7	2.2
No periodic inspection and maintenance	2.5	3.0	2.0	2.5

also depends on the manufacturing conditions, i.e., in a controlled factory environment, or on site, where the control of the process is more difficult, and defects are more likely to occur.

The partial factors for fatigue and adhesive joint resistance thus double, or almost double, from the most to the least favourable overall condition, i.e., from (i) a fail-safe design, with factory application of adhesive joints, and fully accessible details or joints during periodic inspection and maintenance, to (ii) a non-fail-safe design, with on-site application of adhesive joints, without periodic inspection and maintenance. Since the required material quantity and thus cost both increase with the partial factor, this large difference is thought to promote a safer design. It should be noted, however, that the values given in Table 5.1 are based on engineering judgement and not on statistical analyses, as explained in [2].

5.3.2 Fatigue resistance verification

Fatigue resistance in structural engineering related to composite materials typically concerns vehicular bridges and wind turbine blades, and more recently carbon fibre hangers of railway arch bridges [7]. The corresponding fatigue verification for vehicular bridges is specified in [1] and performed according to the ULS condition shown in Fig. 5.2, but with a single partial factor (γ_M), as listed in Table 5.1. The verification occurs at the member or joint level based on internal forces, since stress-based verifications in fatigue critical details, e.g., in a web-flange junction, are difficult to perform due to stress singularities or required labour-intensive testing [1, 2].

To perform the fatigue verification, the action effects must be known (i.e., the left side of the ULS equation in Fig. 5.2); they result from the application of a fatigue load model (FLM). However, as mentioned in [1] and Section 3.8.9, a FLM calibrated for the fatigue damage rates that occur in composite bridges does not yet exist, and engineering judgement is thus required in the selection of such a model. A procedure outlining the method by which a FLM, calibrated for composites, can be derived is presented in Section 3.8.9. The procedure is applied to the bridge deck of the Avançon

5. Composites – How to Design

Bridge, assuming a two-axle fatigue load model (Fig. 3.101). The axle loads are calibrated for different damage rates and number of cycles, assumed to occur during a design service life of 100 years (Fig. 3.104). The combinations of damage rates and number of cycles always produce the same damage, as real traffic does. The results in Section 3.8.9 demonstrate that the axle loads of the FLM depend on the damage rate, which is variable for composites, in contrast to the FLMs for steel, where the damage rate is constant.

Assuming that a calibrated FLM for composites exists, for which the axle loads depend on the damage rate and number of cycles during the design service life, fatigue verification can be performed by combining the provisions given in [1] and the results of Section 3.8.9, taking the following steps:

1) *Determination of the damage rate (slope coefficient of the internal force vs number of cycle relationship)*
The damage rate can be obtained from testing or based on engineering judgement.

2) *Selection of the number of test cycles to be performed (≥ 2 million cycles)*
Depending on the applicable test frequency and possible test duration, the number of test cycles can be selected. A minimum of two million cycles should be performed, however, so as not to overestimate the fatigue resistance by underestimating creep-fatigue interaction effects (see Section 3.8.4).

3) *Derivation of the axle loads of the FLM*
Based on the damage rate and selected number of test cycles, the axle load can be derived from the calibrated FLM. A lower number of test cycles will result in a higher axle load.

4) *Calculation of the action effects (internal forces, Ed) at the fatigue critical location*
Based on influence lines or surfaces for the critical internal forces (e.g., a bending moment) at the critical location (e.g., in a bonded member joint, see example in Section 3.8.9), the action effects (e.g., the moment range), caused by the FLM, can be calculated.

5) *Set-up of the fatigue tests at the member or joint level according to [1]*
The test specimen type and dimensions, boundary conditions, and fatigue test load can be determined, in a way that the set-up produces the same action effects (internal forces) as the fatigue load model does, at the critical location in the actual structural member. The test specimen should thus include the critical detail (e.g., a bonded member joint).

6) *Static and fatigue tests according to [1], and derivation of the characteristic value of the fatigue resistance (R_k)*
A procedure is specified in [1] that applies to qualification testing for new products, e.g., a novel bridge deck system to obtain the fatigue resistance. For proof testing of existing systems, the testing protocols can be agreed on by the relevant parties [1].

7) *Comparison of the obtained fatigue resistance (R_k) with the critical action effect (E_d) obtained from the application of the FLM, i.e., $E_d \leq \eta_c \cdot R_k / \gamma_M$.*

Further elements that may affect the resulting fatigue resistance, such as the surfacing layer and the test loading device configuration, are also specified in [1]. The surfacing layer may assist in distributing the tyre load to a larger surface inside the member and thus decrease the fatigue stresses, see Section 2.6.2. However, the interface bond strength between deck and surfacing must be designed to not cause disintegration of the surfacing and associated fatigue damage, see Section 3.8.8. Test loading devices should represent the tyre load, including avoidance of detrimental contact pressure concentrations at the device edges.

5.3.3 Adhesive joint resistance verification

In addition to the fail-safe design principle introduced in Section 5.3.1, a second principle concerns the failure mode of adhesive joints, i.e., the failure mode should be either cohesive failure in the adhesive layer or fibre-tear failure in the adherend (see Section 3.7.2 regarding the latter). Failure in the adherend-adhesive interface should be prevented through appropriate material selection and surface preparation (e.g., by applying peel-plies, see Fig. 2.14, and/or primers) [1]. This principle was set since interface failure is normally unpredictable and not quantifiable and cannot thus serve as a basis for a safe design.

Resistance verification is again performed according to Fig. 5.2, using a single partial factor (γ_M), see Table 5.1. Three methods of verification can be applied according to [1], i.e., design (i) assisted by testing, (ii) based on stress analysis, and (iii) based on fracture mechanics. Since both the stress-based and fracture mechanics-based verifications include numerical methods, the experimental validation of the results is mandatory [1].

Design assisted by testing should be performed according to [6], ideally on full-scale specimens of the same material, and under the same manufacturing and loading conditions, as in the case of the actual joint.

Design based on stress analysis can be performed, in the case of fibre-tear failure, by using the shear-tensile stress interaction failure criterion, shown in Eq. 5.1:

$$\left(\frac{\sigma_{z,t,Ed}}{\kappa_\sigma \cdot f_{z,t,d}} \right)^2 + \left(\frac{\tau_{xy,Ed}}{\kappa_\tau \cdot f_{xy,v,d}} \right)^2 \leq 1.0 \tag{5.1}$$

where $\sigma_{z,t,Ed}$ and $f_{z,t,d}$ are the design values of the out-of-plane tensile (peeling) stress and strength, $\tau_{xy,Ed}$ and $f_{xy,v,d}$ are the design values of the in-plane shear stress and strength, and κ_σ and κ_τ are two correction factors for the out-of-plane tensile and shear strength, which consider the statistical (geometrical) size effect, mentioned in Section 2.4.3. These factors are not comprised in [1] but they are introduced in [2], based on [8, 9]. They consider that larger material volumes exhibit lower strength due to the higher probability of existing critical defects. More uniform stress distributions over larger material volumes, as in the case of flexible adhesives, thus result in lower strength, while the strength at localised stress peaks is higher. Values for these corrections factors are given in **Fig. 5.4** for pultruded glass fibre composite double-lap joints with epoxy, polyurethane, and acrylic adhesives, which produce the indicated stress distributions over a 100 mm overlap length, in the layer of fibre-tear failure, i.e., at about 0.5 mm depth inside the inner adherend. As can be seen, the values of the

5. Composites – How to Design

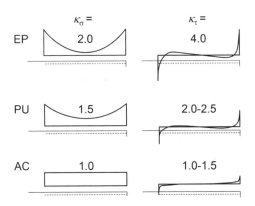

Fig. 5.4 Correction factors for in-plane shear strength **(left)** and out-of-plane tensile strength **(right)** to consider geometrical size effects, derived from double-lap joints with pultruded glass fibre composite adherends and different adhesives (EP = epoxy, PU = polyurethane, AC = acrylic) and respective stress distributions, and 100 mm overlap length, according to [8, 9]

correction factors are higher for stress distributions with narrower peaks while they are 1.0 for uniform stress distribution, which is the basis for their derivation [8].

Design verification based on fracture mechanics is built on a similar interaction criterion, i.e., instead of tensile and shear stresses, the criterion includes Mode-I and Mode-II strain energy release rates, for crack initiation or propagation, in the latter case taking into account fibre bridging. An example of such a fracture-based failure criterion is shown in Fig. 3.62.

The fail-safe condition can generally be met in three ways: (i) by selecting a redundant structural system, (ii) by designing a joint which is itself fail-safe, or (iii) by a combination of the previous two ways. In a redundant structural system, an individual joint failure does not lead to the collapse of the entire structure. Joints can be conceived as fail-safe themselves by, as an example, adding back-up bolts. The bolted connection part should then meet the provisions concerning an accidental design situation (see Section 5.3.1). Examples of fail-safe concepts at the structural or joint levels are given in Section 2.4.3.

5.4 EXECUTION

Execution will be discussed in this section from the perspective of the designer (and not from that of the manufacturer). Standards specifying corresponding execution rules are available for conventional materials, while for composites, the specification of such rules is currently underway, as a complement to CEN/TS 19101 [1]. General technical management measures for design and execution are further provided in prEN 1990-1 [6]. The following overview of aspects relevant for the execution of composite structures is derived from [6] and the execution rules specified for timber structures in [10]; the latter were selected since timber materials are closely related to composites, see Section 1.2.

The task of the designer regarding execution is to provide rules for manufacturing, assembly, and erection of composite structures, to ensure that what is built meets the requirements set in the design, regarding structural safety, serviceability, durability, fatigue, and fire performance, during the entire design service life. Such execution rules should primarily provide guidance on workmanship and inspection during execution.

The execution rules are summarised in a document referred to as the "Execution specification", which is established before the start of the execution. The document mainly includes (i) structural design specifications (calculations, drawings), (ii) consequence classes (CC1-3, according to [6]), (iii) instructions about workmanship, (iv) inspection levels and their specification (IL1-3, according to [6]), (v) permitted geometrical deviations, and (vi) specifications regarding the documentation of the execution.

The instructions regarding workmanship should specify the manufacturing processes, in factory and on site, such as (i) component and member manufacturing (e.g., fibre lay-up, sandwich core placement, resin impregnation, pressure and curing conditions), (ii) mechanical fastening (drilling, moisture protection) and adhesive bonding (environmental conditions, surface preparation, adhesive application, curing), (iii) protection system application (concerning durability, fire), and (iv) erection of the structure (including construction loads, lifting points, assumptions about temporary supports or stability, and sequence of erection).

Inspection during execution is specified according to the consequence classes, i.e., the required sample size for testing and the extent of documentation of test results both increase as the consequence class increases. Inspection should include the experimental verifications or measurements of (i) the assumed basic mechanical properties, (ii) the efficiency of protection measures required for durability or fire performance, (iii) the performance of structural details, such as connections and joints, (iv) the dimensions and permitted deviations, and (v) the assumed environmental conditions (for on-site execution, in particular). Furthermore, corrective measures in cases of deviations from the execution specifications should be defined.

The entire execution process should, furthermore, be documented, to prove that the work performed meets the requirements of the "Execution specification". Such documentation includes, in addition to the results of the inspection, (i) material documentation (e.g., delivery notes), (ii) a list of persons responsible and involved, and (iii) the time/dates of manufacturing and assembly.

5.5 MAINTENANCE AND REHABILITATION

Maintenance of composite structures is indispensable to ensure that the initial physical and mechanical properties are maintained within the assumed limits in the design, during the entire design service life. Composite structures should therefore be delivered with a "Maintenance plan". This plan is based on regular inspections and should comprise, at least, (i) as-built drawings of the structure and specification of the used materials, (ii) regular maintenance procedures, (iii) inspection procedures

5. Composites – How to Design

and intervals, (iv) members and joints of the structure to be inspected, (v) acceptance criteria, and (vi) instructions for how to rehabilitate or replace components. As a basis for future inspections, it is recommended to perform a benchmark assessment directly after installation, which can also include load tests, e.g., in the case of bridges.

Inspection type and intervals of public works should be specified in alignment with inspection plans of authorities, e.g., for their bridge inventory. Inspectors should, however, be qualified to detect anomalies which are typical for composites. Inspectors qualified for the inspection of steel and concrete bridges, for instance, need further training in this respect.

Inspections are chiefly visual but can also include non-destructive diagnosis methods if more quantitative information is required. Visual inspections essentially detect anomalies such as (i) discoloration, fibre blooming, blistering, (ii) cracks, local crushing, delamination, (iii) permanent deformations, and (iv) leakage at joints. Non-destructive diagnosis methods may include, (i) hammer tapping, (ii) acoustic emission logs, (iii) infrared thermography, (iv) laser shearography, (v) ultrasonic measurements, and (vi) strain measurements (e.g., by embedded optical fibre sensors). Further information about anomalies and non-destructive diagnosis can be found in [11].

If damage occurs, which affects the levels of serviceability or structural safety, rehabilitation measures must be taken to re-establish the initial levels of structural integrity. Practical methods of rehabilitation are summarised in **Table 5.2** and shown in **Figs. 5.5–5.6** and 3.97. Anomalies such as discoloration, fibre blooming, and blistering normally only concern serviceability requirements and not structural safety. After an appropriate cleaning and surface treatment of the substrate, the application of a coating is often sufficient.

Cracks, local crushing, or delamination can lower the level of structural safety. In such cases, a structural "bypass" can be applied, which can re-establish the level of safety. Depending on the extent of the damage pattern, such bypasses can consist of large-scale over-laminated layers (see Section 3.10.3) or can be locally applied,

Table 5.2 Practical methods of rehabilitation

Anomaly or damage	Rehabilitation method	Examples
Discoloration, fibre blooming, blistering	Coating	Fig. 3.157 (right)
Cracks, local crushing, delamination	Bypass - laminated layer - bonded profile Replacement	Section 3.10.3 Figs. 5.5, 3.97
Permanent deformations	Additional supporting elements	Figs. 5.6, 3.97
Leakage of joints	Sealing or joint replacement	Fig. 3.161

Fig. 5.5 Pontresina Bridge: flange crushing in bottom chord profile **(left)**; rehabilitation by adhesively bonding upper and lower bypass profiles **(middle and right)**

Fig. 5.6 Pontresina Bridge: cracks and permanent deformation in open-section outstand flange at bridge support **(left)**; rehabilitation by supporting flange with high-stiffness polymer filler block and coating cracks with resin **(middle and right)**

e.g., by sandwiching and thus bridging the damage by adhesively bonded structural profiles, see Figs. 5.5 and 3.97. This rehabilitation technique is straightforward if the local increase of the member thickness is possible. More difficult is an in-plane rehabilitation, i.e., within the existing thickness; in such situations, solutions must be found on a case-by-case basis, in the worst case by replacing an entire member.

Permanent or intolerable deformations can be corrected or prevented by applying additional supporting elements, as shown in Figs. 5.6 and 3.97. In the latter case, high-density foam inserted into the cavities of the web-core deck support the newly added face panel bypass plates. The stiffness of such supporting elements should be sufficiently high, however, to become effective. Furthermore, leakages in joints, e.g., expansion joints, should be removed to avoid penetration of humidity into the composite structure; an example is shown in Fig. 3.161.

5.6   REFERENCES

[1]   *CEN/TS 19101: Design of fibre-polymer composite structures*. European Committee for Standardization, Brussels, 2022.

[2]   J. R. Correia, T. Keller, J. Knippers, J. T. Mottram, C. Paulotto, J. Sena-Cruz, and L. Ascione, *Design of fibre-polymer composite structures: Commentary to European Technical Specification CEN/TS 19101:2022*. CRC Press, London, 2025.

5. Composites – How to Design

[3] L. Ascione, T. Keller, J. Knippers, J. T. Mottram, C. Paulotto, and J. Sena-Cruz, *Design of fibre-polymer composite structures: Worked examples for European Technical Specification CEN/TS 19101: 2022*. CRC Press, London, 2024.

[4] T. Vallée, "Adhesively bonded lap joints of pultruded GFRP shapes", PhD thesis, EPFL, Lausanne, 2004. [Online]. Available: 10.5075/epfl-thesis-2964.

[5] J. de Castro San Román, "System ductility and redundancy of FRP structures with ductile adhesively-bonded joints", PhD thesis, EPFL, Lausanne, 2004. [Online]. Available: 10.5075/epfl-thesis-3214.

[6] *prEN 1990-1: Basis of structural and geotechnical design – Part 1: New structures*. European Committee for Standardization, Brussels, 2023.

[7] U. O. Meier, A. U. Winistörger, and L. Haspel, "World's first large bridge fully relying on carbon fiber reinforced polymer hangers", in *Sampe Europe Conference Amsterdam*, 2020.

[8] T. Keller and T. Vallée, "Adhesively bonded lap joints from pultruded GFRP profiles. Part II: Joint strength prediction", *Compos. Part B Eng.*, vol. 36, no. 4, pp. 341–350, 2005, doi: 10.1016/j.compositesb.2004.11.002.

[9] J. De Castro and T. Keller, "Ductile double-lap joints from brittle GFRP laminates and ductile adhesives, Part II: Numerical investigation and joint strength prediction", *Compos. Part B Eng.*, vol. 39, no. 2, pp. 282–291, 2008, doi: 10.1016/j.compositesb.2007.02.016.

[10] *prEN 1995-3: Design of timber structures – Part 3: Execution rules*. European Committee for Standardization, Brussels, 2023.

[11] A. Castelo, J. R. Correia, S. Cabral-Fonseca, and J. de Brito, "Inspection, diagnosis and rehabilitation system for all-fibre-reinforced polymer constructions", *Constr. Build. Mater.*, vol. 253, 2020, doi: 10.1016/j.conbuildmat.2020.119160.

CHAPTER 6

COMPOSITES – HOW TO PROCEED

6.1   INTRODUCTION

This final chapter summarises the lessons learnt and conclusions drawn from the previous Chapters 2 to 5. It also identifies research needs and addresses the issue of education, which are both considered essential to further encourage the application of composites in structural engineering and architecture.

6.2   LESSONS LEARNT AND CONCLUSIONS

6.2.1   Composites – Compact

Lessons learnt and conclusions drawn from the state-of-the-art overview provided in Chapter 2 are summarised below, concerning composite materials, components and members, connections and joints, manufacturing, and composites in bridge construction and architecture.

At the material level, a large diversity of fibre, resin, sandwich core, and adhesive materials are available, which allow adaptation to almost every indoor or outdoor design situation. Critical design aspects that should be considered are, in particular, (i) the glass transition temperature of the polymeric material components (resin, adhesive, core materials), in respect to the maximum service temperature, and (ii) the resistance to environmental conditions. A sufficient temperature difference should exist in the former case, while material protection systems may be necessary in the latter case, to ensure durability during the whole design service life.

In terms of components and members, a similarly large variety of linear and two-dimensional design options exist, from monofunctional profiles up to multifunctional sandwich panels in complex shapes. However, a close interaction exists between component and member properties, manufacturing process, and joining technology. The selection of an appropriate combination of these ingredients will be decisive for the quality of the project, in terms of robustness, durability, aesthetics, sustainability, and economy.

Concerning connections and joints, adhesive solutions have proven to be reliable and have thus been taken into consideration in design standards. Their detailing, manufacturing, and environmental protection require specific attention, however, to ensure fatigue resistance and durability, in particular.

Manufacturing is often the most limiting parameter regarding the versatility of composite material applications. A clear idea about the manufacturing process should guide the project development from the first concepts, to obtain convincing results, as mentioned above.

Vehicular bridges have always been one of the main foci of composite material applications. Looking back at the "journey" of composites in vehicular bridge construction suggests that bridge decks, in particular, can offer significant advantages compared to decks made of conventional materials. However, the fatigue resistance of composite decks is still not yet fully proven and, therefore, their application is recommended to be limited to bridge categories for less traffic, until more long-term data are available. Such limitations do not exist, however, for pedestrian bridges, where fatigue is not normally involved. Pedestrian bridges are thus among the most interesting applications of composites, since they can take advantage of the excellent mechanical properties and may offer unique architectural features.

The use of composites in architecture represents the main field of potential innovation, based on the freedom of form and possible integration of functions. However, since a close collaboration between architect, structural engineer, building physicist, and manufacturer is required from the beginning of the project, and difficulties such as fire reaction response and resistance may need to be considered, progress in this field is not easy and largely depends on the designers involved. If such projects are successful, however, they will demonstrate the real potential of composites compared to conventional materials.

6.2.2 Composites – Overcoming limitations

As discussed in Chapter 3, certain aspects of composites may cause limitations in properties, design, and use of composite materials in structural engineering and architecture, when compared to conventional materials. The aspects that are considered relevant are listed in Table 3.1 of Section 3.1, together with examples of limitations and possible measures for how to prevent or overcome such limitations.

Some of these aspects are related to each other since they can entail similar limitations, and, furthermore, similar measures can solve potential problems caused by different aspects. This discussion across the aspects, which began in Section 3.1, continues here. Analysing the measures derived for the individual aspects in Table 3.1 allows for their regrouping into basically five groups of measures to overcome limitations, as defined in **Table 6.1**. More details about the listed measures can be found in the individual aspect discussion, in Sections 3.2–11.

Concerning the first group of measures "Structural design and application", nonlinear analysis can take into account effects of geometrical imperfections on the internal forces, while appropriate testing at the component, member, or joint levels can consider effects of residual stresses and manufacturing defects on the mechanical properties. Lacking composite material ductility can be (i) compensated for by system ductility through either pseudo-ductile progressive failure in redundant composite systems or (ii) provided by hybrid systems involving ductile components, e.g., made of steel. Redundant structural systems can also limit fire damage by preventing complete structural collapse. Life cycle analysis and an appropriate application of composites

6. Composites – How to Proceed 353

Table 6.1 Measure groups and potential measures to overcome limitations of composites and related aspect categories (Sections 3.2–11) [1]

Measure groups	Potential measures	Related aspects (Sections)
Structural design and application	Nonlinear analysis Appropriate testing	Imperfections and tolerances (3.3)
	System redundancy and ductility	Ductility (3.6), Fire (3.9)
	Life cycle analysis Appropriate application	Sustainability (3.11)
Material selection (including fibre architecture)	Fibre-dominated response	Anisotropy (3.2), Viscoelasticity (3.4), Curing and physical ageing (3.5), Fire (3.9)
	Activation of fibre bridging	Ductility (3.6), Fracture (3.7)
	Appropriate material system Surface veil	Durability (3.10)
Detailing	Compensation of tolerances	Imperfections and tolerances (3.3)
	Reduction of through-thickness tensile stresses	Ductility (3.6), Fracture (3.7)
	Reduction of stress concentrations	Fatigue (3.8)
	Closed (cellular) sections	Fire (3.9), Durability (3.10)
	Gravity driven dewatering	Durability (3.10)
Manufacturing	Appropriate manufacturing process	Anisotropy (3.2)
	Post curing	Viscoelasticity (3.4), Curing and physical ageing (3.5)
Protection systems and maintenance	Fire protection	Fire (3.9)
	Protective coating	Durability (3.10)
	Sealing of cut member ends, borehole surfaces, and adhesive joint edges	
	Regular inspections and maintenance	

can ensure sustainability. Evaluating the complete life cycle makes it possible to take advantage of the "beneficial" stages regarding sustainability during the design service life of the structure. Since the product and end-of-life stages are normally not beneficial, significant advantages should exist during the construction and use stages to provide an overall sustainable solution, i.e., an appropriate application of composites.

Appropriate "Material selection" (second group, including fibre architecture) may have significant positive effects on the structural response. If a relevant fibre-dominated behaviour can be achieved for the critical design limit states, negative effects of time and temperature dependence on deformations and fire resistance can be prevented or diminished. Activating fibre-bridging can improve ductility and fracture resistance. Selecting an appropriate fibre and resin system, and implementing surface veils, can enhance durability.

Appropriate "Detailing" (third group) may provide major advantages in several respects. Member tolerances can be well compensated for in adhesive joints and connections, for instance. Reducing through-thickness tensile stresses and stress concentrations by minimising eccentricities, through appropriate curvatures, and transitions in changes of materials and member and joint geometries, may effectively increase ductility, fatigue life, and fracture resistance. Using closed (cellular) instead of open cross sections may (i) increase fire resistance through the insulation capacity of the entrapped air and (ii) increase durability through the absence of outstand flanges, which may crack at the web-flange junctions at already low transverse impact. Protecting exposed surfaces via appropriate slopes and drip edges from stagnant and flowing water can also considerably improve durability.

Concerning "Manufacturing" (fourth group), the selected manufacturing process should be compatible with the required degree of anisotropy. To give an example, a quasi-isotropic architecture cannot be produced by pultrusion, which normally requires a significant number of fibres in the pulling direction. Post curing (if possible) may attenuate unwanted matrix-dominated responses caused by viscoelasticity or incomplete curing.

The subject of the fifth and last group of measures are "Protection systems and maintenance". Depending on the composite application and exposure, active or passive fire protection measures can provide the required fire resistance and fire reaction behaviour. If durability cannot be ensured by appropriate material selection and detailing, protective coatings can be applied. They should be renewed periodically, however, to remain effective. Similarly, cut member ends or borehole surfaces should be sealed to prevent humidity ingress along the exposed fibre ends. Regular inspections and maintenance also effectively contribute to ensure durability.

In addition to these measures, opportunities of composites, as addressed in Chapter 4 and in the following Section 6.2.3, may also contribute to overcoming potential limitations. Worth mentioning in this respect are material-tailored structural forms and systems (Section 4.3) and hybrid and modular construction (Sections 4.6 and 4.7).

6.2.3   Composites – Exploring opportunities

The unique material and system properties offered by composites allow new design options to be explored in structural engineering and architecture. These options are either unattainable or achievable only with significant difficulty when using conventional structural materials. Six design options were derived from these material and system properties and discussed in the sections of Chapter 4; they were related to opportunities in architecture, sustainability, and energy, as listed in Table 4.1 of Chapter 4, whereby energy and structural engineering aspects were subsumed under sustainability and economy. In the following, the results from Chapter 4 are summarised according to the opportunities, which emerge from the specified design options, as shown in **Table 6.2**.

Table 6.2 Opportunities offered by design options of composites ($\checkmark$ = applicable, - = not applicable) [1]

Opportunities	Design options					
	Lessons learnt from nature	Structural form	Transparency	Function integration	Hybrid construction	Modular construction
Architecture	$\checkmark$	$\checkmark$	$\checkmark$	$\checkmark$	-	$\checkmark$
Sustainability	$\checkmark$	$\checkmark$	-	$\checkmark$	$\checkmark$	$\checkmark$
Economy	$\checkmark$	$\checkmark$	-	$\checkmark$	$\checkmark$	$\checkmark$

Opportunities in architecture mainly comprise new possible ways of realising targeted architectural concepts and modes of expression. The specific language of composites and associated material-tailored structural forms and systems clearly differ from those of conventional materials. Double-curved and other complex shapes are possible, for instance, which are normally not achievable with reasonable efforts with conventional materials. If the thermal insulation function is integrated into such shapes, single-layer structural building envelopes can be designed, which can directly separate interior from exterior spaces without any additional layers. The lightweight and large-scale prefabrication of such function-integrated members allows modular construction, which can also support architectural concepts and expression. Furthermore, composite laminates can be designed with different degrees of transparency, thus fulfilling, in addition to structural functions, daylit buildings or solar cell encapsulation requirements. In the latter case, the generally unpleasant cladding of buildings with solar panels can be avoided. Finally, colour, light, and surface texture can be integrated into structural or function-integrated members.

Opportunities in sustainability can help to make composite applications sustainable by compensating for the often high carbon footprints of manufacturing, for instance. Design options that increase sustainability are mainly those which minimise material consumption, i.e., material-tailored structural forms and systems, function-integration, and modular construction. The latter also reduces waste, and the rapid on-site assembly of lightweight and large-scale prefabricated elements can, in the case of bridge construction, for instance, significantly reduce traffic congestion or detours and thus social cost. Design options such as hydronic thermal activation or solar cell encapsulation can further contribute to energy savings or harvesting during the whole use stage, and thus increase the sustainability of composite construction. In addition to the above-mentioned design options, designs which lead to more durable infrastructure, with less maintenance requirements during service life, also increase the sustainability.

Opportunities in economy can help to compensate for the relatively high material cost of composites. As was the case for sustainability, the reduction of material consumption and waste increases economy in composite construction. The same design options as mentioned above for sustainability, complemented by hybrid construction, may thus contribute to the reduction of cost and make composite construction economical.

The realisation of all the design options presented in this chapter requires a close collaboration between architects, structural engineers, and manufacturers, from the very beginning of the design process, since all the related aspects are interconnected. Only collaborative or integrated design, and not sequential design, will make it possible to fully benefit from the novel design options offered by composites.

6.2.4   Composites – How to design

Conceptual design is the most critical phase in the design process to achieve a successful composite project. The work to be accomplished in this phase is much less linear than in the case of conventional materials (if the target is to conceive a fully material-tailored project and not just a material substitution in a concept developed for conventional materials). In composites, in addition to the member, the material also needs to be designed, and both designs, furthermore, depend on the manufacturing process, which is often the most limiting factor. If function integration is targeted, the preliminary design process becomes even more complex and highly iterative. It is therefore crucial that the required expertise and experience in composite construction are already represented in the design team in this initial phase.

On the other hand, the preliminary dimensioning of a composite structure is not much more complicated than that of a structure made of conventional materials. Thanks to the design standards available today, such as CEN/TS 19101 [2], the preliminary dimensioning can be performed by engineers without previous profound knowledge and experience in composite construction. CEN/TS 19101 is built up in the same way as the Eurocodes for conventional materials. The preliminary dimensioning procedures for composites and conventional materials are thus not so different. The differences from conventional materials are basically limited to the resistance side of the ultimate state verification, as discussed in Chapter 5. Differences in the detailed design are more pronounced; however, the available "Commentary" to CEN/TS 19101 [3] and the collection of "Worked Examples" [4] can safely guide the engineer through these design verifications. A certain investment in time is, nevertheless, required if a detailed design is being performed for the first time.

Designers also have tasks to accomplish for the execution, maintenance, and rehabilitation of composite structures. Of particular importance is the preparation of substantiated "Execution specification" and "Maintenance plan" documents, as described in Chapter 5.

6.3   RESEARCH NEEDS

The survey I made in 2001 about the future application of composites in bridge construction, as described in the Preface, also included a question about research needs [5]. The responses were that research (or development) was needed in the following fields (with percentage of entries): durability (23%), design standards (18%), material-tailored construction (15%), sensors and smart structures (13%), more cost-effective manufacturing (8%), cable anchoring systems (8%), pilot projects (5%), others (single responses, 10%).

Although significant knowledge has been acquired in the time since, this list still seems applicable, including the order of priority, i.e., with durability in the first position, but without design standards. Durability, if subsuming fatigue resistance, is still the field with most unknowns and particular needs for research (also see below). Concerning design standards, with the completion of CEN/TS 19101 [2] in 2022, engineers now have access to a unique document that is founded on an almost 1 000-page background document [3], where all the decisions made and numbers selected are justified, based on detailed statistical analyses of experimental data or transparent engineering judgement where databases were insufficient.

Based on my practical experience and the reflections made during the writing of this monograph, I would place the focus for research needs on an additional six points, where progress in knowledge could significantly support the application of composites, as follows:

1) Targeted acquisition of further experimental data for the calibration of partial and conversion factors of CEN/TS 19101

Ideally, partial and conversion factors are calibrated based on statistical analyses of representative experimental data. Experimental data in literature are numerous, however, they cannot be used for this purpose in many cases, since the studies reported are incomplete or results are not applicable (e.g., because not following test standards). While for certain resistance models of CEN/TS 19101 limited experimental databases were available for a statistical derivation of the partial factors for the model uncertainty, sufficient data could not be found to statistically determine the conversion factors for temperature and moisture, and their final selection was thus based on engineering judgement [3]. The same applies to the single partial factors for the fatigue and adhesive connection resistances, whose assumptions are based on engineering judgement. While these selections and assumptions seem plausible based on the available information, their quantitative proof would improve design confidence significantly.

2) Damage accumulation in composites

As has been explained in Section 3.8, damage accumulation knowledge is required to determine fatigue resistances based on variable amplitude data or to calibrate fatigue load models. Due to missing alternatives, the Palmgren-Miner linear damage accumulation assumption is normally used, although it is well documented that the assumption is not reliably applicable for composites. Nonlinear extensions of the assumption do not really improve the situation since they are based on fitting of experimental data and not on mechanical background. Not to have reliable basic knowledge in a field as important as fatigue represents a real problem, and priority should thus be given to research in this field. An approach to separate fibre- from matrix-dominated behaviour, as outlined in Section 3.8.5, may help to derive more mechanically based models of damage accumulation.

3) Strategies to create fibre bridging

Delamination is the most critical failure mode in composites, particularly if an only small fracture energy is required to produce sudden and unstable crack propagation, which can lead to a brittle collapse of an entire structure. The fracture energy can be significantly increased to provide stable and limited crack propagation by conceiving a fibre architecture that produces large-scale fibre bridging, see Section 3.7.2. Activating significant fibre bridging can fundamentally change the structural response from brittle to pseudo-ductile. Comprehensive strategies to implement such fibre bridging and reliably quantify its contribution in design are, however, still unavailable.

4) Design requirements tailored for reduced design service life

The typical design service life of buildings and bridges is 50 and 100 years, respectively. However, as discussed in Section 4.2.3, bridges and buildings today often become obsolete at their location, or need heavy transformation to fulfil changing requirements, after only two to three decades. Downsizing the lifetime requirements regarding durability thus seems reasonable from economic and sustainability points of view. Concerning sustainability, shorter lifetimes may, furthermore, allow the use of bio-based composites. However, a comprehensive relationship between requirements regarding durability and effective lifetime does not yet exist.

5) Design for reuse

Full recycling of composite materials is difficult, and downcycling, e.g., the use as shredded filler granulate, appears to be nothing but a stop-gap solution for such high-performance materials, see Section 3.11.3. A much more reasonable solution for the end-of-life stage is the reuse or repurposing of composite members. However, a substantial and economic reuse should also take advantage of the excellent material properties. If pursuing this approach, a clear idea of possible reuses should therefore already exist at the beginning of the member design, and this design should consider potential second life phases. It is clear that this approach involves the introduction of a certain redundancy into the design and thus increases the initial cost. This additional initial cost seems marginal, however, when taking into account the possible cost savings offered by a second phase of life. Without such an approach, a reasonable reuse might only be possible in rare cases and remain a stop-gap, as discussed in Section 3.11.3. The development of concepts of "design for reuse", in conjunction with modular construction (see example in Section 4.7.3), could thus significantly contribute to a sustainable use of composites.

6) Possibilities to detach adhesive connections

Adhesive connections may offer significant advantages in design, as discussed at various locations in the monograph, but they also have the disadvantage that they are not detachable, which becomes critical at the end-of-life stage or if reuse is planned. Detachable adhesive connections can be conceived in some cases, as shown in Fig. 4.70 for a detachable adhesive bridge deck-to-girder joint; however, the basic adherends cannot also be separated in this case. Cutting the connection in the adhesive layer plane may present another solution, but, depending on the geometry, this is not always practicable. Systematic research, which also includes possible manipulations of the adhesive material itself, may lead to more general practical solutions.

6.4 EDUCATION

Just as important as progress in critical research fields is education. Barriers to considering these materials in new projects will exist as long as structural engineers and architects are not educated in composite technology, going hand in hand with a lack of courage when faced with the unknown, or lack of time to invest in knowledge acquisition.

Composites should therefore be added to and covered by basic courses on construction materials, structural engineering, and architecture, at undergraduate level, and taught in the same way as conventional materials are. At the graduate level, specific courses on composite materials and their application in structural engineering and architecture can be taught. Teaching units and joint seminars with students of civil engineering, architecture, building physics, and environmental engineering can draw attention to the multifunctional potential of the materials (see Section 4.3.1). Preliminary designs of bridge and building structures, complemented by detailed designs for critical members, can be undertaken in master's dissertations.

Ideally, structural engineers and architects should be ready to design composite structures, as they can currently do so for steel and concrete structures, when they enter professional life. This would represent a significant step forward in converting the perception of composites from novel and still unknown to conventional materials. Composites will become "mature" in structural engineering and architecture when they are regarded as conventional materials.

6.5 REFERENCES

[1]   Original contribution.

[2]   *CEN/TS 19101: Design of fibre-polymer composite structures.* European Committee for Standardization, Brussels, 2022.

[3]   J. R. Correia *et al.*, *Design of fibre-polymer composite structures: Commentary to European Technical Specification CEN/TS 19101:2022.* CRC Press, London, 2025.

[4]   L. Ascione *et al.*, *Design of fibre-polymer composite structures: Worked examples for European Technical Specification CEN/TS 19101: 2022.* CRC Press, London, 2024.

[5]   T. Keller, *Use of fibre reinforced polymers in bridge construction.* Structural Engineering Documents, SED 7, International Association for Bridge and Structural Engineering, IABSE, Zurich, 2003.

POSTSCRIPT

COMPOSITES – OR WHY THINGS (DON'T) GET BUILT

The inspiration for the title of this Postscript (and of the Preface) came from one of my favourite books: J. E. Gordon's *Structures: Or why things don't fall down* [1]. I have had the opportunity to work on several composite bridge and building projects and on the development of several composite products over the past two decades. Some of them were built and implemented and others were not, with the number of the latter exceeding that of the former. This poses the question of "why things (don't) get built?" I will go through some of the projects and products and try to find a response to the question. In all the examples, the composite materials offered significant advantages over similar solutions built with conventional materials, except initial cost, which was always higher.

1) Pontresina Bridge – built, 1997 (Appendix A.2): Based on exclusive use during winter and a critical floodwater situation during summer, the Pontresina Municipality wanted to construct a lightweight temporary bridge that could easily be installed in autumn and removed in spring. Three similar two-span truss bridges were designed, composed of timber, steel, or pultruded composite profiles. The weights of the timber and steel bridges were 2.3 and 2.7 t per span, respectively, while the composite bridge weighed 1.7 t per span. Despite fierce opposition from forest offices of the surrounding municipalities, a composite bridge was finally built for the following reasons: (i) it had the lowest weight in view of the numerous lifting procedures, (ii) the absence of maintenance (as assumed at the time), (iii) the profiles were donated by the manufacturer and the assembly was carried out by students, and (iv) the mayor of the municipality supported a pioneering project.

2) Eyecatcher Building – built, 1998 (Appendix A.3): The architect of the building became aware of the Pontresina Bridge and was impressed by the smooth and shiny appearance and the tactile sensation of pultruded composite profiles. Due to the low thermal conductivity, he could position the load-bearing structure into the planes of the envelope, without creating significant thermal bridges, and thus use the visible structure as an architectonic element. Furthermore, the low weight facilitates the disassembly and re-installation of the mobile five-storey building. Since the architect was the co-owner of the building, he was able to make the decision to switch from the initially planned steel to composite profiles.

3) Roof of the Novartis Campus Entrance Building – built, 2006 (Appendix A.4): The architect's intention was to create a "floating roof". To achieve this, the building was to be fully transparent, i.e., consist only of glass walls, and the roof was not to contain visible joints or supports and would exhibit a double-curved wing form. A lightweight and function-integrated composite sandwich construction best fulfilled those requirements; in fact, no other material could satisfactorily achieve the architect's intentions.

4) Avançon Bridge – built, 2012 (Appendix A.5): The Canton's (owner's) intention was to replace the existing one-lane concrete bridge, without interrupting the traffic, with a new, two-lane bridge that would not need maintenance, i.e., not comprise any steel. An all-composite bridge was targeted since it was associated with "no maintenance". After many iterations of the project and comparisons with conventional concrete deck-steel girder options, the decision was taken in favour of the hybrid composite sandwich deck-steel girder solution, as presented in Appendix A.5, although the initial cost was approximately 40% higher than that of a conventional bridge. Since an all-composite solution was even more expensive, the two steel girders were accepted, because they are protected from critical environmental impact under the slab, as is the case for the adhesive deck-to-girder connections. A complete traffic interruption was not able to be avoided, but the interruption was minimised. The decision to proceed with a hybrid composite bridge largely depended on the Cantonal bridge engineer, who was in support of this solution. The manufacturer, 3A Composites, developed the balsa-based composite deck system under the brand "Colevo" (see Section 2.6.2), with the aim of entering the construction market. Several decks for pedestrian bridges were subsequently built, but in the end the Swiss market was too small for the business and the return on investment required to extend the market into Europe would have taken too long. The business was thus stopped, and key members of the company left.

5) TSCB Bridge – not built (Appendix B.4): The TSCB Pedestrian Bridge was a project developed by an architect, who was inspired by the optimised and complex forms of diatoms and their possible modular construction, and the appearance of carbon fibre composites with their shiny black surface and the visible underlying texture of the fabrics. The project development suffered from the limited skills of the engineers and manufacturers involved. With my direction, qualified engineers and manufacturers were found and a structurally sound project was developed; however, the cost increased significantly compared to early estimations. The main reason for the failure of the project was, in the end, a change in the Cantonal government and the associated loss of support by the new person in charge of the bridge from the owner's side.

6) Tape/filament winding products – not implemented: Three composite products were developed in conjunction with an industry partner to replace certain own company products made of conventional materials, i.e., (i) a composite thermal break component (Fig. 4.52), (ii) carbon fibre strap strengthening components for concrete flat slabs (Fig. 3.7), and (iii) carbon fibre ground and rock anchors (Fig. 3.6). In all cases, the estimated initial cost was higher than that of the conventional products, but since the new products also offered added value, the cost differences were considered acceptable. The products were developed with public funding, patented, and almost

ready for implementation. However, they eventually failed due to (i) a necessary significant short-term investment in a tape-winding machine for the thermal break component, (ii) a certain resistance to acquire and invest in (expensive) pilot applications, (iii) ultimately, a fear that the successful conventional products would be cannibalised, (iv) some mistrust between those involved, and (v) a change in management who no longer supported the new products.

7) Adhesive joints – implemented: Adhesive joints were one of CCLab's research axes from the beginning, see Preface. Despite significant resistance – adhesive joints were explicitly not recommended in some countries at the time – research was pursued, and adhesive joints were implemented in pilot projects, as documented in the monograph, and finally found entry in design standards, such as CEN/TS 19101 [2]. Reasons for this development were persistence based on the conviction to do the "right thing", and the scientific proof of the feasibility, which finally also convinced owners of built projects.

These experiences demonstrate that the realisation of composite projects and implementation of products so far basically depended on (i) representatives of involved authorities/owners, i.e., their openness to new technologies and level of risk tolerance, (ii) industry, i.e., their willingness to invest in products with a relatively long-term prospect of return and associated persistence of support, (iii) team spirit and trust within the design team, and (iv) architects, who in many cases made the material selection. If the first three "ingredients" were in favour of composites, the elevated initial cost was normally not project-critical, because the long-term return on investment during the "use" phase was acknowledged.

By looking back on the "journey" of today's conventional materials, it can be seen that the development has been similar to the one outlined for composites (also see Section 4.3.3 and [3]). The more composites will be perceived as conventional materials, the more the criteria for selecting a composite project will approach those of conventional projects, i.e., the decisions will be less "human-driven" and more measurable.

REFERENCES

[1]   J. E. Gordon, *Structures: Or why things don't fall down*. Dacapo, 1981.

[2]   *CEN/TS 19101: Design of fibre-polymer composite structures*. European Committee for Standardization, Brussels, 2022.

[3]   S. Dooley, "The development of material-adapted structural form", PhD thesis, EPFL, Lausanne, 2004. [Online]. Available: 10.5075/epfl-thesis-2986.

APPENDIX

A.0 INTRODUCTION

This appendix comprises additional information regarding the bridge and building projects referred to in Chapters 1 to 5, and for which I was involved in the design. As the information given in Chapters 1 to 5 is not repeated but only complemented to give an overall view of the projects, the appendices are not independent. The projects are split into two groups, i.e., "A-projects", which were built, and "B-projects", which were not. Reference is also made to the Postscript, which provides information about why some of these projects were built or not.

The structure of the appendices is always similar and consists of three sections: (i) basic information, including an Allocation Table, indicating project information already comprised in Chapters 1 to 5, (2) additional images and information, and (iii) a list of papers giving further details or relating to the project.

In the section on basic information, the following items are addressed: concept of the project, dimensions, materials, loads, owner of the construction, architectural and structural designers (also specifying my contribution), execution (manufacturers and methods), and years of completion (A-projects) or design (B-projects).

The following projects are included:

A.1 Verdasio Bridge
A.2 Pontresina Bridge
A.3 Eyecatcher Building
A.4 Novartis Building
A.5 Avançon Bridge
A.6 Flower Sculpture
A.7 TCy Sculpture

B.1 Dock Tower
B.2 RLC Building
B.3 CLP Building
B.4 TSCB Bridge
B.5 1K Bridge

A.1   VERDASIO BRIDGE

A.1.1   Basic information

Background   Two-span prestressed concrete bridge

Loss of structural safety due to significant pitting corrosion in one 2 350 kN prestressed steel cable after only 14 years, caused by chloride-contaminated water, running from surface dewatering tubes down along outer side of one web

Strengthening concept   Strengthening of two-span prestressed concrete box girder with four external prestressed carbon fibre-polymer cables, in polygonal path over entire bridge length

Cables are accessible at any time for examination or replacement

Continuous monitoring of load relaxation in new cables

Dimensions   Length of new cables of 70.0 m

Materials   Carbon fibre-epoxy composite wires of 5 mm in diameter, pultruded, 70 vol% fibres

19 wires per cable, cables prestressed to 600 kN each, i.e., 65% of nominal ultimate load (912 kN), minimum radius of curvature of 3.00 m

High density polyethylene sheaths of 32 mm in diameter, steel cradle deviators

Owner   Canton of Tessin

Structural design   Thomas Keller, WKP Bauingenieure AG Zurich (concept and design of strengthening project, I was a co-owner of this company)

Urs Meier, EMPA Dübendorf (cable and anchor design based on required cable force)

Execution   BBR Systems Ltd. and EMPA Dübendorf: cable installation and prestress

Completion   1999

(Allocation) Table A.1   Further information in Chapters 1 to 5

Section	Figure	Subject
3.4.5.4	3.28	Strengthening of Verdasio Bridge with four external carbon fibre cables inside concrete box girder, and cable force and temperature vs time (relaxation), over 21 years

A.1.2 Additional images and information

Fig. A.1 Prestressed concrete Verdasio Bridge with corrosion potential distributions measured on damaged web areas indicating corrosion of steel reinforcement **(left)**; bottom view of damaged zones through chloride contaminated water flow **(right)**

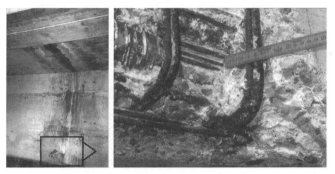

Fig. A.2 Detail of chloride contaminated water flow along web from upper surface dewatering tube with brown signs of steel corrosion **(left)**; local removal of concrete (in green window) shows pitting corrosion of steel wires of one prestressed steel cable (and rebars and stirrups) **(right)**

Fig. A.3 Installed four carbon fibre-polymer cables inside concrete box **(left)**; new concrete cable deviator (poured from top through hole in deck slab) **(right)**

Fig. A.4 Prestressing of new carbon fibre-polymer cables in abutment chamber behind end cross-beam **(left)**; feeding lightweight cables through drilled hole in deck slab into box girder **(right)**

A.1.3 Papers related to Verdasio Bridge

[1] T. Keller, "FRP applications in bridge and building construction – case studies", in *IABSE Technical Conference on Composite Structures*, Madrid 2002.

[2] T. Keller, "Strengthening of concrete bridges with carbon cables and strips", in *6th International Symposium on FRP Reinforcement for Concrete Structures (FRPRCS-6)*, Singapore, 2003.

A.2   PONTRESINA BRIDGE

A.2.1   Basic information

Concept — Temporary lightweight pedestrian bridge, installed each year in autumn and removed in spring due to exclusive use during winter and critical floodwater situation in summer

Two truss girders, adhesively bonded joints in one span (fully load-bearing), bolted joints in the other span

Structural redundancy through crossed diagonals and back-up bolts in bonded joints

Dimensions — Spans of 2×12.50 m, 1.5 (1.93) m interior (exterior) width

Materials — Pultruded glass fibre-polyester profiles, 50 vol% fibres, natural white colour (initially planned to be light green), unprotected

Two-component epoxy adhesive from Degussa, uncontrolled thickness of ~ 0.5-1.0 mm, stainless steel bolts

Loads — Permanent load, pedestrian load 4 kN/m^2 together with 1.2 kN/m^2 snow load (30 cm wet snow), plus 10 kN snow removal vehicle, lateral wind, and lateral impact on top chord handrail

Alternatively, permanent load plus 8.7 kN/m^2 snow (4 m high) with load factor 1.0

Owner   Pontresina Municipality

Architectural design   Thomas Keller, ETH Zurich

Structural design   Thomas Keller, ETH Zurich

     Design based on Fiberline Design Manual (1995), and full-scale testing of bolted and bonded joints

Execution   Fiberline Building Profiles, Denmark (pultruded profiles)

     ETH Zurich, 20 architecture students (assembled in two weeks)

Completion   December 1997

(Allocation) Table A.2   Further information in Chapters 1 to 5

Section	Figure	Subject
2.4.3	2.18 (left)	Manufacturing of adhesive joints
2.5.5	2.26 (left)	Installation by helicopter
2.6.3	2.33 (middle)	View of Pontresina Bridge in winter
3.10.2	-	Results of long-term survey of Pontresina Bridge
3.10.2.1	3.140	Installation during winter, storage on riverbanks during summer, and on river with removed gratings during floodwater
	3.141	Destruction of Pontresina Bridge in 2023, caused by all-time record floodwater on 28 August 2023
3.10.2.2	Table 3.8	Climate data of Pontresina
	Table 3.9	Intensity of sun irradiation in Pontresina
3.10.2.3	3.143	Full-scale serviceability (SLS) loading during periodic inspections
	3.144	Coupon tensile strength and elastic modulus development over 25 years
	3.145	Load vs mid-span deflection of bolted and bonded spans over 25 years
	3.146	Full-scale experiments on bolted and bonded joints in 1997
3.10.2.4	3.148 (left)	Adhesively bonded connections after 25 years (no detachment)
3.10.2.5	3.149	Cracks in outstand flanges and rehabilitation by supporting flanges with high-stiffness polymer filler block
	3.150	Flange crushing in bottom chord profile and rehabilitation with adhesively bonded bypass/bridging profiles
3.10.2.6	3.151	Fibre blooming levels and effect of missing surface veil
	3.153 (left)	Progress of fibre blooming levels over 25 years
	3.154	Effect of Level-2 fibre blooming on tensile strength and elastic modulus

Section	Figure	Subject
3.10.2.7	3.155	Transverse beam end with lost sealing and fungi growth after 25 years
3.10.2.8	3.157	Colour change between white-coated top chord after 18 years and uncoated profiles
	3.158	Level of yellowing due to UV radiation after 25 years
3.10.4	3.164 (left)	Spacer tubes to prevent cracking of closed-section profiles due to bolt tightening
4.3.3.1	4.16 (bottom left)	Material substitution by profile-truss structural form
5.5	5.5	Rehabilitation procedure to bypass/bridge crushed flange with adhesively bonded profiles
	5.6	Rehabilitation procedure to support cracked and deformed flange with high-stiffness polymer filler block
Postscript		Why was this bridge built?

A.2.2 Additional images and information

Fig. A.5 Pontresina Bridge in 1997

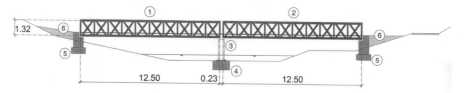

Fig. A.6 Original (1997) elevation drawing (1/2 = bonded/bolted spans, 3 = removable steel support (was changed to permanent concrete support), 4 and 5 = concrete foundation, 6 = filled ground), dimensions in [m]

Appendix

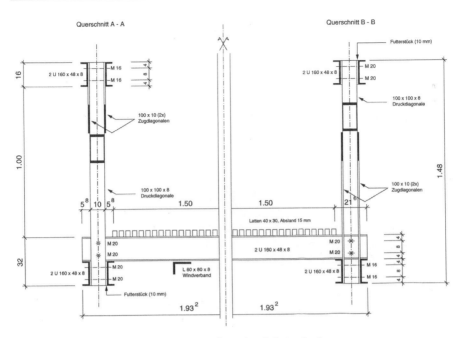

Fig. A.7 Original (1997) plan of cross section, dimensions in [m] or [cm]

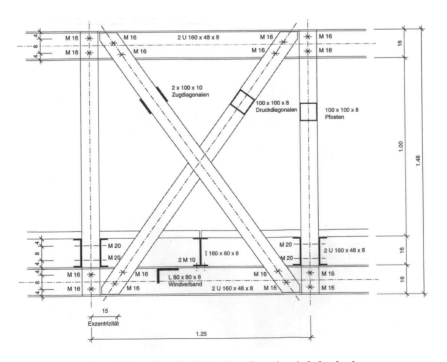

Fig. A.8 Original (1997) plan of longitudinal section, dimensions in [m] or [cm]

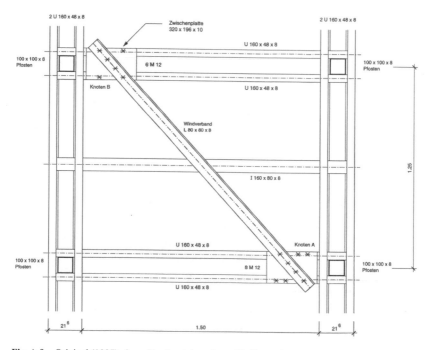

Fig. A.9 Original (1997) plan of horizontal section with diagonal wind bracing fixed on bottom flanges of transverse beams via intermediate plates, dimensions in [m] or [cm]

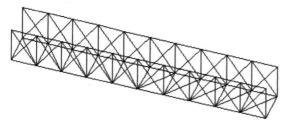

Fig. A.10 Structural system used for calculating internal forces (top and bottom chords continuous, remaining joints hinged, without considering joint eccentricities)

Table A.3 Eigenfrequencies and mode shapes resulting from impact excitation, showing that adhesive bonding has a significant effect on eigenfrequencies in lateral, but less in vertical mode shapes [1]

Mode	Eigenfrequency [Hz]		Damping [%]		Mode shape
	Bolted	Bonded	Bolted	Bonded	
1	4.0	5.2	3.9	3.1	lateral
2	8.2	10.9	1.4	4.2	lateral
3	12.5	13.0	-	5.2	vertical

Appendix

Fig. A.11 Construction team on top of 4 kN/m² brick loaded adhesively bonded span without back-up bolts installed, December 1997

A.2.3 Papers related to Pontresina Bridge

[1] Y. Bai and T. Keller, "Modal parameter identification for a GFRP pedestrian bridge", *Compos. Struct.*, vol. 82, pp. 90–100, 2008, doi: 10.1016/j.compstruct.2006.12.008.

[2] T. Keller, "Towards structural forms for composite fibre materials", *Struct. Eng. Int. J. Int. Assoc. Bridg. Struct. Eng.*, vol. 9, pp. 297–300, 1999, doi: 10.2749/101686699780481673.

[3] T. Keller, "Influences of advances composite materials on structural concepts for bridges and buildings", in *The 3rd International Conference on Composites in Infrastructure (ICCI'02)*, San Francisco, 2002.

[4] T. Keller, Y. Bai, and T. Valĺe, "Long-term performance of a glass fiber-reinforced polymer truss bridge", *J. Compos. Constr.*, vol. 11, pp. 99–108, 2007, doi: 10.1061/(ASCE)1090-0268(2007)11:1(99).

[5] T. Keller, N. A. Theodorou, A. P. Vassilopoulos, and J. De Castro, "Effect of natural weathering on durability of pultruded glass fiber-reinforced bridge and building structures", *J. Compos. Constr.*, vol. 20, 2016, doi: 10.1061/(ASCE)CC.1943-5614.0000589.

A.3 EYECATCHER BUILDING

A.3.1 Basic information

Concept   Mobile lightweight five-storey building, still tallest multi-storey building worldwide with primary all-composite structure

First use at a building fair ("Swissbau", 1999, Basel Messe Zentrum), subsequently disassembled and reassembled at Basel Dreispitz

	Envelope with integration of structural, building physics, and architectural functions
	Three composite frames composed of adhesively bonded built-up profiles
	Translucent aerogel-filled composite sandwich envelope on lateral sides
Dimensions	12.7×10.0×14.6 m (length×width×height)
Materials	Pultruded glass fibre-polyester profiles, 50 vol% fibres, grey colour, unprotected
	Two-component epoxy adhesive Sikadur-330, uncontrolled thickness of ~ 1.0 mm, stainless-steel bolts
Loads	Permanent loads plus live loads 3 (4) kN/m^2 (terrace on top), together with wind on envelope
Owner in 1998	I-catcher GmbH, Basel, Ruedi Tobler and Felix Knobel, Basel
Architectural design	Felix Knobel, Artevetro Architekten AG
Structural design	Thomas Keller, ETH Zurich / EPFL Lausanne
	Design based on Fiberline Design Manual (1995), and testing of full-scale built-up beams
Execution	Fiberline Building Profiles, Denmark (pultruded profiles)
	Scobalit AG, Winterthur (translucent panels on lateral sides)
	Jakem Steel Construction, Münchwilen (frame assembly and installation)
Completion	December 1998

(Allocation) Table A.4   Further information in Chapters 1 to 5

Section	Figure	Subject
1.2	1.1 (left)	Channel profile section from Eyecatcher showing fibre architecture, taken during pultrusion
2.4.3	2.18 (middle left)	Applying adhesive to flat profile to connect channel sections of built-up beams
2.5.5	2.26 (middle left)	Installation of frame by crane
2.7.3	Table 2.16b	Summary of project information
	2.43	Eyecatcher at night with illuminated translucent lateral side envelope and during installation

Appendix

Section	Figure	Subject
3.9.1	3.106 (left)	Interior view on third floor with fire detector and sprinkler tube
3.10.2	-	Results of long-term survey of Eyecatcher Building
3.10.2.1	3.142	Installation of timber slabs, Eyecatcher at different locations in 1998 and 2023
3.10.2.2	Table 3.8	Climate data of Basel
	Table 3.9	Intensity of sun irradiation in Basel
3.10.2.3	3.147	Built-up beam and column sections, full-scale experiment on built-up column in 1998, and failure mode outside of adhesive connection
3.10.2.4	3.148 (right)	Adhesively bonded connections after 25 years (no detachment)
3.10.2.6	3.152	Fibre blooming on columns profiles after 25 years
	3.153 (right)	Progress of fibre blooming levels over 25 years
3.10.2.7	3.156	Moisture migration around bolted joints
3.10.4	3.164 (right)	Spacer tubes to prevent cracking of closed-section profiles due to bolt tightening
4.3.3.1	4.17	Channel profile section from Eyecatcher showing fibre architecture, taken during pultrusion
4.3.4.2	4.23 (right)	Section-active system of Eyecatcher frame
Postscript		Why was this building built?

A.3.2 Additional images and information

Fig. A.12 Eyecatcher Building in 1998

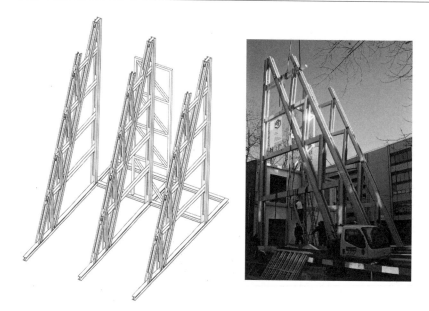

Fig. A.13 Primary load-bearing structure consisting of three parallel frames on steel profile foundation, each frame composed of pultruded composite built-up beams and columns arranged in three layers to decrease joint forces, transverse timber truss in the back acting as wind bracing together with timber slabs (not shown) **(left)**; timber staircase boxes in the back during installation **(right)**

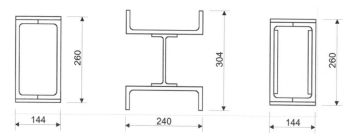

Fig. A.14 Adhesively bonded built-up profiles: horizontal beam **(left)**; column with central (not continuous) I-profile to prevent buckling of channel profiles **(middle)**; horizonal beam with web reinforcement plates at bolted joint locations **(right)**, dimensions in [mm]

Appendix

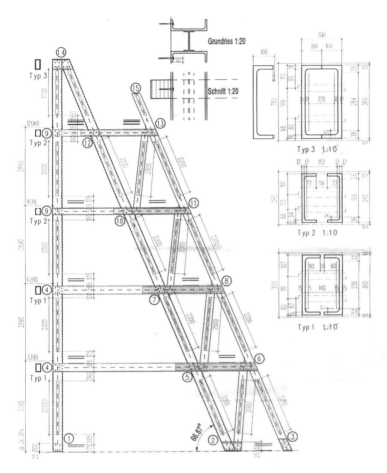

Fig. A.15 Original (1998) construction plan of frame composed of pultruded composite built-up profiles, with location of discontinuous column I-profiles and web reinforcements (grey shaded), dimensions in [mm]

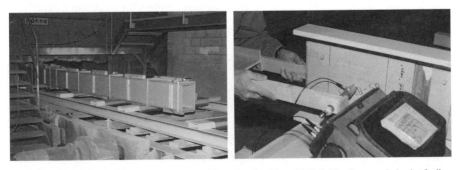

Fig. A.16 Bonded built-up beam exits oven after curing for 2 h at 60°C **(left)**; ultrasound check of adhesive connection of web reinforcement plate **(right)**

Fig. A.17 Interior views during first use in 1998, unprotected profiles **(left and right)**; (white) fire detector on bottom of ceiling beam and sprinkler tube fixed to timber slab **(right)**

Fig. A.18 Details of inclined glass front envelope and lateral frame with closed/protected beam ends **(left)**; columns approaching steel plates of foundation for joining **(right)**

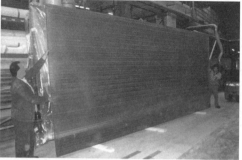

Fig. A.19 Web-core glass fibre composite panel for lateral envelope, 50 mm thick, 30 vol% fibres result in 70% transparency, manufactured by continuous lamination, subsequently filled with translucent aerogel thermal insulation, $U = 0.40$ W/(m^{2}K): cross section **(left)**; full panel **(right)** (1998, Scobalit)

Appendix

Fig. A.20 Reconstruction at new location in Basel Dreispitz, 1998

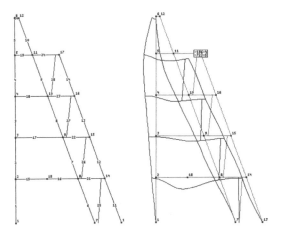

Fig. A.21 Structural system used for calculating internal forces, considering joint eccentricities, beams and columns are continuous, diagonals are hinged **(left)**; vertical and horizontal deformation under vertical live load on each storey, max. 19 mm **(right)**

A.3.3 Papers related to Eyecatcher Building

[1] T. Keller, "Towards structural forms for composite fibre materials", *Struct. Eng. Int. J. Int. Assoc. Bridg. Struct. Eng.*, no. 4, pp. 297–300, 1999, doi: 10.2749/101686699780481673.

[2] T. Keller, "Fibre reinforced polymer materials in building construction", in *IABSE Symposium Towards a Better Built Environment*, Melbourne, 2002.

[3] T. Keller, "Influences of advances composite materials on structural concepts for bridges and buildings", in *The 3rd International Conference on Composites in Infrastructure (ICCI'02)*, San Francisco, 2002.

[4] T. Keller, N. A. Theodorou, A. P. Vassilopoulos, and J. De Castro, "Effect of natural weathering on durability of pultruded glass fiber-reinforced bridge and building structures", *J. Compos. Constr.*, vol. 20, 2016, doi: 10.1061/(ASCE)CC.1943-5614.0000589.

[5] Y. Bai, T. Keller, and C. Wu, "Pre-buckling and post-buckling failure at web-flange junction of pultruded GFRP beams", *Mater. Struct. Constr.*, vol. 46, pp. 1143–1154, 2013, doi: 10.1617/s11527-012-9960-9.

A.4 NOVARTIS BUILDING

A.4.1 Basic information

Concept
Lightweight composite foam-filled web-core sandwich roof, on load-bearing glass envelope, double curvature, seamless, wing shape

Function integration: static, building physics, and architectural functions

Prefabrication in four segments and rapid installation, adhesively bonded connections

Built up from 460 CNC-machined PUR-foam blocks of different geometries, no moulds required

Secured against wind uplift by stainless steel rods between glass envelope double-stiffeners

Design service life of 50 years

Dimensions
Roof 21.6×18.5 m (400 m^2), thickness of 70–620 mm

Materials
Glass fibre-polyester face sheets and webs, 3–6 mm in thickness, 29 vol% fibres

PUR foam of 60–145 kg/m^3 density (depending on shear loading)

Two-component filled epoxy adhesive Sikadur-30

White PU-enamel silk gloss coating

Loads
Permanent loads, prestress in steel rods securing roof from uplift, creep, shrinkage of laminates, temperature (global changes and gradients), wind (horizontal pressure and uplift), snow, walking on roof for cleaning, impact from overturning tree, failure of two adjacent supports or a corner support (due to partial collapse of glass envelope), earthquake, fire (2×2 m burn through without collapse), and construction stages

Owner
Novartis Pharma AG, Basel

Architectural design
Marco Serra, Basel

Structural design
Ernst Basler & Partner AG, Zurich (final design)

Swissfiber AG, Zurich & ZHAW Winterthur (preliminary design)

Thomas Keller, EPFL-CCLab (technical and scientific support and supervision)

EPFL-CCLab (material characterisation and full-scale testing)

Design based on Eurocomp Design Code and Handbook (1996), and testing

Execution (roof)
Scobalit AG, Winterthur (CNC cutting of foam blocks, hand lay-up, and installation, incl. adhesive bonding)

Completion
November 2006

Appendix

(Allocation) Table A.5 Further information in Chapters 1 to 5

Section	Figure	Subject
2.4.3	2.18 (middle right)	On-site application of adhesive on web to join sandwich panel segments
2.5.1	2.20	Hand lay-up and assembly of PUR web-core sandwich panel segment
2.5.5	2.26 (middle right)	Installation of sandwich panel segment on scaffolding
2.7.4	Table 2.16b	Summary of project information
	2.49	Different view on installation of segment on scaffolding
3.4.5.2	3.27	Effect of creep on shear force distribution between webs and foam core
3.9.1	3.106 (right)	Interior view
3.10.3	3.159	Blistering and cracks along adhesive joints after 12 years
4.3.2	4.14	Roof support on glass envelope, detail section and view during installation
	4.15	CNC milling of foam blocks
4.3.3.1	4.16 (top right)	View on illuminated building during night
4.4.4	4.44	Study on potential integration of skylights into roof
4.5.2.3	4.48	Study on thermal transmittance of roof
4.7.2	4.67	Roof segment during fabrication, transportation, installation (top), and adhesive joining of segments (bottom)
Postscript		Why was this building built?

A.4.2 Additional images and information

Fig. A.22 Construction of Novartis Building in 2006

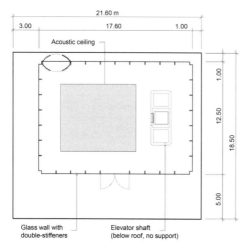

Fig. A.23 Plan view with glass envelope and glass double-stiffeners, roof cantilevers, recess for acoustic ceiling, and elevator shafts (from underlying parking, not connected to roof), dimensions in [m]

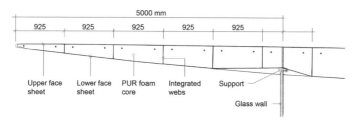

Fig. A.24 Longitudinal section through 5.0 m long overhang with 900 mm foam blocks and internal webs, glass envelope and support (made invisible according to architect's requirement), dimensions in [mm]

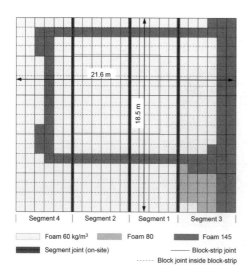

Fig. A.25 Plan view with internal web grid, core density distribution (according to shear loading), and arrangement of blocks, four-block strips, and segments

Appendix

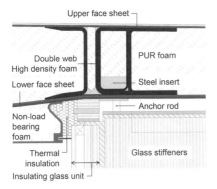

Fig. A.26 Cross section with steel insert and associated suspension laminates for rod anchorage against uplift

Fig. A.27 Glass double-stiffeners (**left**) with anchorage rod in between (**right**)

Fig. A.28 Separate hand lay-up manufacturing of 70 mm thick roof edge with water drip: corner mould (**left**); edge element integrated into sandwich panel manufacturing (downside up) (**right**)

Fig. A.29 Complex shape of shortest 1.0 m long overhang during manufacturing (segment downside up)

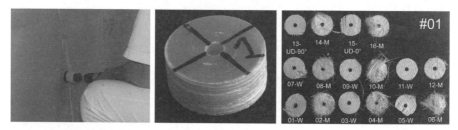

Fig. A.30 Quality control: 40 mm diameter cutouts taken from laminates **(left)**; cutout for laminate thickness control **(middle)**; control of fibre layers after resin burn-off (01–16 = layer number, M = chopped strand mat, W = woven fabric 0/90°, UD = unidirectional binding warp fabric) **(right)**

Fig. A.31 Full-scale four-point-bending experimental design verification: beam failure in adhesive joint at mid-span **(left)**; beam with bottom recess for acoustic ceiling during loading **(right)** (beams composed of eight blocks, length of 7.20 m, span of 5.4 m, thickness of 300–460 mm)

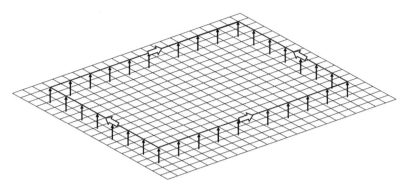

Fig. A.32 Structural system composed of internal girder grid (webs plus effective width of face sheets), vertical supports at location of glass stiffeners (vertical arrows), and horizontal supports in plane directions of glass envelop (horizontal arrows)

A.4.3 Papers related to Novartis Building

[1] T. Keller, C. Haas, and T. Vallée, "Structural concept, design, and experimental verification of a glass fiber-reinforced polymer sandwich roof structure", *J. Compos. Constr.*, vol. 12, pp. 454–468, 2008, doi: 10.1061/(ASCE)1090-0268(2008)12:4(454).

[2] S. Yanes-Armas, J. de Castro, and T. Keller, "Long-term design of FRP-PUR web-core sandwich structures in building construction", *Compos. Struct.*, vol. 181, pp. 214–228, 2017, doi: 10.1016/j.compstruct.2017.08.089.

[3] B. D. Manshadi, A. P. Vassilopoulos, J. De Castro, and T. Keller, "Instability of thin-walled GFRP webs in cell-core sandwiches under combined bending and shear loads", *Thin-Walled Struct.*, vol. 53, pp. 200–210, 2012, doi: 10.1016/j.tws.2011.12.021.

[4] B. D. Manshadi, A. P. Vassilopoulos, and T. Keller, "Post-wrinkling behavior of webs in GFRP cell-core sandwich structures", *Compos. Struct.*, vol. 138, pp. 276–284, 2016, doi: 10.1016/j.compstruct.2015.11.025.

[5] B. D. Manshadi, A. P. Vassilopoulos, J. De Castro, and T. Keller, "Contribution to shear wrinkling of GFRP webs in cell-core sandwiches", *J. Compos. Constr.*, vol. 15, 2011, doi: 10.1061/(ASCE)CC.1943-5614.0000212.

[6] B. D. Manshadi, A. P. Vassilopoulos, and T. Keller, "Shear buckling resistance of GFRP plate girders", *J. Compos. Constr.*, vol. 15, 2011, doi: 10.1061/(ASCE)CC.1943-5614.0000167.

A.5 AVANÇON BRIDGE

A.5.1 Basic information

Concept   Lightweight composite-balsa sandwich bridge deck adhesively bonded onto two steel girders

                   Semi-integrated bridge without expansion joints and no waterproofing layer

                   Replacement of deteriorated one-lane concrete bridge by new two-lane bridge without increasing total load on old brick abutments

                   Reduction of bridge closure time from 50 days (conventional replacement) to ten days

                   Design service life of 100 years

Dimensions   Deck 11.45×7.50 m (length×width), 65° skew angle

                   Deck (face sheet) thickness of 285 (22) mm, steel girder depth of 650 mm

                   Deck-girder adhesive layer thickness of 10 mm (ensured by spacers)

                   Surfacing layer thickness of 60 mm

Materials   Glass fibre-vinylester face sheets, 55 vol% fibres, 25 (75) % in longitudinal (transverse) directions

                   Balsa LVL 0°/0° core of 250 kg/m^3 nominal density

                   Two-component epoxy adhesives Sikadur-330 for deck-girder connections, and Sikadur-300 for injection of transverse panel-panel joints

                   PU coating of apparent/exposed composite surfaces

                   Steel S355 girders, medium-temperature asphalt applied at 120°C

Loads   Permanent loads, vehicular traffic loads (including braking and centrifugal forces), temperature (global changes and gradients), impact on crash barrier, earthquake (wind and snow are negligible)

Owner   Canton of Vaud

Structural design Suisse Technology Partners, Neuhausen am Rheinfall

                   3A Composites, Sins (material characterisation)

                   Thomas Keller, EPFL-CCLab (expert from owner side)

                   EPFL-CCLab (full-scale static and fatigue testing)

                   Design based on German BÜF Guideline (2014), Eurocomp Design Code and Handbook (1996), and testing

Execution   3A Composites, Sins (vacuum infusion of deck and on-site adhesive joints)

Completion   November 2012

Appendix

(Allocation) Table A.6   Further information in Chapters 1 to 5, including balsa wood

Section	Figure	Subject
1.2	1.1 (right)	Sandwich panel specimen of Avançon Bridge deck
2.2.2.1	2.2	Microstructure of balsa wood with longitudinal tracheid and radial ray cells
	2.3	Cell wall thickness vs density of balsa wood
	2.4	Top view on balsa wood end-grain panel with blocks and adhesive bond lines
2.2.2.2	Table 2.5	Relevant properties and application fields of balsa wood
	2.5	Normalised shear modulus vs temperature of balsa wood
2.3.3	2.12 (bottom left)	Sandwich panel section for pedestrian bridges with applied sanding for non-slip surface, and dewatering channel (spruce core)
2.4.3	2.14 (right)	Peel-ply removal on transverse sandwich panel joint before bonding
	2.17	Transverse scarf joints in sandwich panel face sheets and epoxy adhesive injection into panel-to-panel joint
	2.18 (right)	Adhesive application on top of steel girder flange
2.5.5	2.26 (right)	Lifting into place of new bridge superstructure
2.6.2.3	2.32	Experimental verification of crash barrier post fixation in sandwich deck
2.6.3.1	2.34	Pedestrian bridges with balsa sandwich decks
3.2.2.2	3.3	Anisotropy in end-grain balsa panels
	3.4	Crack patterns in end-grain balsa panels
	3.5	Shear failure modes of laminated veneer lumber (LVL) 0/90° balsa
3.3.3	3.13	Handling of tolerances in deck-to-girder adhesive connections
3.7.3.2	3.70 (middle and right)	Mode-I 2D face sheet debonding from balsa core
3.7.3.3	3.74	2D Mode-I load and total radial crack length vs opening displacement, and compliance vs total crack area
	3.75	Effect of R-ratio on fatigue crack (area) growth rate dA/dN of 2D Mode-I sandwich panel face sheet debonding
3.8.8	3.95	Effect of balsa core joint type on stress concentrations
	3.96	Continuous support of upper face sheet provided by balsa core to reduce stress concentrations
	3.98	Full-scale composite-balsa sandwich bridge girder with adhesive joint at mid-span under four-point bending fatigue loading
	3.99	Debonding of glass fibre composite face sheet from balsa core under mixed-mode bending fatigue
	3.100	Mixed-mode static failure criterion for debonding of face sheet from balsa core, and crack growth rate vs fracture energy comparison between face sheet debonding from balsa core and fibre-tear failure in adhesive connection

Section	Figure	Subject
3.8.9	3.101	Two-axle fatigue load model on Avançon deck and fatigue critical location
	3.104	Calibrated fatigue load model for Avançon deck
3.9.2.2	3.107 (bottom)	Remaining mass fraction and mass loss rate vs temperature for balsa wood
	3.108	Balsa wood structure at 20°C and after cooling down from 250°C
3.9.2.3	3.109 (right)	Effective thermal conductivity vs temperature for balsa wood
3.9.2.4	3.110 (right)	Effective specific heat capacity vs temperature for balsa wood
3.9.2.5	3.112	Storage modulus and tan-δ vs temperature at different moisture levels
	3.113	Glass transition temperatures of partially and fully cured polyurethane adhesive used in balsa wood panels between blocks
3.9.2.7	3.116	Normalised compressive strength vs temperature for balsa wood
3.9.2.8	3.117	Coefficient of thermal expansion vs temperature for balsa wood
3.9.3.2	Table 3.7	Balsa-core sandwich beams subjected to simultaneous SLS load and one-sided ISO 834 fire
	3.119	Span/mid-span deflection vs time relationships of balsa-core sandwich beams under simultaneous SLS load and one-sided ISO 834 fire
3.9.3.3	3.127	Progress of char depth into balsa core under fire vs time in case of through-thickness core air gaps or steel inserts
	3.129	Heating and cooling effects of air gaps and steel inserts in balsa core under one-sided fire
	3.130	Failure modes of composite-balsa core sandwich beam under three-point bending and one-sided fire
	3.131	Charring of balsa core under three-point bending and one-sided fire
3.10.3	3.160	View on Avançon Bridge in 2012 and after 11 years of use
	3.161	Rehabilitation of deformed elastomer plug joint, and cracks in protective coating
3.10.4	3.162	Concrete vs composite bridge kerb, simple kerb detail of Avançon Bridge
4.6.2	4.61	Hybrid concepts of composite-balsa slab bridges and bridge decks with carbon fibre composite arch and high-density timber and steel inserts in core
4.7.3	4.69	Construction and installation of Avançon Bridge
	4.70	Modular bridge concept with composite deck panels detachable from steel girders
Postscript		Why was this bridge built?

A.5.2 Additional images and information

Fig. A.33 Construction of Avançon Bridge in 2012

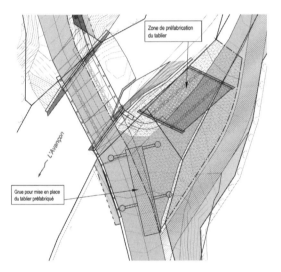

Fig. A.34 Plan view of new bridge with skew angle over Avançon River, and installation place with crane location for bridge lifting

Fig. A.35 Former one-lane concrete bridge on stone abutment with foundations under river level **(left)**; demounted former bridge **(middle)**; new two-lane bridge in service **(right)**

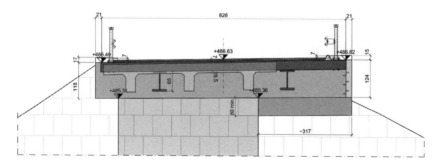

Fig. A.36 Former one-lane three-girder concrete bridge section, and new two-lane hybrid-composite bridge on only slightly modified stone abutment wall thanks to identical total weight (permanent plus traffic load)

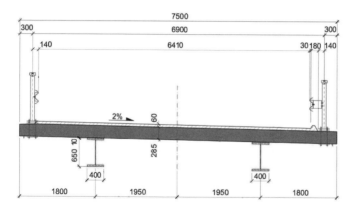

Fig. A.37 Cross section of new hybrid-composite bridge with dimensions in [mm]

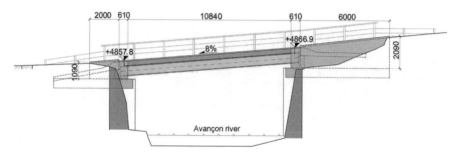

Fig. A.38 Longitudinal section of new hybrid-composite bridge with dimensions in [mm]

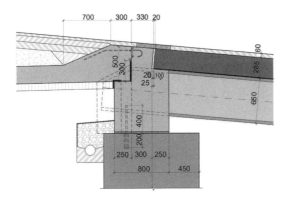

Fig. A.39 Detail of abutment with dimensions in [mm] (concrete transition slab behind bridge, on left side)

Fig. A.40 Mixing of two-component epoxy adhesives: viscous adhesive for deck-girder connections **(left)**; fluid adhesive for panel-panel injections **(middle)**; on-site test specimen fabrication for quality control **(right)**

Fig. A.41 Steel girder end with applied adhesive on steel flange and rubber supports on concrete crossbeam, ready for deck panel installation **(left)**; kerb with inserted crash barrier post fixation, and infused diagonal panel-panel joint in the back through asphalt adherend layer **(right)**

Fig. A.42 Installation of new superstructure using outstanding steel girders for chain suspension

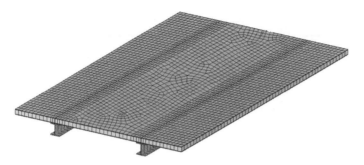

Fig. A.43 Finite element model of bridge superstructure for structural design, adhesive layer and steel girders included in model

A.5.3 Papers related to Avançon Bridge and balsa wood

[1] T. Keller, J. Rothe, J. De Castro, and M. Osei-Antwi, "GFRP-balsa sandwich bridge deck: Concept, design, and experimental validation", *J. Compos. Constr.*, vol. 18, 2014, doi: 10.1061/(ASCE)CC.1943-5614.0000423.

[2] M. Osei-Antwi, J. De Castro, A. P. Vassilopoulos, and T. Keller, "Structural limits of FRP-balsa sandwich decks in bridge construction", *Compos. Part B Eng.*, vol. 63, pp. 77–84, 2014, doi: 10.1016/j.compositesb.2014.03.027.

[3] M. Osei-Antwi, J. De Castro, A. P. Vassilopoulos, and T. Keller, "FRP-balsa composite sandwich bridge deck with complex core assembly", *J. Compos. Constr.*, vol. 17, 2013, doi: 10.1061/(ASCE)CC.1943-5614.0000435.

[4] M. Osei-Antwi, J. De Castro, A. P. Vassilopoulos, and T. Keller, "Fracture in complex balsa cores of fiber-reinforced polymer sandwich structures", *Constr. Build. Mater.*, vol. 71, pp. 194–201, 2014, doi: 10.1016/j.conbuildmat.2014.08.029.

[5] M. Osei-Antwi, J. de Castro, A. P. Vassilopoulos, and T. Keller, "Analytical modeling of local stresses at Balsa/timber core joints of FRP sandwich structures", *Compos. Struct.*, vol. 116, pp. 501–508, 2014, doi: 10.1016/j.compstruct.2014.05.050.

Appendix

[6] M. Osei-Antwi, J. de Castro, A. P. Vassilopoulos, and T. Keller, "Modeling of axial and shear stresses in multilayer sandwich beams with stiff core layers", *Compos. Struct.*, vol. 116, pp. 453–460, 2014, doi: 10.1016/j.compstruct.2014.05.049.

[7] M. Osei-Antwi, J. De Castro, A. P. Vassilopoulos, and T. Keller, "Shear mechanical characterization of balsa wood as core material of composite sandwich panels", *Constr. Build. Mater.*, vol. 41, pp. 231–238, 2013, doi: 10.1016/j.conbuildmat.2012.11.009.

[8] A. Cameselle-Molares, A. P. Vassilopoulos, and T. Keller, "Two-dimensional quasi-static debonding in GFRP/balsa sandwich panels", *Compos. Struct.*, vol. 215, pp. 391–401, 2019, doi: 10.1016/j.compstruct.2019.02.077.

[9] A. Cameselle-Molares, A. P. Vassilopoulos, and T. Keller, "Two-dimensional fatigue debonding in GFRP/balsa sandwich panels", *Int. J. Fatigue*, vol. 125, pp. 72–84, 2019, doi: 10.1016/j.ijfatigue.2019.03.032.

[10] N. Vahedi, C. Tiago, A. P. Vassilopoulos, J. R. Correia, and T. Keller, "Thermophysical properties of balsa wood used as core of sandwich composite bridge decks exposed to external fire", *Constr. Build. Mater.*, vol. 329, 2022, doi: 10.1016/j.conbuildmat.2022.127164.

[11] N. Vahedi, J. R. Correia, A. P. Vassilopoulos, and T. Keller, "Effects of core air gaps and steel inserts on thermal response of GFRP-balsa sandwich panels subjected to fire", *Fire Saf. J.*, vol. 134, 2022, doi: 10.1016/j.firesaf.2022.103703.

[12] N. Vahedi, J. R. Correia, A. P. Vassilopoulos, and T. Keller, "Effects of core air gaps and steel inserts on thermomechanical response of GFRP-balsa sandwich panels subjected to fire", *Compos. Struct.*, vol. 313, 2023, doi: 10.1016/j.compstruct.2023.116924.

[13] C. Wu, N. Vahedi, A. P. Vassilopoulos, and T. Keller, "Mechanical properties of a balsa wood veneer structural sandwich core material", *Constr. Build. Mater.*, vol. 265, 2020, doi: 10.1016/j.conbuildmat.2020.120193.

[14] N. Vahedi, C. Wu, A. P. Vassilopoulos, and T. Keller, "Thermomechanical characterization of a balsa-wood-veneer structural sandwich core material at elevated temperatures", *Constr. Build. Mater.*, vol. 230, 2020, doi: 10.1016/j.conbuildmat.2019.117037.

A.6 FLOWER SCULPTURE

A.6.1 Basic information

Concept	Sculpture composed of three flowers, anchored in the ground
	Translucent flowers with integrated lighting
	Flowers move slightly in the wind
Dimensions	Flower heights of 6.2, 5.9, and 5.7 m (0.8 cm of which are in the ground)
	Blooms of ~180×95×50 cm, stem diameter of 11–12 cm, laminate thickness of ~6 mm
Materials	Glass fibre-polyester composites
Location	Hospital, Brig
Artists	Carlo Schmidt (Guttet-Feschel, Canton of Wallis) and Pascal Seiler (Gampel-Bratsch, Canton of Wallis)

Structural design	Thomas Keller, EPFL-CCLab
	Max. horizontal displacements under wind: 50, 40, and 35 cm
Execution	Artists (shaped bloom and stem in foam)
	Scobalit AG, Winterthur (produced mould of bloom by wrapping shaped foam, subsequently hand-lay-up of blooms in mould, and wrapping of stems)
Completed	2002

(Allocation) Table A.7 Further information in Chapters 1 to 5

Section	Figure	Subject
2.5.1	2.19	Hand lay-up manufacturing of stem and bloom

A.6.2 Additional images and information

Fig. A.44 Composite Flower Sculpture with illuminated blooms **(left)**; detail **(right)**

Fig. A.45 Shaped foam by artists, to manufacture bloom mould and stems

A.7 TCy SCULPTURE

A.7.1 Basic information

Concept	"TCy" refers to Tensegrity Cylinder
	Deployable indoor sculpture for Hermès pavilion at Watches & Wonders Show, Geneva, 2022
	Suspended tensegrity structure delimited by external and internal cylinder
	12 cable helices per cylinder, six of which are arranged clockwise and the other six counter-clockwise
	12 helices of tubes and 12 rings of radial cables located between cylinders
	360 cables, 156 tubes, and 156 joints
	Complex joints with connection of 2 tubes, 4 helical cables, and 1 radial cable each
	Application of prestress to stabilise sculpture by shortening radial cables
Dimensions	Height of 6.2 m, outer (inner) cylinder diameter of 3.3 (1.5) m
	External (internal) diameter of tubes 40 (36) mm, cable diameters of 3 and 4 mm
Materials	Tubes of carbon fibre-epoxy composites, elastic modulus of 70 GPa, compressive strength of 470 MPa
	Stainless steel cables and aluminium/steel joints
Owner	Hermès, Paris
Artist	Clément Vieille, Saint Alban Leysse, France
Architectural design	Filippo Broggini AEA & BlueOffice Architecture, Genève-Bellinzona
Structural design	Landolf Rhode-Barbarigos, University of Miami
	Tara Habibi, Thomas Keller, EPFL-CCLab
	Luca Diviani, University of Applied Sciences of Italian Switzerland (SUPSI), Lugano-Viganello (joint design)
Execution	Pre-assembly and experimental work at EPFL-CCLab, subsequent folding and then deployment in Geneva
Completed	2022

(Allocation) Table A.8 Further information in Chapters 1 to 5

Section	Figure	Subject
4.2.4.1	4.5 (right)	View from top on TCy (Tensegrity Cylinder) structure, Geneva, 2022

A.7.2 Additional images and information

Fig. A.46 Pre-assembly of TCy structure at EPFL-CCLab and 3D scan to check geometry after prestress

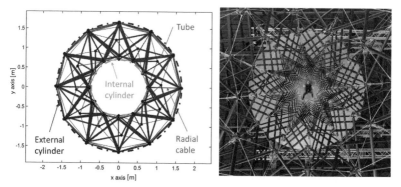

Fig. A.47 Composition of structure in plan and image from top into assembled structure

Fig. A.48 Joint composed of 12 aluminium and steel components: connection of two carbon fibre-epoxy tubes (**left**); installed joint with support fixation, two carbon tubes, and radial and one visible external cable connected (**right**)

Fig. A.49 Deployment and installation of folded sculpture at Geneva exhibition site

A.7.3 Papers related to TCy Tensegrity Sculpture

[1] T. Habibi, L. Rhode-Barbarigos, and T. Keller, "Effects of prestress implementation on self-stress state in large-scale tensegrity structure", *Eng. Struct.*, vol. 288, no. November 2022, p. 116222, 2023, doi: 10.1016/j.engstruct.2023.116222.

[2] T. Habibi, L. Rhode-Barbarigos, F. Broggini, L. Diviani, and T. Keller, "Self-stress distribution in large-scale cylindrical tensegrity structure", in *Proceedings of the IASS 2022 Symposium affiliated with APCS 2022 conference*, Beijing, 2022.

B.1 DOCK TOWER

B.1.1 Basic information

Background	High-rise building designed for "Urban visions" exhibition at Swiss Building Fair "Swissbau" in 2002
Concept	Energetically autonomous composite high-rise building of 36 storeys, comprising living and office spaces in the south and north, respectively
	Light weight, about 40% of that of a conventional steel building with glass envelope, provides significant advantages in seismic region
	Function integration: static, building physics, and architectural functions
	Circular cross section derived from plant stems offers optimised structural, protection, and supply functions
	Structural function provided by (i) an exterior Vierendeel system integrated into envelope, consisting of five elevator/stairwell shafts and connecting translucent two/six-storey envelope segments (of living spaces with circular windows to reduce stress concentrations), and (ii) five interior supply shafts
	Cross-sectional effective depth of 27 m results in small stresses caused by horizontal wind and seismic actions

	Entry of natural light into building through alternating two-storey sky gardens with transparent envelope, also creating an interior envelope enclosing the central (non-load bearing) stairwell shaft
	Water-circulation system integrated into cells of web-core sandwich panel walls and slabs provides heating and cooling, and fire resistance for up to two hours
	Central and northern stairwell shafts serve as fire escapes
	Encapsulated photovoltaic cells into exterior shaft panels and five vertical wind turbines on roof provide energy harvesting and make the building autonomous
Dimensions	Diameter of 27 m and height of 108 m
Materials	Glass fibre-polymer composites
	Transparent aerogel and opaque PUR thermal insulation in cells of web-core sandwich panel members
Location	Basel
Architectural design	Thomas Keller, EPFL-CCLab
Structural design	Thomas Keller, EPFL-CCLab
Year of design	2001, complemented by water-cooling system in 2004, and photovoltaic cell encapsulation in 2006

(Allocation) Table B.1 Further information in Chapters 1 to 5

Section	Figure	Subject
4.2.3	4.4	Derivation of cross section from stem of Lolium perenne
4.5.3.2	4.54	Thermally activated fibre-polymer composite building slab

B.1.2 Additional images and information

Fig. B.1 Rendering of composite Dock Tower during day, with vertical axis wind turbines on top **(left and middle)**, and at night, showing transparency of double-storey sky gardens, translucency of living spaces with transparent windows, and opacity of elevator shafts with encapsulated photovoltaic cells **(right)**

Appendix

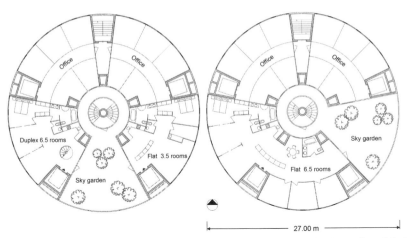

Fig. B.2 Cross section of 27 m diameter: office spaces in the north, flats in the south, and double-storey sky garden in the south **(left)**, and southeast **(right)**

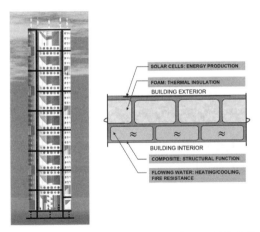

Fig. B.3 Vierendeel structural system of tower, 2001 **(left)**; multifunctional wall element of elevator shaft providing structural function, thermal insulation, heating/cooling and fire resistance through cellular water flow, and energy harvesting through encapsulated photovoltaic cells **(right)**, concept developed in 2004 (water flow integration) and extended in 2006 (photovoltaic cell integration)

B.1.3 Papers related to Dock Tower

[1] T. Keller, "Fibre reinforced polymer materials in building construction", in *IABSE Symposium Towards a Better Built Environment*, Melbourne, 2002.

[2] T. Keller, "Influences of advances composite materials on structural concepts for bridges and buildings", in *The 3rd International Conference on Composites in Infrastructure (ICCI'02)*, San Francisco, 2002.

[3] D. Khovalyg, A. Mudry, and T. Keller, "Prefabricated thermally-activated fiber-polymer composite building slab P-TACS: Toward the multifunctional and pre-fabricated structural elements in buildings", *J. Build. Phys.*, 2023, doi: 10.1177/17442591221150257.

[4]   T. Keller, J. De Castro, and M. Schollmayer, "Adhesively bonded and translucent glass fiber reinforced polymer sandwich girders", *J. Compos. Constr.*, vol. 8, 2004, doi: 10.1061/(ASCE)1090-0268(2004)8:5(461).

[5]   T. Keller, C. Tracy, and E. Hugi, "Fire endurance of loaded and liquid-cooled GFRP slabs for construction", *Compos. Part A Appl. Sci. Manuf.*, vol. 37, pp. 1055–1067, 2006, doi: 10.1016/j.compositesa.2005.03.030.

[6]   T. Keller, A. P. Vassilopoulos, and B. D. Manshadi, "Thermomechanical behavior of multifunctional GFRP sandwich structures with encapsulated photovoltaic cells", *J. Compos. Constr.*, vol. 14, pp. 470–478, 2010, doi: 10.1061/(ASCE)CC.1943-5614.0000101.

B.2   RLC BUILDING

B.2.1   Basic information

Concept	"RLC" refers to Rolex Learning Center
	Alternative project to planned conventional foil-covered roof 700 mm in thickness
	Double-curved composite web-core sandwich roof of complex shape and monolithic seamless appearance
	Minimised constant thickness of 350 mm thanks to function integration
Dimensions	Roof of 166.5×121.5 m, composed of (i) 12 sections of thermal dilatation and (ii) about 350 composite sandwich panels of 3×18 m size
	127 mm diameter steel columns in grid of 9×9 m
Materials	Glass fibre-polymer face sheets and webs
	PUR foam, two-component epoxy adhesive
Loads	Permanent loads, snow, wind, walking on the roof
	Fire: no structural collapse in case of single column failure
Location	EPFL Lausanne
Architectural design	Building: SANAA Architects, Japan
Structural design	Composite sandwich roof: Ernst Basler & Partner AG, Zurich
	Thomas Keller, EPFL-CCLab
	Scobalit AG, Winterthur (manufacturer of Novartis roof, cost estimate)
Manufacturing	Sandwich panels: lay-up and adhesive bonding of prefabricated face sheets and composite laminate-wrapped foam blocks on shape-adjustable "bed" of jacks, adhesive bonding of panels on site
Year of design	2006–2007

Appendix 401

Project subsequently abandoned since (i) estimated cost of 1 100 CHF/m² was too high compared to foil-covered roof (530.- CHF/m²) and (ii) risk of upscaling from 400 m² (Novartis Building, A.4, just completed in 2006) to 15 000 m² (without courtyards) was considered too high

(Allocation) Table B.2 Further information in Chapters 1 to 5

Section	Figure	Subject
4.5.2.3	4.49	Thickness comparison of function-integrated composite sandwich roof (350 mm) vs built foil-covered roof (700 mm)

B.2.2 Additional images and information

Fig. B.4 Rolex Learning Center (RLC), EPFL Lausanne, with conventional function-separated steel-timber insulation-foil roof

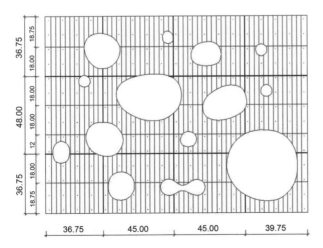

Fig. B.5 Plan view with about 350 composite panels of 3×18 m size in most cases, and 12 sections of thermal dilatation, dimensions in [m]

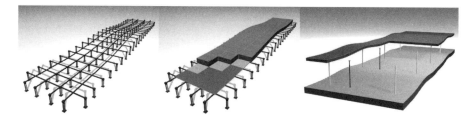

Fig. B.6 Manufacturing and installation of 3×18 m composite panels: shape-adjustable "bed" of jacks **(left)**; lay-up of prefabricated (deformable) bottom face sheet, adhesive bonding of 1.0×1.0×0.35 m laminate-wrapped foam blocks (compensation of small angles between blocks within adhesive layer), bonding of prefabricated (deformable) top face sheet (not shown) **(middle)**; on-site adhesive bonding of panels on final (white) and auxiliary (light grey) columns **(right)**

B.2.3 Paper related to RLC Building

[1] T. Keller and S. Yanes-Armas, "FRP sandwich structures in bridge and building construction", in *11th International Conference on Sandwich Structures (ICSS-11)*, Fort Lauderdale, USA, 2016.

B.3 CLP BUILDING

B.3.1 Basic information

Background	Replacement of deteriorated timber roof on indoor space with swimming pool
	Minimum closure time through use of existing columns of timber roof
	Existing columns with limited load-bearing capacity
	Indoor temperature up to 34°C and relative humidity up to 80%
Concept	Lightweight composite web-core sandwich roof with integration of structural, building physics, and architectural functions
	Integrated thermal insulation through use of PUR foam
	Daylight entry through three half-waves
	Two roof options: roof with simple top coating or added green roof to fulfil urban planning requirements
	Stepwise installation of new roof in parallel to dismantling of old roof
Dimensions	Roof area of 1 040 m^2, composed of ten prefabricated panels of 5.0 m width, up to 22.4 m length, and 90–600 mm thickness
	Maximum span of 8.80 m and 1.70 m overhang
	Three half-waves of identical double-curved geometry, each half-wave comprised in one panel

	Pre-camber to compensate for creep deformations
Materials	Glass fibre-polymer face sheets and webs, 55 vol% fibres
	PUR foam of 60 kg/m^3, two-component epoxy adhesive
Loads	Permanent loads and snow
	Green roof of 170 kg/m^2, with increase of permanent loads by a factor of 3.4 and of sandwich roof weight by 60%
Location	Cannot be disclosed
Architectural design	Brönnimann & Gottreux Architectes SA, Vevey
Structural design	Thomas Keller, Sonia Yanes-Armas, EPFL-CCLab
	Design based on Eurocomp Design Code and Handbook (1996) and German BÜF Guideline (2014)
Manufacturing	Resin infusion of modules, adhesive bonding of joints
Year of design	2015–2016
	Project subsequently abandoned since estimated cost of 2 500 CHF/m^2 was too high

(Allocation) Table B.3 Further information in Chapters 1 to 5

Section	Figure	Subject
3.4.5.2	-	Effect of green roof weight on dimensions

B.3.2 Additional images and information

Fig. B.7 Rendering of new composite sandwich roof with half-waves for daylighting of indoor space and cover technical facilities: roof with simple white topcoat on exterior face sheet **(left)** or with green roof on top of exterior face sheet **(right)**

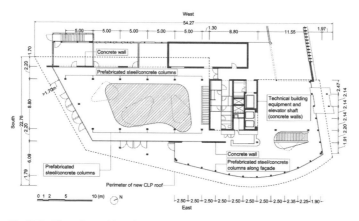

Fig. B.8 Plan view with swimming pool in centre, existing columns from previous timber roof, and perimeter of new composite roof on existing columns (dashed), dimensions in [m]

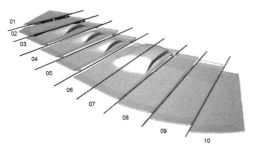

Fig. B.9 Subdivision of new composite roof into 10 panels of sizes up to 5×23 m, three of the four half-waves with identical geometry

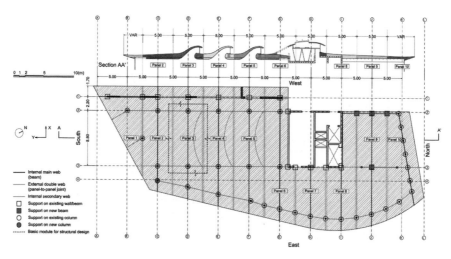

Fig. B.10 Longitudinal section **(top)** and plan view **(bottom)** of new roof, showing panels with double-web panel joints (green), internal primary transverse webs (red), and secondary longitudinal webs (blue) over existing columns, dimensions in [m]

Appendix

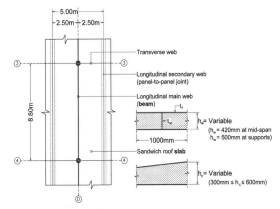

Fig. B.11 Structural system for preliminary design: simple I-beam with overhangs on existing columns, with cross section consisting of primary web and 1.0 m wide face sheets. (t_f (t_w)) = face sheet (web) thickness, h_w (h_c) = web (core) thickness)

B.3.3 Paper related to CLP Building

[1] S. Yanes-Armas and T. Keller, "Structural concept and design of a GFRP-polyurethane sandwich roof structure", in *Proceedings of the Eighth International Conference on Fibre-Reinforced Polymer (FRP) Composites in Civil Engineering (CICE2016)*, Hong Kong, 2016.

B.4 TSCB BRIDGE

B.4.1 Basic information

Concept   "TSCB" refers to <u>T</u>win-<u>S</u>haped <u>C</u>omposite <u>B</u>eam

Overhead spatial frame, structural concept of radiolarians and diatoms

Side frames undulated to form three waves of 6 m in length each

Undulating width of parallel roof frame and walkway slab

Double-curved edges between horizontal and undulated side frames

Modular construction, six modules of 3 m in length, composed of two half-modules each, with walkway slabs integrated in bottom half-modules

Adhesive connections, with back-up bolts in vertical module joints to provide redundancy

Carbon fibre-PET foam sandwich construction

Black surface texture of woven carbon fibre fabrics as architectural feature

Twill-weave fabrics due to high drapability over complex shape

	Quasi-isotropic fibre lay-up due to complex geometry
	Design service life of 50 years
Dimensions	18.00×4.20×2.40 m (length×width×height)
	Sandwich panel member thickness of 30–40 mm
Tolerances	10 mm misalignment between upper and lower half-module (see loads below)
	3 mm misalignment between module joints, considered in testing
Materials	Carbon fibre-epoxy face sheets, of 0.75–1.5 mm thickness, 63 vol% fibres, 1/3 in 0°/30°/60° directions each
	PET foam of 135 and 320 kg/m^3, solid PUR material of 1 380 kg/m^3 at supports and joints with mechanical back-up
	Two-component epoxy adhesive, stainless steel bolts
Loads	Permanent loads, pedestrian loads, lateral load on handrail, snow, wind, temperature (global changes and gradients), vandalism (failure of one face sheet of one diagonal, accidental design situation), installation
	Constraint forces in joints to compensate 10 mm tolerances between upper and lower half-modules
Location	Bissone, Canton of Tessin
Architectural design	Filippo Broggini, BlueOffice Architecture, Bellinzona
Structural design	Ernst Baser & Partner AG, Zurich
	University of Applied Sciences of Italian Switzerland, Mendrisio (material characterisation)
	Thomas Keller, EPFL-CCLab (expert from owner side)
	EPFL-CCLab (DMA experiments)
	Design based on Eurocomp Design Code and Handbook (1996) and testing
	First (second) eigenfrequency of 3.82 (5.25) Hz, lateral (vertical) mode
	Temperature increases up to 84°C due to black colour (measured at 1 000 W/m^2 irradiation, as present in Bissone)
	Max. allowable service temperature of $T_{g,\text{onset (DMA)}}$ -10°C
	Required compromise after temperature analysis: exterior surfaces painted with light grey colour to prevent overheating; interior surfaces left black with transparent UV coating
	Hail tests performed successfully
Manufacturing	Resin infusion of modules
	Module joints infused or adhesively bonded
Year of design	2012–2016 (including testing and prototyping)

(Allocation) Table B.4 Further information in Chapters 1 to 5

Section	Figure	Subject
2.5.3	2.24	Prototype of basic carbon fibre-PET sandwich module and assembly of two modules
4.2.1	4.1 (left)	Model of twin-shaped modular composite bridge inspired by diatoms
4.3.3.1	4.16 (bottom right)	Structural form development: example of third phase material-tailored frame-sandwich form
Postscript		Why was this bridge not built?

B.4.2 Additional images and information

Fig. B.12 Rendering with bridge (small in centre) over connection between northern and southern part of Lake Lugano in Bissone

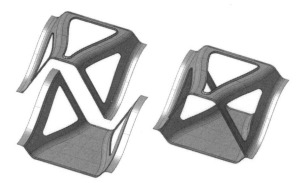

Fig. B.13 Two bridge half-modules identical in shape **(left)**; connected to one module **(right)** (walkway slab integrated in bottom half-module)

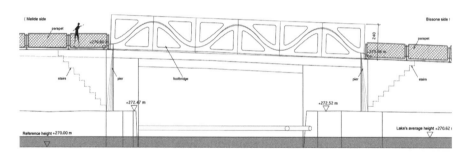

Fig. B.14 Bridge elevation with six 3 m long modules (twelve half-modules) forming three waves

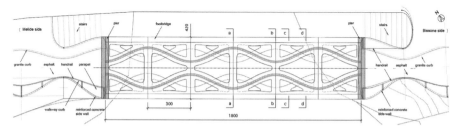

Fig. B.15 Bridge plan **(top)** view with three waves in roof frame and three opposite waves in walkway slab, dimensions in [cm]

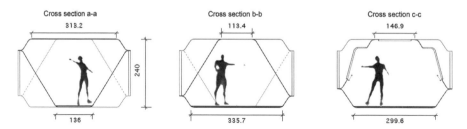

Fig. B.16 Three typical cross sections, as indicated in Fig. B.15, dimensions in [cm]

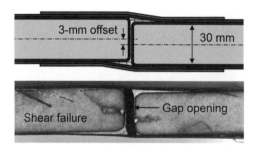

Fig. B.17 Assumed misalignment in adhesively bonded vertical module joints: test joint geometry with 3 mm offset **(top)**; experimental shear failure mode in PET foam under axial tension (leading to gap between panels) **(bottom)**

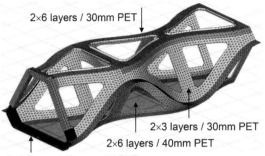

Fig. B.18 Finite element model of half of bridge superstructure, indicating number of fabric layers, core type, and thickness

B.4.3 Paper related to TSCB Bridge

[1] F. Broggini, L. Diviani, D. Rüegg, M. Ambrosini, and T. Keller, "Twin-shaped CFRP-sandwich pedestrian bridge", in *Proceedings of the Eighth International Conference on Fibre-Reinforced Polymer (FRP) Composites in Civil Engineering (CICE2016)*, Hong Kong, 2016.

B.5   1K BRIDGE

B.5.1   Basic information

Concept	"1K" refers to 1 km bridge length
	Suspension bridge for pedestrians in Swiss Alps
	Bridge composed of 50 bending-active, self-supporting modules
	Modules assembled on-site by bending straight profiles into arches and fixing them to the deck
	Arches with flat-rectangular hollow cross section to facilitate bending and provide lateral stability
	Short wider platforms between modules to enable crossing of bicycles
	Modules suspended to two bearing cables and stabilised by two counter-cables
	Lateral wind action taken by curvature of the four main cables in ground plan
	Minimised transportation effort due to lightweight and straight members
Dimensions	Length of 1 065 m
	Module length (including one platform) of 21.30 m
	Walkway width of 1.0 m (crossing platform 1.5 m)

Materials	Arches and four main cables: carbon fibre-polymer composites
	Deck: glass fibre-polymer composites
Loads	Permanent loads, pedestrian loads, wind, snow (conceptual design stage)
Location	Lumino, Canton of Tessin
Architectural design	Filippo Broggini AEA & BlueOffice Architecture, Genève-Bellinzona
Structural design	Thomas Keller, EPFL-CCLab (conceptual design)
	Design based on CEN/TS 19101
Manufacturing	Pultruded profiles, bent into arch shape, assembled with deck to basic module
Year of design	In progress (conceptual design completed in 2023)

(Allocation) Table B.5   Further information in Chapters 1 to 5

Section	Figure	Subject
4.3.4.3	4.33	Rendering of 1K Bridge and suspended module

B.5.2   Additional images and information

Fig. B.19   Aerial view of bridge **(left)**; rendering of suspended bending-active modules **(right)**

IMAGE & VIDEO CREDITS

Images with © Thomas Keller / CCLab, licenced under CC BY NC licence

P.1	2.33 middle	3.40 left
P.2	2.43 right	3.41
P.3 right	2.47 left	3.42 right
1.2 all	2.49	3.43 all
1.3 all	3.1 right	3.44 all
1.4 top all	3.2 all	3.45
1.5 all	3.7 top left	3.46
1.6 left	3.9 all	3.47 all
1.7 all	3.10	3.48 all
1.9 all	3.12 left	3.49 right
1.10	3.13 all	3.50 left & middle
1.11	3.14 all	3.53 right
2.1	3.15 left & top middle	3.54 all
2.2 all	3.16 bottom left & right	3.56 all
2.3 all	3.18 all	3.57 left
2.4	3.19 all	3.59 right & middle
2.5	3.20 all	3.60
2.7	3.22 all	3.61
2.8 right	3.23 all	3.65 left
2.10 all	3.24 all	3.67 all
2.13	3.25	3.69 left & right
2.14	3.26 all	3.70 middle & right
2.16 all	3.27 all	3.71 all
2.17 all	3.28 left	3.72 all
2.18 all	3.30	3.73 right
2.19 all	3.31	3.74 all
2.20 all	3.32 all	3.79 right
2.21 right	3.34 all	3.81 right
2.22 all	3.35 all	3.82
2.26 left & right	3.36 all	3.83 all
2.26 middle left	3.37 left	3.84 right
2.27 all	3.38 left	3.85
2.28 all	3.39 left	3.88

3.91	3.163 all	5.6 middle & right
3.92	3.164 right	A.1 all
3.93	4.2	A.2 all
3.95 left	4.3	A.3 all
3.96 all	4.5 right	A.4 all
3.98 bottom & right	4.8	A.5
3.99 all	4.11	A.6
3.100 all	4.12 all	A.7
3.105	4.14 right	A.9
3.106 all	4.17 right	A.10
3.107 all	4.19	A.11
3.109 all	4.21 all	A.12
3.110 all	4.22 all	A.13 right
3.111 all	4.23 right	A.14
3.117 all	4.24	A.16 all
3.119	4.36	A.17 all
3.121 all	4.37	A.18 all
3.124 all	4.38 all	A.19 all
3.126 all	4.39 all	A.21 all
3.129	4.41 all	A.22
3.130 top	4.42	A.27 all
3.140 left & right	4.43 all	A.28 all
3.141 all	4.46	A.29
3.142 all	4.48	A.30 all
3.143 all	4.49	A.31 all
3.144 top left & right	4.51 all	A.32
3.145 all	4.52 middle	A.33
3.146 all	4.54 right	A.35 all
3.147 middle right & right	4.56	A.40 all
3.148 all	4.62 top left & right	A.41 all
3.148 all	4.62 bottom left	A.42 all
3.149 left & middle	4.63 top right	A.44 all
3.151 right	4.63 bottom left	A.45 all
3.152 all	4.64 right	A.46 left
3.153 all	4.65 all	A.47 left
3.154	4.67 top left & right	A.48 all
3.155 all	4.67 bottom, all	B.3
3.156 all	4.68 left	B.8
3.157 all	4.69 bottom, all	B.10
3.158 all	4.70	B.11
3.159 all	5.1	
3.160 middle & right	5.2	
3.161 all	5.3	
3.162 all	5.5	

Image & Video Credits 413

Images from different rightsholders, licenced under different types of CC licences

P.3 left	Wikimedia Commons, CC BY-SA 3.0, Gregory Zeier
1.1 all	Alain Herzog, EPFL, CC BY
1.8 all	Alain Herzog, EPFL, CC BY
2.9 left	Alain Herzog, EPFL, CC BY
2.12 top left	Alain Herzog, EPFL, CC BY
2.12 bottom left	Alain Herzog, EPFL, CC BY
2.14 left	Alain Herzog, EPFL, CC BY
2.33 left	Wikimedia Commons, CC BY-SA 2.0, Andrew Curtis, bottom and right side slightly cropped
2.35 all	Wikimedia Commons: Photographs in the Carol M. Highsmith Archive, Library of Congress, Prints and Photographs Division (according to library, there are no known copyright restrictions)
2.39 left	ETH-Bibliothek Zürich, Bildarchiv / Fotograf: Comet Photo AG (Zürich) / Com_BC25-004-017 / CC BY-SA 4.0, top and bottom cropped
2.50	Wikimedia Commons, CC BY-SA 4.0, Justin Ormont
2.52 left	Wikimedia Commons, CC BY-SA 3.0, Henk Monster
3.1 left	Wikimedia Commons, CC BY-SA 3.0, Achim Hering, photo slightly cropped
3.8 all	© 2020 The Authors. Published by Elsevier Ltd., CC BY-NC-ND
3.15 bottom, middle & right	© 2021 The Authors. Published by Elsevier Ltd., CC BY-NC-ND
3.16 top left	© 2022 The Authors. Published by Elsevier Ltd., CC BY-NC-ND
3.17	© 2022 The Authors. Published by Elsevier Ltd., CC BY-NC-ND
3.55 all	© 2020 The Authors. Published by Elsevier Ltd., CC BY-NC-ND
3.63 left	© 2023 The Authors. Published by Elsevier Ltd., CC BY-NC-ND
3.65 right	© 2023 The Authors. Published by Elsevier Ltd., CC BY-NC-ND
3.68 right	© 2023 The Authors. Published by Elsevier Ltd., CC BY-NC-ND
3.73 left	© 2021 The Authors. Published by Elsevier Ltd., CC BY-NC-ND
3.76 all	© 2023 The Authors. Published by Elsevier Ltd., CC BY-NC-ND
3.77 all	© 2023 The Authors. Published by Elsevier Ltd., CC BY-NC-ND
3.78 all	© 2023 The Authors. Published by Elsevier Ltd., CC BY-NC-ND
3.86	© 2022 The Authors. Published by Elsevier Ltd., CC BY
3.89 all	© 2022 The Authors. Published by Elsevier Ltd., CC BY
3.90	© 2022 The Authors. Published by Elsevier Ltd., CC BY
4.5 left	Flickr, CC BY-NC 2.0, Robin Capper
4.9 left	Wikimedia Commons, CC BY-SA 4.0, Paris Orlando, top and bottom slightly cropped
4.10	Alain Herzog, EPFL, CC BY
4.13	Wikimedia Commons, CC BY-SA 3.0, Holger Ellgaard, photo slightly cropped

4.16 top left	Wikimedia Commons, CC BY-SA 4.0, Henning Schlottmann, photo slightly cropped
4.17 left	Alain Herzog, EPFL, CC BY
4.18 top left	Wikimedia Commons, CC BY-SA 3.0, Jorune, photo printed in black and white
4.18 bottom right	Wikimedia Commons, CC BY-SA 3.0, Christian Kleis, photo slightly cropped
4.26 right	Wikimedia Commons, CC CC0 1.0 Universal, S. Juhl, photo slightly cropped
4.27	Wikimedia Commons, CC BY-SA 3.0, Hubert Berberich
4.28 all	Alain Herzog, EPFL, CC BY
4.29 all	Alain Herzog, EPFL, CC BY
4.34 left	Alain Herzog, EPFL, CC BY
4.47 left	Wikimedia Commons, CC BY-SA 4.0, Gunnar Klack, photo slightly cropped
4.47 right	Wikimedia Commons, CC BY-SA 2.0, Gunnar Klack, photo slightly cropped
4.57	Wikimedia Commons, CC BY-SA 4.0, Roland Fischer
4.58 left	Wikimedia Commons, CC BY-SA 3.0, Michielverbeek, right edge cropped
4.58 right	Wikimedia Commons, Free Art License, NOX / Lars Spuybroek
4.66 top left	Flickr, CC BY-NC-SY 2.0, Ralph Crane, bottom cropped
4.68 right	Flickr, CC BY 2.0, 師事務所
B.4	Wikimedia Commons, CC BY-SA 4.0, Mediacom EPFL

Images from different rightsholders, excluded from CC BY NC licences

Book cover	© Tara Habibi, EPFL-CCLab
1.4 bottom left & middle	© Christian de Portzamparc
1.4 right	© Nicolas Borel
1.6 right	© Fiberline Building Profiles
2.9 right	© Courtesy of Strongwell
2.11	© FiberCore Europe
2.12 top right	© Fiberline Building Profiles
2.12 middle right	© Fiberline Building Profiles
2.12 bottom right	© Courtesy of Strongwell
2.21 left	© Fiberline Building Profiles
2.23 all	© Multiplast Groupe Carboman
2.24 all	© Filippo Broggini, Bellinzona
2.25 all	© AM Structures Ltd.
2.29 all	© FiberCore Europe
2.30 all	© FiberCore Europe
2.31	© Creative Composites Group
2.32 all	© 3A Composites

2.33 right	© AM Structures Ltd.
2.34 all	© 3A Composites
2.36 all	© gta Archiv / ETH Zürich, Heinz Isler
2.37 all	© Jonathan Quarmby
2.38 all	© gta Archiv / ETH Zürich, Heinz Isler
2.39 right	© Schnetzer Puskas Ingenieure AG
2.40 all	© Courtesy of Fondazione Renzo Piano
2.41 all	© Roselli Architetti, Roma
2.42 left	© Nicolas Borel
2.42 right	© Christian de Portzamparc
2.44 all	© Holland Composites
2.45	© British Antarctic Survey, photographer unknown
2.46 all	© NIO Architecten, Rotterdam
2.47 right	© Multiplast Groupe Carboman
2.48 all	© Holland Composites
2.51 left	© Jean Dubuffet / 2023, ProLitteris, Zurich / Archives Fondation Dubuffet / Photo: Kurt Wyss
2.51 right	© Archives Fondation Dubuffet / 2023, ProLitteris, Zurich / Photo: Zingher
2.52 middle	© Archives Fondation Dubuffet / 2023, ProLitteris, Zurich / Photo: Atelier Dhoedt
2.52 right	© Archives Fondation Dubuffet / 2023, ProLitteris, Zurich / photo: DR
2.53 left	© Jean Dubuffet / 2023, ProLitteris, Zurich / Archives Fondation Dubuffet / Photo: Centre National des Arts Plastiques (CNAP), Paris
2.53 middle	© Archives Fondation Dubuffet / 2023, ProLitteris, Zurich / Photo: Atelier Dhoedt
2.53 right	© Archives Fondation Dubuffet / 2023, ProLitteris, Zurich / Photo: Atelier Dhoedt
2.54 left & right	© Hans Hammarskiöld Heritage © Niki Charitable Art Foundation / 2023, ProLitteris, Zurich 2023
2.54 middle	© Moderna Museet, Stockholm © Niki Charitable Art Foundation / 2023, ProLitteris, Zurich 2023
2.55 all	© Hans Hammarskiöld Heritage © Niki Charitable Art Foundation / 2023, ProLitteris, Zurich 2023
2.56 all	© gta Archiv / ETH Zürich, Heinz Isler
2.57 all	© Yayoi Kusama and Benesse Art Site Naoshima
2.58 left	© 2013 / Comité International Olympique (CIO) / Leutenegger, Catherine
2.58 right	© 2009 / Comité International Olympique (CIO) / Widmer, Cédric
2.59 left & right	© Jonathan Borofsky
2.59 middle left & right	© Jonathan Borofsky / Munich Re
2.60 all	© Jonathan Borofsky
3.7 bottom left & right	© American Concrete Institute, Authors, DOI: 10.14359/51686148
3.28 right	© Data with courtesy of prof. Urs Meier
3.94	© CCLab research ongoing

3.97 all	© Prof Wendel Sebastian
3.101	© CCLab research ongoing
3.102	© CCLab research ongoing
3.103	© CCLab research ongoing
3.104	© CCLab research ongoing
3.116	© Goodrich T, Nawaz N, Feih S, Lattimer BY, Mouritz AP. Derived from Fig. 3 in "High-temperature mechanical properties and thermal recovery of balsa wood." *J Wood Sci.* 2010;56(6):437–43.
4.1 left	© Filippo Broggini, Bellinzona
4.1 right	© Reproduced with permission of Cambridge University Press through PLSclear
4.4 left	© Antonia Kesel, Hochschule Bremen
4.4 right	© Pierre Cagna, EPFL
4.6 all	© Jonathan Quarmby
4.7 all	© itke, Universität Stuttgart
4.9 right	© F.L.C. / 2023, ProLitteris, Zurich
4.15 left	© Richard Steger
4.16 bottom right	© Filippo Broggini, Bellinzona
4.18 top right	© gta Archiv / ETH Zürich, Heinz Isler
4.18 bottom left	© Fonds Bétons armés Hennebique, CNAM/SIAF/Cité de l'architecture et du patrimoine/Archives d'architecture contemporaine
4.20 left & middle	© Stadtmuseum Oldenburg
4.20 right	© Niedersächsische Landesmuseen Oldenburg, Landesmuseum für Kunst & Kultur, Oldenburg
4.23 left	© gta Archiv / ETH Zürich, Heinz Isler
4.25	© Dr. Bodo Rasch
4.26 left	© Courtesy of Strongwell
4.30 all	© CCLab research ongoing
4.31	© CCLab research ongoing
4.32	© CCLab research ongoing
4.33 all	© AEA & BlueOffice Architecture
4.35 all	© Authors, CClab, DOI: 10.1177/0021998313511653
4.40 all	© Authors, CClab, DOI: 10.1177/0021998313511655
4.44	© Authors, CClab, DOI: 10.1177/0021998314567696
4.45	© Authors, CClab, DOI: 10.1177/0021998316656393
4.54 left	© Authors, CClab, DOI: 10.1177/17442591221150257
4.55 right	© Authors, CClab, DOI: 10.1177/0021998316656393
4.59 left	© HR Giger Museum
4.59 right	© Henrik Kam
4.60 all	© Christian de Portzamparc
4.66 top right	© Jonathan Quarmby
4.66 bottom left & right	© Eastman Chemical Company
4.67 top middle	© Richard Steger

Image & Video Credits

4.69 top middle & right	© 3A Composites
A.15	© René Clausen
A.20 all	© Felix Knobel
A.34	© Canton of Vaud, Monod-Piguet+Associés Ingénieurs Conseils SA
A.36	© Canton of Vaud, Monod-Piguet+Associés Ingénieurs Conseils SA
A.43	© Suisse Technology Partners, Neuhausen am Rheinfall
A.46 right	© AEA & BlueOffice Architecture
A.47 right	© AEA & BlueOffice Architecture
A.49 all	© AEA & BlueOffice Architecture
B.1 right	© Georges Abou Jaoudé, EPFL
B.2	© Pierre Cagna, EPFL
B.5	© Ernst Basler & Partner AG, Zurich
B.6	© Ernst Basler & Partner AG, Zurich
B.7 all	© Brönnimann & Gottreux Architectes
B.9	© Brönnimann & Gottreux Architectes
B.12	© AEA & BlueOffice Architecture
B.13	© AEA & BlueOffice Architecture
B.14	© AEA & BlueOffice Architecture
B.15	© AEA & BlueOffice Architecture
B.16	© AEA & BlueOffice Architecture
B.17 top	© Ernst Basler & Partner AG, Zurich
B.17 bottom	© University of Applied Sciences of Italian Switzerland
B.18	© Ernst Basler & Partner AG, Zurich
B.19 all	© AEA & BlueOffice Architecture

Images with © Reprinted with permission from Elsevier, excluded from CC BY licences

(The data shown in these figures resulted from CCLab research and have already been published in Elsevier papers, as indicated in the figure captions and reference lists.)

2.6 all	3.40 right	3.68 left
2.8 left	3.42 left	3.69 middle
2.15 all	3.49 left	3.70 left
3.3 all	3.50 right	3.75 all
3.4 all	3.51 all	3.79 left
3.5 all	3.52 all	3.80 all
3.6 all	3.53 left & middle	3.81 left
3.11 all	3.57 right	3.84 all
3.21 all	3.58 all	3.87
3.29	3.59 left	3.95 right
3.33 all	3.62 all	3.108 all
3.37 right	3.63 right	3.112 all
3.38 right	3.64	3.113
3.39 right	3.66 all	3.114

3.115 all	3.130 bottom-3	4.61 top right & bottom
3.118 top all	3.131	4.62 bottom right
3.120 all	3.133 all	4.63 bottom right
3.122 all	3.147 left & middle left	4.64 left
3.123 all	4.50 all	5.4
3.127 all	4.52 left & right	
3.128	4.53 all	

Images with © Used with permission of American Society of Civil Engineers, permission conveyed through Copyright Clearance Center, Inc., excluded from CC BY licences

(The data shown in these figures resulted from CCLab research and have already been published in ASCE papers, as indicated in the figure captions and reference lists.)

2.26 middle right	3.144 bottom left	4.69 top left
2.43 left	3.149 right	5.6 left
3.12 middle & right	3.150 all	A.8
3.98 top left	3.151 left	A.13 left
3.118 bottom all	3.160 left	A.23
3.125 all	3.164 left	A.24
3.132 all	4.14 left	A.25
3.134 all	4.15 right	A.26
3.135 all	4.16 top right & bottom left	A.37
3.136 all		A.38
3.137 all	4.34 right	A.39
3.138 all	4.55 left	B.1 left
3.139 all	4.61 top left	
3.140 middle	4.63 top left	

Videos in ebook, excluded from CC BY licences

3.70 left	© Thomas Keller / CCLab, 09 August 2018
3.99 left	© Thomas Keller / CCLab, 03 June 2013
4.28	© Thomas Keller / CCLab, 28 June 2023
4.29 middle	© Thomas Keller / CCLab, 28 June 2023
4.29 right	© Thomas Keller / CCLab, 28 June 2023
4.34 left	© Thomas Keller / CCLab, 15 December 1998
4.63 top left	© "Menschen-Technik-Wissenschaft: Kunststoff statt Beton im Brückenbau", SRF television channel (Switzerland), 27 February 2003 (English)
A.5	© "Schweiz Aktuell: Kunststoff-Brücke" telecast, broadcast station SRF (Switzerland), 28 October 1998 (Romansh and German)
A.12	© "Menschen-Technik-Wissenschaft: Eyecatcher – Gebäude aus Kunststoff", SRF television channel (Switzerland), 28 January 1999 (English)

A.22	© "Menschen-Technik-Wissenschaft: Glashaus mit Flügeldach", SRF television channel (Switzerland), produced by Offroad Reports Gmbh, Zürich, 26 October 2006 (English)
A.33	© "Nano: First road bridge in Switzerland with composite-sandwich deck", 3sat television channel (Austria, Germany, Switzerland), produced by Offroad Reports Gmbh, Zürich, 21 January 2013 (German)

INDEX

This index indicates the sections (and figures) where the below-listed, alphabetically ordered terms are defined.

Term	Section
Anisotropic	3.2.1
Balanced laminate	2.3.1 (Fig. 2.7)
Bending-active system	4.3.4.3 (Fig. 4.28)
Bio-based composites	3.11.4
Blistering	3.10.3
Brittle response	3.6.2 (Fig. 3.48)
CFM: Continuous filament mat	2.2.1
Char, fire	3.9.2.2
Cohesive failure	3.4.2 (Fig. 3.15)
Cohesive zone model (CZM)	3.7.2.2
Cold cured	3.5.1
Component	2.3.1
Composite action	4.6.2-3
Conceptual design	5.1-2 (Fig. 5.1)
Connection	2.4.1
Constant life diagram, fatigue	3.8.3 (Fig. 3.84)
Constructible form	4.3.3.3 (Fig. 4.19)
Convertible structure	4.2.4.2
Crack arrest	3.7.3.3, 3.7.5 (Table 3.6)
Crack front (2D)	3.7.3.2 (Fig. 3.70), 3.7.3.4 (Fig. 3.78)
Crack growth rate, fatigue (1D-2D)	3.7.3.3, 3.7.5 (Table 3.6)
Crack opening displacement	3.7.2.1 (Fig. 3.64)
Crack tip	3.7.2.1 (Fig. 3.64)
Crash barrier	2.6.2.3 (Fig. 2.32)
Creep	3.4.5.1 (Fig. 3.24)
Creep rupture	3.4.5.2 (Fig. 3.26 right)
Critical SERR	3.7.1

Cross-link network, density, reaction	3.5.1-2
CSM: Chopped strand mat	2.2.1
Curing degree	3.5.2
Cyclic stiffness, fatigue	3.8.4 (Fig. 3.86)
Damage energy	3.6.2 (Fig. 3.48)
Daylit building	4.4.1-2
Decomposition, fire	3.9.2.2
Detailed design	5.1 (Fig. 5.1)
Diffusion, moisture	3.10.2.7
Discolouration	3.10.2.8
Dissipated energy, fatigue	3.8.4 (Fig. 3.86)
Dissipated energy, quasi static	3.6.2 (Fig. 3.48)
DMA: Dynamic Mechanical Analysis	3.4.3
Drip edge	3.10.4 (Fig. 3.162)
DSC: Differential Scanning Calorimetry	3.4.3
Ductile response	3.6.2 (Fig. 3.48)
Ductility index	3.6.2 (Eqs. 3.4-5)
Effective coefficient of thermal expansion	3.9.2.8
Effective specific heat capacity	3.9.2.4
Effective thermal conductivity	3.9.2.3
Elastic energy	3.6.2 (Fig. 3.48)
Elastomer	2.2.1
End-grain balsa	2.2.2.1 (Fig. 2.4)
Execution specification	5.4
Experimental compliance method	3.7.2.1
Fatigue load model	3.8.9
Fibre blooming	3.10.2.6 (Fig. 3.151)
Fibre bridging	3.7.2.1 (Fig. 3.63)
Fibre bridging length	3.7.2.1 (Fig. 3.64)
Fibre-dominated	3.2.3
Fibreglass	2.8
Fibre-tear failure	3.7.2.1 (Fig. 3.66)
Filled adhesive	3.5.3.1
Fire curve (ISO 834, external, hydrocarbon)	3.9.1 (Fig. 3.105)
Fire dynamic simulation	3.9.4
Fire reaction behaviour	3.9.1
Fire resistance	3.9.3.1
Flexible adhesive	2.4.3
Flexible core	2.2.2.2
Form-active structural system	4.3.4.2
Fractal dimension	4.2.4.3
Fracture energy	3.7.1 (Table 3.4)

Fracture process zone (FPZ)	3.7.2.1
Fracture resistance (G_c)	3.7.1
Fracture toughness	3.7.1
Gel time	4.4.3
Glass transition temperature	3.4.3 (Fig. 3.18)
Glassy state	3.4.3 (Fig. 3.18)
Hand lay-up	2.5.1
Homogeneous core	2.3.3
Hybrid	1.1
Hybrid (bonded and bolted) joints	3.6.3 (*Example-3*)
Hydrolysis	3.10.2.7
Hydronic radiant system	4.5.3.2
Hysteretic energy	3.6.2 (Fig. 3.48)
Ideal form	4.3.3.3 (Fig. 4.19)
Imperfections	3.3.2
Implemented form	4.3.3.3 (Fig. 4.19)
Inelastic energy	3.6.2 (Fig. 3.48)
Interlaminar fracture	3.7.1 (Fig. 3.60)
Intralaminar fracture	3.7.1 (Fig. 3.60)
Isotropic	3.2.1
Joint	2.4.1
Joint efficiency	3.6.3 (*Example-3*)
Kerb (bridge)	3.10.4 (Fig. 3.162)
Kink band failure	3.9.2.7 (Fig. 3.115)
Laminate	2.3.1
Lap shear connection/joint	2.4.1 (Fig. 2.13)
Life cycle assessment (LCA)	3.11.2
Life cycle costing (LCC)	3.11.2
Linear elastic fracture mechanics	3.7.2.2
Loss factor (tan-δ)	3.4.3
Loss modulus	3.4.3 (Fig. 3.18)
Maintenance plan	5.5
Material substitution	4.3.3.1
Material-tailored structural form/system	4.3.3.2 / 4.3.4.1
Matrix-dominated	3.2.3
Mean strain (fatigue)	3.8.4 (Fig. 3.86)
Mean stress (fatigue)	3.8.2
Member	2.3.2, 2.3.3

Microcrack	3.4.2 (Fig. 3.14)
Mixed-mode fracture	3.7.2.1
Mode-I, -II, -III fracture	3.7.2.1 (Fig. 3.61)
Modular construction	4.7.1
Moment redistribution	3.6.3 (*Example-4*)
Multiple cracks	3.7.4 (Fig.3.79)
Nonlinear fracture mechanics	3.7.2.2
Onset value of storage modulus	3.4.3 (Fig. 3.18)
Orthotropic	3.2.1
Palmgren-Miner linear damage accumulation	3.8.1, 3.8.5
Peel-ply fabric	2.4.3 (Fig. 2.14)
Physical ageing	3.5.2
Pigment	4.5.5.2
Pin-bearing failure	3.2.3 (Fig. 3.8)
Plasticisation	3.4.6, 3.5.2
Ply	2.3.1
Post cured	3.5.3.1
Preliminary dimensioning	5.2.6
Prepreg	2.5.4
Prestressed arch	4.3.4.3 (Fig. 4.28)
Pseudo-ductile response	3.6.2 (Fig. 3.48)
Pultrusion, pultruded	2.5.2
R-curve, fracture resistance	3.7.2.1 (Fig. 3.65)
R-ratio, fatigue	3.8.2-3
Recovery, after creep	3.4.5.1, 3.4.5.3 (Fig. 3.24)
Recovery, after exceeding T_g	3.9.3.4
Recycling	3.11.3
Redundant structural system	3.6.2
Reflectance	4.4.2
Refractive index	4.4.2
Rehabilitation	5.5
Relaxation	3.4.5.1, 3.4.5.4 (Fig. 3.24)
Relaxation enthalpy	3.5.2 (Fig. 3.31)
Residual heat	3.5.2 (Fig. 3.31)
Reuse	3.11.3, 6.3
Rigid core	2.2.2.2
Roving bridging	3.7.2.1 (Fig. 3.65)
Rubbery state	3.4.3 (Fig. 3.18)
Sandwich panel	1.2 (Fig. 1.11), 2.3.3
Scarf joint	2.4.1 (Fig. 2.17)
Secondary bond	3.4.2 (Fig. 3.14)

Section-active structural system	4.3.4.2
Shear-out failure	3.2.3 (Fig. 3.9)
Sizing	2.2.1
Slope coefficient, fatigue	3.8.2
S-N curve, fatigue	3.8.2 (Fig. 3.82)
Softening 1D (2D)	3.7.2.1 (Fig. 3.67), 3.7.3.2 (Fig. 3.71)
Spacer tube	3.10.4 (Fig. 3.164)
Specific strength and modulus	1.2 (Fig. 1.10)
Statically in/determinate system	3.6.2
Stiff adhesive	2.4.3
Strain energy release rate (SERR)	3.7.1
Strain hardening	3.4.2 (Fig. 3.15)
Strain rate	3.4.4
Stress concentration	3.8.8
Stress intensity factor	3.7.1
Stretching, polymer chains	3.4.2 (Fig. 3.14)
Stretching, 2D crack propagation	3.7.3.1 (Fig. 3.69)
Storage modulus	3.4.3 (Fig. 3.18)
Surfacing	2.6.2.3
Sustainable development	3.11.2
Symmetric laminate	2.3.1 (Fig. 2.7)
System ductility	3.6.1
Tensegrity structure	4.2.4.1 (Fig. 4.5)
TGA: Thermogravimetric Analysis	3.9.2.2 (Fig. 3.107)
Thermal conductivity	4.5.2.2
Thermoplastic resin	2.2.1
Thermoset resin	2.2.1
Tolerance	3.3.3
Total energy	3.6.2 (Eq. 3.4-5)
Transition point, 2D crack propagation	3.7.3.2 (Fig. 3.71)
Translaminar fracture	3.7.1 (Fig. 3.60)
Translucency	4.4.2
Transmittance, optical (regular, diffuse, total)	4.4.2
Transmittance, thermal (linear)	4.5.2.3
Transparency, optical	4.4.2
Transparency, thermal	4.5.3.3
Vacuum infusion, vacuum-infused	2.5.3
Variable amplitude loading	3.8.5
Vibration	2.6.3.2
Vierendeel frame	3.6.3 (*Example-1*)
Virtual crack closure technique (VCCT)	3.7.2.2
Vitrification	3.5.2
Viscosity	3.9.2.6

Web-core	2.3.3
Web-crippling failure	3.6.3 (Fig. 3.59)
Wicking mechanism	3.10.2.7
Zero-emission building	4.5.2.4